de Gruyter Studies in Mathematics 22

de Gruyter Studies in Mathematics

1 Riemannian Geometry, 2nd rev. ed., *Wilhelm P. A. Klingenberg*
2 Semimartingales, *Michel Métivier*
3 Holomorphic Functions of Several Variables, *Ludger Kaup and Burchard Kaup*
4 Spaces of Measures, *Corneliu Constantinescu*
5 Knots, *Gerhard Burde and Heiner Zieschang*
6 Ergodic Theorems, *Ulrich Krengel*
7 Mathematical Theory of Statistics, *Helmut Strasser*
8 Transformation Groups, *Tammo tom Dieck*
9 Gibbs Measures and Phase Transitions, *Hans-Otto Georgii*
10 Analyticity in Infinite Dimensional Spaces, *Michel Hervé*
11 Elementary Geometry in Hyperbolic Space, *Werner Fenchel*
12 Transcendental Numbers, *Andrei B. Shidlovskii*
13 Ordinary Differential Equations, *Herbert Amann*
14 Dirichlet Forms and Analysis on Wiener Space, *Nicolas Bouleau and Francis Hirsch*
15 Nevanlinna Theory and Complex Differential Equations, *Ilpo Laine*
16 Rational Iteration, *Norbert Steinmetz*
17 Korovkin-type Approximation Theory and its Applications, *Francesco Altomare and Michele Campiti*
18 Quantum Invariants of Knots and 3-Manifolds, *Vladimir G. Turaev*
19 Dirichlet Forms and Symmetric Markov Processes, *Masatoshi Fukushima, Yoichi Oshima, Masayoshi Takeda*
20 Harmonic Analysis of Probability Measures on Hypergroups, *Walter R. Bloom and Herbert Heyer*
21 Potential Theory on Infinite-Dimensional Abelian Groups, *Alexander Bendikov*

Vladimir E. Nazaikinskii ·
Victor E. Shatalov · Boris Yu. Sternin

Methods of Noncommutative Analysis

Theory and Applications

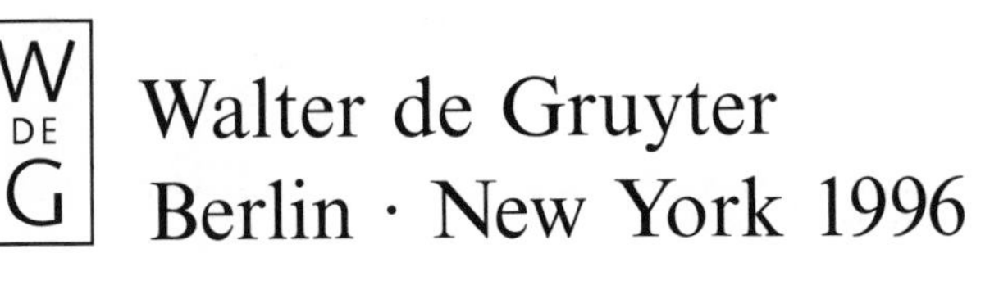

Walter de Gruyter
Berlin · New York 1996

Authors:

Vladimir E. Nazaikinskii
Moscow State Institute of
Electronics and Mathematics
Technical University
3/12 B. Vuzovskii per
Moscow 109028, Russia

Victor E. Shatalov, Boris Yu. Sternin
Department of Computational
Mathematics and Cybernetics
Moscow State University
Vorob'evy Gory
Moscow 119899, Russia

1991 Mathematics Subject Classification: 35-02; 22 Exx, 44-XX, 47-XX, 81R50

Keywords: Functional analysis, partial differential equations, Fourier integral operators, asymptotic theory, representation theory, Yang-Baxter equations, quantum groups

♾ Printed on acid-free paper which falls within the guidelines of the ANSI to ensure permanence and durability.

Library of Congress Cataloging-in-Publication Data

Methods of noncommutative analysis : theory and applications / Vladimir E. Nazaikinskii, Viktor E. Šhatalov, Boris Yu. Sternin
p. cm. – (De Gruyter studies in mathematics ; 22)
Includes bibliographical references and index.
1. Geometry, Differential. 2. Noncommutative algebras. 3. Mathematical physics. I. Šhatalov, V. E. (Viktor Evgen'evich) II. Sternin, B. Yu. III. Title. IV. Series.
QC20.7.G44N39 1996 95-39641
515'.72–dc20 CIP

Die Deutsche Bibliothek – Cataloging-in-Publication Data

Nazajkinskij, Vladimir E.:
Methods of noncommutative analysis : theory and applications / Vladimir E. Nazaikinskii ; Victor E. Shatalov ; Boris Yu. Sternin. – Berlin ; New York : de Gruyter, 1996
(De Gruyter studies in mathematics ; 22)
ISBN 3-11-014632-0
NE: Šatalov, Viktor E.:; Sternin, Boris J.:; GT

Printed in Germany.
Typeset using the authors'TeX files: I. Zimmermann, Freiburg. – Printing: Gerike GmbH, Berlin.
Binding: Dieter Mikolai, Berlin. – Cover design: Rudolf Hübler, Berlin.

Preface

Noncommutative analysis, that is, the calculus of noncommuting operators, is one of the main tools in contemporary mathematics. Indeed, the theory of differential and pseudodifferential operators, various problems of algebra, functional analysis, and theoretical physics deal with functions of noncommuting operators. This was clearly understood by such outstanding scientists as H. Weyl, I. Schur, R. Feynman and many others. It is therefore not surprising that the development of mathematics required creating noncommutative analysis clear and convenient in applications. R. Feynman was clearly a pioneer in the field. As early as 1951, he noticed in his paper "An operator calculus having applications in quantum electrodynamics", that noncommutativity of operators can be accounted for by introducing numbers or indices showing the order of action of the operators. This apparently simple remark served as a starting point for the noncommutative operational calculus created by V. Maslov in the early 70's.

Maslov's ideas have been developed in numerous papers, and very deep and important results have been obtained. In particular, they show that noncommutative operator calculus is intimately related not only to traditional mathematical and physical theories but also to rapidly developing new ones. Thus, noncommutative analysis occurs to be useful in such mathematical fields as the theory of geometric and asymptotic quantization, representation theory, the theory of quantum groups, and so on. In this book the reader will find a lot of examples from functional analysis, algebra, representation theory, and the theory of differential equations where non commutative analysis is involved.

Unfortunately, up to now there does not exist a sufficiently simple exposition of noncommutative analysis which might serve as an introduction to the subject for scientists who are just beginning to get acquainted with this area and that is why this book has been written. It is primarily addressed to those who are not specialists in noncommutative analysis.

At the same time, even the experienced mathematician can find in this book many new and interesting topics. Noncommutative analysis gives a new outlook both on quite traditional and modern mathematical topics such as representation theory, operator theory, the theory of (pseudo)differential operators, Yang–Baxter equations and others.

Acknowledgements. We express our kind gratitude to Professor Victor P. Maslov, whose great influence inspired our work and whose advice was of much use for us.

This book has been written under the support of the Chair of Nonlinear Dynamic Systems and Control Processes, Moscow State University, and of Max-Planck-Arbeitsgruppe "Partielle Differentialgleichungen und Komplexe Analysis", Institut für Mathematik, Universität Potsdam. We are thankful to the heads of these Departments – Professor Stanislav V. Emel'yanov and Professor Bert-Wolfgang Schulze.

We are also grateful to Dr. Manfred Karbe, whose advice was truly invaluable. Finally, we are cordially thankful to Mrs. Helena R. Shashurina, who prepared the manuscript for the publishers.

Moscow – Potsdam, 1994 *The Authors*

Contents

Chapter I
Elementary Notions of Noncommutative Analysis

1 Some Situations where Functions of Noncommuting Operators Arise

In this section we consider a few examples of problems from different areas of mathematics and physics whose study requires the usage of functions of noncommuting operators. In fact, each of these problems can be studied by its own inherent methods, but if considered as a whole they suggest that there should be a universal apparatus that would permit us to consider them all from a common point of view. Indeed, such an apparatus is already developed. It is called *noncommutative analysis*; the aim of his section is to illustrate its main features by simple examples and, in particular, to introduce the so-called Feynman indices, which play the main role in the machinery of noncommutative analysis.

It should be clearly noted that the examples given in the following were not chosen at random. In the course of our exposition we will return to them repeatedly and show how the ideas and methods of the theory developed work in simple situations.

1.1 Nonautonomous Linear Differential Equations of First Order. T-Exponentials

The equation

$$\dot{x} = A(t)x$$

can easily be integrated in quadratures if $x \in \mathbb{R}^1$ and $A(t)$ is a given continuous function:

$$x(t) = \exp\left(\int_0^t A(t)\,dt\right) x(0). \tag{I.1}$$

No more complicated is the case in which $x \in \mathbb{R}^n$ but the $n \times n$ matrices $A(t)$ and $A(t')$ commute with each other:

$$[A(t), A(t')] \overset{\text{def}}{=} A(t)A(t') - A(t')A(t) = 0, \tag{I.2}$$

for any t, t'.

The solution is expressed by the same formula, where exp stands for the *matrix exponential,* defined as the sum of the convergent series

$$\exp(B) = \sum_{n=0}^{\infty} \frac{B^n}{n!}.$$

Indeed, under condition (I.2) we have

$$\exp\left(\int_0^{t+\Delta t} A(t)\,dt\right) = \exp\left(\int_0^{t} A(t)\,dt\right) \exp\left(\int_t^{t+\Delta t} A(t)\,dt\right)$$

(since the arguments of the exponentials commute with each other, one can expand the exponentials into Taylor series and repeat verbatim the proof of the identity $e^a e^b = e^{a+b}$ valid for the case in which a and b are numbers). Since

$$\exp\left(\int_0^{t+\Delta t} A(t)\,dt\right) = 1 + \int_t^{t+\Delta t} A(t)\,dt + O(\Delta t^2),$$

it is easy to differentiate $x(t)$ with respect to t and prove that $x(t)$ satisfies the original equation.

However, the simple formula fails if we do not require that the commutator (I.2) is equal to zero. In general, we can only write down the solution in the form of the following limit:

$$x(t) = \lim_{\substack{N\to\infty \\ \max\limits_i \Delta t_i \to 0}} \exp(A_N \Delta t_N) \exp(A_{N-1} \Delta t_{N-1}) \dots \exp(A_1 \Delta t_1)\, x(0),$$

where

$$0 = t_0 < t_1 < \cdots < t_n = t, \quad \Delta t_i = t_i = t_i - t_{i-1}$$

and $A_i = A(\theta_i)$ for some $\theta_i \in [t_{i-1}, t_i]$. In physics literature the limit on the right-hand side of the latter equality is called a *T-exponential* (see, e.g., [55]) and is sometimes denoted by T-exp $\left(\int_0^t A(\tau)\,d\tau\right) x(0)$. Generally speaking,

$$T\text{-}\exp\left(\int_0^t A(\tau)\,d\tau\right) x(0) \neq \exp\left(\int_0^t A(\tau)\,d\tau\right) x(0),$$

since the identity

$$\exp(A)\,\exp(B) = \exp(A + B),$$

which could be used to represent the expression under the limit sign as the exponential of an integral sum, is not valid if $[A, B] \neq 0$, that is, if A does not commute with B. However, Feynman invented a fairly simple notation enabling us to sidestep this difficulty and, in a sense, pay no attention to the fact that the matrices $A(t)$ do not commute with each other for different values of t.

Namely, we equip matrices with indices prescribing their arrangement in products: the greater the index, the further to the left stands the corresponding matrix. In writing, these indices will be placed over the letters denoting matrices; for example,

$$\overset{1}{A}\overset{2}{B} = BA,$$
$$(\overset{1}{A} + \overset{2}{B})^2 = \overset{1}{A}{}^2 + 2\overset{1}{A}\overset{2}{B} + \overset{2}{B}{}^2 = A^2 + B^2 + 2BA,$$
$$(\overset{1}{A} + \overset{2}{B})^3 = A^3 + 3BA^2 + 3B^2A + B^3,$$

etc. In this notation, we see that

$$\exp(\overset{1}{A} + \overset{2}{B}) = \sum \frac{(\overset{1}{A} + \overset{2}{B})^n}{n!} = \exp(\overset{1}{A})\exp(\overset{2}{B}) = \exp(B)\,\exp(A).$$

Note that the order of the factors $\exp(A_j\,\Delta t_j)$ in the product coincides with the order of the points θ_j on the real axis. Hence we can rewrite $x(t)$ as

$$x(t) = \lim_{\substack{N\to\infty \\ \max_i \Delta t_i \to 0}} \exp(\overset{\theta_1}{A_1}\Delta t_1 + \cdots + \overset{\theta_N}{A_N}\Delta t_N)\,x(0).$$

Formally, the argument of the exponential is the integral sum of some integral; let us write down this integral explicitly:

$$x(t) = \exp\left(\int_0^t \overset{\theta}{A}(\theta)\,d\theta\right) x(0).$$

Note that equation (I.1) is a special case of the last equation. Indeed, if $\{A(t)\}$ is a family of commuting matrices then their arrangement in products is irrelevant, and the index θ over $A(\theta)$ may as well be omitted. As a result, we obtain formula (I.1).

At first it may seem that the expression on the right-hand side of the last equation is no more than yet another notation for the T-exponential. But this is a false impression: its meaning can be defined *directly* as the result of substituting the family $\{\overset{t}{A}(t)\}$ of matrices ordered by the parameter t for the argument $y(t)$ into the functional

$$F[y] = \exp\left(\int_0^t y(\tau)\,d\tau\right).$$

Thus we have introduced *indices over operators* (in the example considered these operators are matrices, which, in fact, does not matter). They will be referred to as *Feynman indices*, in honor of their inventor.

Let us sum up their simplest (and evident) properties.

(i) Indices over operators indicate their arrangement in products; namely, an operator with lower index always stands to the right of an operator with greater index. Thus, it is the order relation between the indices that counts, not the values of the indices themselves (which may be changed without affecting the operator expression).

(ii) The mutual order of indices over commuting operators plays no role at all. In particular, if all operators involved in an expression commute with each other, then the indices can be omitted.

1.2 Operators of Quantum Mechanics. Creation and Annihilation Operators

In quantum mechanics, the state of a system of particles is described by the *wave function* Ψ, which is a normalized element of a Hilbert *state space* $\mathcal{H}$; observables correspond to Hermitian operators in $\mathcal{H}$. The quantity

$$\bar{A} = (\Psi, A\Psi),$$

where $(\cdot\,,\,\cdot)$ is the inner product in $\mathcal{H}$, is called the *expectation value* of the observable A in the state Ψ, and

$$(\Delta A)^2 = (\Psi, (A - \bar{A})^2\Psi)$$

is called the *variance* of A in the state Ψ. One says that A has a definite value a in a state Ψ if $A\,\Psi = a\,\Psi$, that is to say, Ψ is an eigenvector of A with eigenvalue a. It is easy to see that in this case $\bar{A} = a$ and $(\Delta A)^2 = 0$. The converse is also true: if $(\Delta A)^2 = 0$, then

$$||A\,\Psi - \bar{A}\Psi||^2 = ((A - \bar{A})\Psi, (A - \bar{A})\Psi) = (\Psi, (A - \bar{A})^2\Psi) = 0$$

(recall that A is a Hermitian operator), so that $A\,\Psi = \bar{A}\Psi$.

Heisenberg's Uncertainty Principle claims that the variances of two observables A and B corresponding to canonically conjugate variables in classical mechanics (such as the position q^i and the dual momentum p_i) satisfy the *uncertainty relation*

$$(\Delta A)\,(\Delta B) \geq h,$$

where h is Planck's constant. In particular, the observables A and B cannot be measured "simultaneously", that is to say, they cannot have definite values in some state Ψ simultaneously. Indeed, the Uncertainty Principle implies that if $\Delta A = 0$, then $\Delta B = \infty$. Thus, A and B have no common eigenvectors. On the other hand, commuting Hermitian operators always have a complete system of common (generalized) eigenvectors. We conclude that A and B do not commute with each other, and the algebra of observables is not commutative.

In the Schrödinger representation the states Ψ are functions of the physical coordinates $x = (x^1, \dots, x^{3N})$, where N is the number of particles in the system, $\Psi = \Psi(x) \in \mathcal{H} = L_2(\mathbb{R}^{3N})$ (we assume that the particles have no internal degrees of freedom). The observables X^i associated with the coordinates x^i are merely the operators of multiplication by x^i, $X^i = x^i$, whereas the observables P_i associated with the momenta p_i are differentiation operators, $p_i = -ih\,\partial/\partial x^i$. Moreover, if some classical variable F has the form of a function $F = F(x, p)$, then the associated quantum observable is given by

$$\hat{F} = F(X, P) \stackrel{\text{def}}{=} F\left(x, -ih\frac{\partial}{\partial x^i}\right). \tag{I.3}$$

In fact, the *quantization procedure* $F \mapsto \hat{F}$ is ambiguous and the last formula is only valid modulo lower-order terms with respect to h. We point out that this formula itself is ambiguous: since the operators x and $-ih\,\partial/\partial x^i$ do not commute with each other, one should fix their ordering, e.g., by equipping them with Feynman indices.

However, the said ambiguity can be "hidden" in some concrete problems. For example, if H is a Hamiltonian of the form

$$H(x, p) = \sum_{i=1}^{N} \frac{p_i^2}{2m} + V(x),$$

then, due to its additive structure, the choice of Feynman indices of

$$x \quad \text{and} \quad -ih\frac{\partial}{\partial x^i}$$

is unimportant, and the energy operator

$$\hat{H} = H(x, p) = -\sum_{i=1}^{N} \frac{h^2}{2m}\Delta_i + V(x)$$

(where Δ_i is the Laplacian w.r.t. the coordinates of the ith particle) is defined uniquely[1].

Relativistic quantum mechanics deals with systems with a variable number of particles. Here the so-called *occupation-number-representation* is convenient. Let us consider it in the simplest model version. Suppose that there is only one type of particle in the system considered, and each particle can occupy one of n distinct basis states[2]. Let us represent the state space $\mathcal{H}$ as the Hilbert sum of one-dimensional subspaces

$$\mathcal{H} = \bigoplus_{j_1,\dots,j_n \geq 0} \mathcal{H}_{j_1,\dots,j_n},$$

[1]The picture would be much more complicated should we consider the quantization in "generalized" coordinates and momenta.

[2]In realistic systems the number of possible states, as a rule, is infinite.

where $\mathcal{H}_{j_1,\ldots,j_n}$ is the space of states such that exactly j_k particles occupy the kth basis state, $k = 1, \ldots, n$. The numbers $j_1, \ldots, j_n$ are referred to as occupation numbers.

The space $\mathcal{H}_0 = \mathcal{H}_{0,\ldots,0}$ is called the *vacuum subspace* and corresponds to the state with no particles at all. The structure of the state space is known to depend on the spin of the particles. Namely, there are two possibilities: if the spin is integral, then the particles obey *Bose–Einstein* statistics, that is, each state can be occupied by an arbitrary number of particles; if the spin is half-integral, then we have *Fermi–Dirac* statistics, that is, at most one particle can occupy each given state. In the first case, the sum is taken over all nonnegative $j_1, \ldots, j_n$, whereas in the second case the sum is finite and extends over $j_i \in \{0, 1\}$. According to the type of statistics, the particles are referred to as *bosons* or *fermions*.

To each of the n basis states there corresponds a *creation operator* a_k^+ and an *annihilation operator* a_k^-, the adjoint of a_k^+. These operators "create" and "destroy" particles in the kth basis state, that is,

$$a_k^+ \mathcal{H}_{j_1\ldots j_k \ldots j_n} \subset \mathcal{H}_{j_1\ldots j_k+1\ldots j_n},$$

$$a_k^- \mathcal{H}_{j_1\ldots j_k \ldots j_n} \subset \mathcal{H}_{j_1\ldots j_k-1\ldots j_n}$$

(it is assumed that $\mathcal{H}_{j_1\ldots j_n} = \{0\}$ provided that at least one of the indices j_k is negative or, in the case of fermions, greater than 1).

The operators a_k^+ and a_k^- satisfy the following commutation relations:

$$[a_k^+, a_l^+]_\pm = [a_k^-, a_l^-]_\pm = 0,$$

$$[a_k^+, a_l^-]_\pm = \delta_{kl}\, I.$$

Here δ_{kl} is the Kronecker delta, I the identity operator, and

$$[A, B]_\pm = AB \pm BA$$

the (anti)commutator of operators A and B (the upper sign is taken for fermions and the lower for bosons).

The *Wick normal form* of an operator A acting on the space $\mathcal{H}$ is its representation in the form

$$A = \sum_{(\alpha,\beta)} c_{\alpha_1,\ldots,\alpha_n\,\beta_1,\ldots,\beta_n} (a_1^+)^{\alpha_1} \ldots (a_n^+)^{\alpha_n} (a_1^-)^{\beta_1} \ldots (a_n^-)^{\beta_n},$$

that is, the representation in which all creation operators in each monomial stand to the left of all annihilation operators. The Wick normal form is very convenient, e.g. for evaluating vacuum expectations (the expectations in the vacuum state $\Psi_0 \in \mathcal{H}_0$, $\|\Psi_0\| = 1$). Indeed, we have

$$\begin{aligned} (\Psi_0, A\Psi_0) &= \sum_{(\alpha,\beta)} c_{\alpha_1,\ldots,\alpha_n\,\beta_1,\ldots,\beta_n} ((a_1^+)^{\alpha_1} \ldots (a_n^+)^{\alpha_n} \Psi_0, (a_1^-)^{\beta_1} \ldots (a_n^-)^{\beta_n} \Psi_0) \\ &= c_{0\ldots 0}, \end{aligned}$$

since all terms with $(\alpha, \beta) \neq (0, 0)$ vanish.

Clearly, any polynomial (series) of creation and annihilation operators can be reduced to the Wick normal form by permuting all creation operators to the left with the help of the commutation relations.

From the viewpoint of noncommutative analysis, the Wick normal form of an operator A is none other that its representation as

$$A = (f(\overset{2}{a_1^+}, \dots, \overset{2}{a_n^+}, \overset{1}{a_1^-}, \dots, \overset{1}{a_1^-}) \overset{\text{def}}{=} f(\overset{2}{a^+}\overset{1}{a^-})$$

(the Feynman indices were assigned taking into account that the creation operators commute with each other, as do the annihilation operators). The function

$$f(z, w) = f(z_1, \dots, z_n, w_1, \dots, w_n) = \sum_{(\alpha,\beta)} c_{\alpha_1,\dots,\alpha_n\,\beta_1,\dots,\beta_n} z_1^{\alpha_1} \dots z_n^{\alpha_n} w_1^{\beta_1} \dots w_n^{\beta_n}$$

is called the Wick symbol of the operator A [11]. It is easy to check that the Wick symbol is unique.

In the case of fermions the creation operators (and the annihilation operators) no longer commute with each other, so they all should get different Feynman indices:

$$A = f(\overset{2n}{a_1^+}, \dots, \overset{n+1}{a_n^+}, \overset{n}{a_1^-}, \dots, \overset{1}{a_n^-}).$$

The problem of calculating the Wick normal form of an operator acting on $\mathcal{H}$ will be considered in Subsection 1.1 of Chapter II.

1.3 Differential and Integral Operators

The theory of linear partial differential equations deals with differential operators of the form

$$P = \sum_{|\alpha| \leq m} a_\alpha(x) \left(\frac{\partial}{\partial x}\right)^\alpha,$$

where $x \in \mathbb{R}^n$, $\alpha = (\alpha_1, \dots, \alpha_n)$ is a multi-index, and m is the order of P. Obviously, the operator P can be interpreted as a function of the operators x^j and $\partial/\partial x^j$, $j = 1, , \dots, n$:

$$P = f\left(\overset{2}{x}, \overset{1}{\frac{\partial}{\partial x}}\right),$$

where

$$f(x, \xi) = \sum_{|\alpha| \leq m} a_\alpha(x)\, \xi^\alpha.$$

For technical reasons (since $-i\partial/\partial x$ is a self-adjoint operator in $L_2(\mathbb{R}^n)$), the operator P is usually represented in a somewhat different form, namely

$$P = g\left(\overset{2}{x}, -i\overset{1}{\frac{\partial}{\partial x}}\right),$$

where

$$g(x, p) = \sum_{|\alpha|\leq m} a_\alpha(x)\,(ip)^\alpha.$$

Thus, a differential operator is a function of the Feynman-ordered operators $\overset{2}{x}$ and $-i\overset{1}{\partial/\partial x}$ with symbol a polynomial in p.

A solution (or almost-solution) of a differential equation

$$Pu = f$$

can often be represented in the form of some integral operator applied to f. For example, if P is an elliptic operator, then the solution modulo smooth functions can be represented via a pseudodifferential (singular integral) operator,

$$u(x) = \int K(x, x-y)\, f(y)\, dy$$

(see, for example, [79]). If P is a hyperbolic operator, then the solution can be represented via a Fourier integral operator (see [81], [137], and others). Let us show, formally for now, that the pseudodifferential operator can be represented as a function of the Feynman-ordered operators $\overset{2}{x}$ and $-i\overset{1}{\partial/\partial x}$. To this end, we make the change of variables $y = x + \tau$ in the integral and obtain

$$u(x) = \int K(x, -\tau)\, f(x+\tau)\, d\tau = \int K(x, -\tau) e^{\tau\,\partial/\partial x} f(x)\, d\tau,$$

where $e^{\tau\,\partial/\partial x}$ is the operator of translation by τ, which can be defined as the solution U of the operator Cauchy problem

$$\frac{\partial U}{\partial \tau} = \frac{\partial U}{\partial x}, \qquad U|_{\tau=0} = \mathrm{id}.$$

Set

$$H(x, p) = \int K(x, -\tau) e^{i\tau p}\, d\tau.$$

Then

$$u(x) = \left(\frac{1}{2\pi}\right)^n \int \tilde{H}(x,\tau) e^{i\tau(-i\,\partial/\partial x)} d\tau\, f(x) = H\left(\overset{2}{x}, -i\overset{1}{\frac{\partial}{\partial x}}\right) f(x),$$

where $\tilde{H} = (2\pi)^n K(x,-\tau)$ is the Fourier transform of $H(x,p)$ with respect to the variable p. The properties of the kernel $K(x, x-y)$ imply that $H(x,p)$ is a homogeneous function of p smooth for $p \neq 0$. We also have

$$H\left(\overset{2}{x}, -i\overset{1}{\frac{\partial}{\partial x}}\right) f(x) = \left(\frac{1}{2\pi}\right)^n \int H(x,\xi) e^{i\xi(x-y)} f(y)\, dy\, d\xi.$$

Fourier integral operators admit a similar representation.

Let us illustrate this fact on a simple example. Suppose that the Fourier integral operator has the form

$$\hat{\Phi} f = \int e^{iS(x,p)} a(x,p) \tilde{f}(p)\, dp, \tag{I.4}$$

where $\tilde{f}(p)$ is the Fourier transformation of the function $f(x)$ (such operators give, in particular, an almost-solution of the Cauchy problem

$$\begin{cases} i\dfrac{\partial u}{\partial t} = H(\overset{2}{x}, -i\overset{1}{\dfrac{\partial}{\partial x}})\, u, \\ u|_{t=0} = f(x), \end{cases}$$

for small values of t; here $H(\overset{2}{x}, -i\overset{1}{\partial/\partial x})$ is a first-order pseudodifferential operator).

The operator (I.4) can be represented in the form

$$\hat{\Phi} f = \int K(x, x-y)\, f(y)\, dy,$$

where

$$K(x,z) = \bar{F}_{p\to z}\{e^{i\Phi(x,p)} a(x,p)\},$$

$\Phi(x,p) = S(x,p) - xp$, and $\bar{F}_{p\to z}$ is the inverse Fourier transform. Now computations similar to those above show that

$$\hat{\Phi} f = e^{i\Phi(\overset{2}{x}, -i\overset{1}{\partial/\partial x})} a\left(\overset{2}{x}, -i\overset{1}{\frac{\partial}{\partial x}}\right) f.$$

1.4 Problems of Perturbation Theory

Let A and B be two noncommuting operators. We will consider a function

$$f(A+\varepsilon B)$$

of the operator $A+\varepsilon B$, where ε is a small parameter, and try to expand it in powers of ε. This is a typical problem of perturbation theory; the operator A is considered as the operator of the nonperturbed problem and εB as a small perturbation.

For example, consider the operator Cauchy problem

$$\frac{du}{dt} = (A+\varepsilon B)\,u, \quad u|_{t=0} = E \tag{I.5}$$

(here E is the identity operator). Suppose that the solution of the unperturbed problem ($\varepsilon = 0$) is known. Then the "corrections" for small ε can be found by expanding the function $f(A+\varepsilon B)$, where $f(x) = e^{tx}$, in powers of ε.

If the operators A and B commute, the solution of this problem is well known. It is given by the Taylor formula

$$f(A+\varepsilon B) = f(A) + \varepsilon B f'(A) + \frac{\varepsilon^2}{2} B^2 f''(A) + \cdots .$$

However, this equation fails if $[A, B] \neq 0$. Indeed, consider the simplest example in which $f(x) = x^2$. We have

$$f(A+\varepsilon B) = A^2 + \varepsilon(AB + BA) + \varepsilon^2 B^2,$$

whereas

$$f(A) + \varepsilon B f'(A) + \frac{\varepsilon^2}{2} B^2 f''(A) = A^2 + 2\varepsilon BA + \varepsilon^2 B^2,$$

and these two expressions differ by $\varepsilon[A, B]$.

Hence there is the problem of giving a counterpart of the Taylor expansion for noncommuting A and B.

Computing the derivative

$$\frac{d}{d\varepsilon} f(A+\varepsilon B)$$

would be a first step in that direction (this provides the desired expansion with the accuracy $O(\varepsilon^2)$); let us compute this derivative for the particular case in which $f(x) = (\lambda - x)^{-1}$ (that is, we consider the perturbation theory for the resolvent $R_\lambda(A)$).

The following resolvent formula is valid (see, e.g., [197])

$$(\lambda - C)^{-1} - (\lambda - A)^{-1} = (\lambda - C)^{-1}(C - A)(\lambda - A)^{-1}$$

(this can easily be checked by multiplying both sides of the last equation by $(\lambda - C)$ on the left and by $(\lambda - A)$ on the right). By inserting $C = A + \varepsilon B$ in this formula, we obtain

$$f(A+\varepsilon B) - f(A) = (\lambda - A - \varepsilon B)^{-1} \varepsilon B\, (\lambda - A)^{-1}.$$

Hence, it is clear that

$$\left[\frac{d}{d\varepsilon} f(A+\varepsilon B)\right]\bigg|_{\varepsilon=0} = (\lambda - A)^{-1} B (\lambda - A)^{-1} = \frac{\overset{2}{B}}{(\lambda - \overset{3}{A})(\lambda - \overset{1}{A})}.$$

Obviously, the same "naive" method permits us to calculate higher-order derivatives with respect to ε; however, we refrain from doing so, since our computation uses the particular form of the function f heavily and is by no means universal. The general formulas of perturbation theory for $f(A+\varepsilon B)$ will be given in Section 3 and now we consider yet another example.

Let us compute the expansion in powers of ε of the exponential $e^{A+\varepsilon B}$ (recall that this expansion is the perturbation theory series for the solution of the Cauchy problem (I.5) at $t = 1$).

For simplicity, we assume that A and B are bounded operators on a normed space. Thus, we intend to find the coefficients $P_1, P_2, \ldots$ of the expansion

$$e^{A+\varepsilon B} = e^A + \varepsilon P_1 + \varepsilon^2 P_2 + \cdots. \tag{I.6}$$

We will use the *extraction formula* for T-exponentials [54]:

$$T\text{-}\exp\left[\int_0^t (A(\tau) + B(\tau))\,d\tau\right] = T\text{-}\exp\left(\int_0^t A(\tau)\,d\tau\right) T\text{-}\exp\left(\int_0^t C(\tau)\,d\tau\right),$$

where $C(\tau)$ is the operator family given by the formula

$$C(t) = \left[T\text{-}\exp\left(\int_0^t A(\tau)\,d\tau\right)\right]^{-1} B(t)\, T\text{-}\exp\left(\int_0^t A(\tau)\,d\tau\right),$$

that is, $C(t)$ is the conjugate of $B(t)$ by the operator $T\text{-}\exp\left(\int_0^t A(\tau)\,d\tau\right)$.

Proof. The proof is quite simple. For brevity, rewrite the desired formula as

$$U_{A+B}(t) = U_A(t) U_C(t).$$

Both sides are equal to the identity operator if $t = 0$. We can differentiate both sides with respect to t, using the definition of a T-exponential (see Subsection 1.1). We obtain

$$\frac{\partial}{\partial t}(U_{A+B}(t)) = (A(t) + B(t)) U_{A+B}(t)$$

and

$$\frac{\partial}{\partial t}(U_A(t)U_C(t)) = A(t)U_A(t)U_C(t) + U_A(t)C(t)U_C(t) = (A(t)+B(t))U_A(t)U_C(t),$$

since

$$C(t) = U_A(t)^{-1} B(t) U_A(t).$$

This implies immediately that $U_{A+B}(t) = U_A(t)U_C(t)$ for all t. □

We choose $A(t)$ and $B(t)$ to be constant families of operators, $A(t) \equiv A$ and $B(t) \equiv \varepsilon B$, and obtain the formula

$$e^{A+\varepsilon B} = e^A T\text{-}\exp\left(\varepsilon \int_0^1 (C(t)\,dt\right),$$

where

$$C(t) = e^{-tA} B e^{tA}.$$

As in Subsection 1.1, we rewrite the T-exponential using Feynman indices and expand it into the Taylor series in powers of ε:

$$\begin{aligned} T\text{-}\exp\left(\varepsilon \int_0^1 C(\tau)\,d\tau\right) &= \exp\left(\varepsilon \int_0^1 \overset{\tau}{C}(\tau)\,d\tau\right) \\ &= 1 + \varepsilon \int_0^1 \overset{\tau}{C}(\tau)\,d\tau + \frac{\varepsilon^2}{2}\left(\int_0^1 \overset{\tau}{C}(\tau)\,d\tau\right)^2 + \cdots. \end{aligned}$$

The coefficient of ε on the right-hand side is equal to

$$\int_0^1 C(\tau)\,d\tau = \int_0^1 e^{-\tau A} B e^{\tau A}\,d\tau$$

(since in the coefficient of ε all $C(\tau)$ enter additively and are not multiplied by one another, their Feynman indices play no role and can be omitted).

Let us calculate the coefficient of ε^2. We have

$$\left(\int_0^1 \overset{\tau}{C}(\tau)\,d\tau\right)^2 = \int_0^1 \overset{\tau}{C}(\tau)\,d\tau \int_0^1 \overset{\theta}{C}(\theta)\,d\theta = \int_0^1\int_0^1 \overset{\tau}{C}(\tau)\overset{\theta}{C}(\theta)\,d\theta d\tau.$$

The order of factors in the integrand is governed by the sign of the difference $\tau - \theta$: we have

$$\overset{\tau}{C}(\tau)\overset{\theta}{C}(\theta) = C(\tau)\,C(\theta)$$

for $\tau > \theta$, and

$$\overset{\tau}{C}(\tau)\overset{\theta}{C}(\theta) = C(\theta)\,C(\tau)$$

for $\tau \le \theta$, that is, the factor with the smaller argument always stands to the right of the other factor. We divide the integration square $[0, 1] \times [0, 1]$ into two triangles $\tau > \theta$ and $\tau \le \theta$ and note that both of them give the same contribution to the integral. We retain only one of these triangles. Then the factor $1/2$ cancels out, and we obtain the following expression for the coefficient of ε^2:

$$\int_0^1 \left(\int_0^\tau C(\tau)C(\theta)\, d\theta \right) d\tau = \int_0^1 \left(\int_0^\tau e^{-\tau A} B e^{(\tau-\theta)A} B e^{\theta A} d\theta \right) d\tau.$$

Finally, we find that

$$e^{A+\varepsilon B} = e^A + \varepsilon \int_0^1 e^{(1-\tau)A} B e^{\tau A} d\tau + \varepsilon^2 \int_0^1 \left(\int_0^\tau e^{(1-\tau)A} B e^{(\tau-\theta)A} B e^{\theta A} d\theta \right) d\tau + \cdots.$$

Thus, we have calculated the coefficients P_1 and P_2 of the expansion (I.6) for $e^{A+\varepsilon B}$. Using Feynman indices again, we can write

$$P_1 = \overset{2}{B} \int_0^1 e^{(1-\tau)\overset{3}{A}+\tau \overset{1}{A}} d\tau,$$

$$P_2 = \overset{2}{B}\overset{4}{B} \int_0^1 \left(\int_0^\tau e^{(1-\tau)\overset{5}{A}+(\tau-\theta)\overset{3}{A}+\theta \overset{1}{A}} d\theta \right) d\tau.$$

It is easy to predict the formulas for P_3, P_4, etc., but we leave that to the reader.

We remark that the choice of two above examples is not accidental. Actually, the exponent e^{iAt} plays an important role in the definition of functions of noncommuting operators (see Subsection 2.4 below) for tempered symbols. Similarly, as is well known, the resolvent $R_\lambda(A)$ of the operator A can be used for the definition of $f(A)$ for symbols $f(\lambda)$ analytic in a neighborhood of the spectrum of the operator A. These examples are also of use when considering Weyl quantization (see Subsection 2.6 of this Chapter).

1.5 Multiplication Law in Lie Groups

The extraction formula can be read from right to left:

$$T\text{-}\exp\left(\int_0^t A(\tau)\,d\tau\right) T\text{-}\exp\left(\int_0^t B(\tau)\,d\tau\right) T\text{-}\exp\left(\int_0^t C(t)\,dt\right),$$

where

$$C(t) = A(t) + T\text{-}\exp\left(\int_0^t A(\tau)\,d\tau\right) B(t) \left[T\text{-}\exp\left(\int_0^t A(\tau)\,d\tau\right)\right]^{-1}$$

(we have changed the notation slightly so as to show some respect for the alphabetic ordering). In this form it has the following meaning: the product of two T-exponentials is again a T-exponential. Furthermore, if

$$[A(t), B(t')] = A(t)B(t') - B(t')A(t) = 0$$

for all t and t', then the exponent $C(t)$ has the form $C(t) = A(t) + B(t)$ (indeed, we can permute $B(t)$ and $T\text{-}\exp\left(\int_0^t A(\tau)\,d\tau\right)$, which follows easily by considering the passage to the limit (see Subsection 1.1) or the corresponding Cauchy problem).

Now let $A(t)$ and $B(t)$ be constant families of operators, $A(t) \equiv A$ and $B(t) \equiv B$. Then for $t = 1$ we obtain

$$e^A e^B = T\text{-}\exp\left(\int_0^1 C(\tau)\,d\tau\right),$$

where

$$C(t) = A + e^{tA} B e^{-tA}.$$

Thus, the product of two "common" exponentials of operators is represented as a T-exponential. There arises the natural question whether this product can be represented as a "common" exponential, that is to say, whether one can find an operator C such that

$$e^{\overset{2}{A}+\overset{1}{B}} = e^C. \tag{I.7}$$

This is an important problem in the theory of Lie groups. It was partially solved already in the XIX century (see [19]). Namely, it was shown that the element C can be expressed as the series

$$C = A + B + \frac{1}{2}[A, B] + \cdots$$

convergent for small $||A||$ and $||B||$, where the dots stand for the sum of commutators of the operators A and B of order ≥ 2 (the order of a commutator is the number of left (or right) commutation brackets it contains).

Let A and B be elements of some matrix Lie algebra; then e^A and e^B are elements of the corresponding local Lie group realized by matrices, and the above relations mean that the multiplication law in a Lie group is uniquely determined by the commutation operation in its Lie algebra. This is just the statement of the famous Campbell–Hausdorff theorem.

An explicit expression for all terms of the series was found about half a century later by E. B. Dynkin [43], who showed that

$$C = \sum_{m=1}^{\infty} \frac{(-1)^{m-2}}{m} \sum_{k_i+l_i \geq 1, k_i, l_i \geq 0} \frac{[A^{k_1}, B^{l_1}, \ldots, A^{k_m}, B^{l_m}]}{k_1! l_1! \ldots k_m! l_m!},$$

where

$$[S_1, \ldots, S_N] \stackrel{\text{def}}{=} \frac{1}{N}[S_1, [S_2 \ldots [S_{N-1}, S_N] \ldots].$$

In Section 4 we obtain in an elementary way, in the framework of noncommutative analysis, a closed formula expressing the operator C via A, B and the commutation operation.

Let us use the considered problem to make a comment on the notation used. We seek an element C such that "$C = \ln D$, where $D = e^{\overset{2}{A}+\overset{1}{B}}$". It would be desirable to convert the phrase in quotes into a formula. Unfortunately, we cannot write

$$C = \ln(e^{\overset{2}{A}+\overset{1}{B}}),$$

since this means the following: "take the function $f(x, y) = \ln(e^{x+y})$ and substitute $x \mapsto \overset{2}{A}$, $y \mapsto \overset{1}{B}$ into it". Taking into account that $\ln(e^{x+y}) = x + y$, we would obtain $C = A + B$. In order to solve the arising problem we extend our notation by introducing the so-called *autonomous brackets* $[\![\ ]\!]$. We write

$$C = \ln([\![e^{\overset{2}{A}+\overset{1}{B}}]\!]);$$

the meaning of this notation is as follows: "calculate the operator in the brackets $[\![\ ]\!]$ and forget about Feynman indices occurring inside $[\![\ ]\!]$; instead, use the resultant operator in the subsequent computations as an indivisible entity".

The introduction of autonomous brackets is motivated by the simple observation that operator expressions involve, in addition to traditional arithmetic operations and function evaluation, only one new operation, namely, substitution of operators (equipped with Feynman indices) instead of numerical arguments. And we do not have any convenient means to denote recursive application of this operation. Recall that the order of computation in arithmetic expressions is governed by brackets. We just introduce here a *new type* of brackets, responsible for substitution of operators.

1.6 Eigenfunctions and Eigenvalues of the Quantum Oscillator

Let us now use a well-known problem of quantum mechanics to show how the above notation works.

Let $\hat{H}$ be the energy operator of the one-dimensional quantum oscillator,

$$\hat{H} = \frac{1}{2}\left[\left(-ih\frac{\partial}{\partial x}\right)^2 + \omega^2 x^2\right],$$

where h is the Planck constant, $\omega > 0$ the frequency of the oscillator, and $x \in \mathbb{R}$. Consider the eigenvalue problem for $\hat{H}$ in the space $L_2(\mathbb{R}^1)$:

$$\hat{H}\Psi(x) = E\Psi(x), \quad \Psi \in L_2(\mathbb{R}^1).$$

One should find the values of E for which this equation has nontrivial (nonzero) solutions Ψ.

It is easy to check that $\hat{H}$ can be represented in the form

$$\hat{H} = h\omega(a^+a^- + 1/2),$$

where

$$a^{\pm} = (2h\omega)^{-1/2}\left(-ih\frac{\partial}{\partial x} \pm i\omega x\right)$$

are the *creation and annihilation operators* (it is not mere chance that this term coincides with the one already introduced in Subsection 1.2; this is related to the decomposition of a free electromagnetic field into oscillators, frequently used in quantum theory).

Thus, our problem is reduced to the eigenvalue problem for the operator $a^+a^- = \overset{2}{a^+}\overset{1}{a^-}$:

$$\overset{2}{a^+}\overset{1}{a^-}\Psi(x) = \lambda\Psi(x).$$

We seek the solution in the form

$$\Psi(x) = G(\overset{2}{a^+}, \overset{1}{a^-}\overset{1}{a^-})v(x), \tag{I.8}$$

where $v(x)$ is an arbitrary function and $G(\xi, y)$ is an unknown symbol to be defined. It suffices to require that the product of the operators $\overset{2}{a^+}\overset{1}{a^-}$ and $G(\overset{2}{a^+}, \overset{1}{a^-})$ be equal to $\lambda G(\overset{2}{a^+}, \overset{1}{a^-})$. Using the introduced notation, we can rewrite the cited product in any of the following forms:

$$\begin{aligned}
[\![\overset{2}{a^+}\overset{1}{a^-}]\!]\,[\![G(\overset{2}{a^+}, \overset{1}{a^-})]\!] &= \overset{4}{a^+}\overset{3}{a^-}G(\overset{2}{a^+}, \overset{1}{a^-}) = \overset{2}{a^+}\overset{1}{a^-}[\![G(\overset{2}{a^+}, \overset{1}{a^-})]\!] \\
&= \overset{3}{a^+}\overset{2}{a^-}{}_{1}[\![G(\overset{2}{a^+}, \overset{1}{a^-})]\!] = BA,
\end{aligned}$$

where $B = a^+a^-$ and $A = G(\overset{2}{a^+}, \overset{1}{a^-})$. (The index over the left autonomous bracket in the last case is the Feynman index to be assigned to the operator obtained by computing the expression inside the autonomous brackets).

We obtain the equation

$$a^+a^-[\![G(\overset{2}{a^+}, \overset{1}{a^-})]\!] = \lambda G(\overset{2}{a^+}, \overset{1}{a^-}) \tag{I.9}$$

for the operator $G(\overset{2}{a^+}, \overset{1}{a^-})$.

Our plan is to reduce this equation to an equation with respect to the symbol $G(\xi, y)$. To this end, we should represent the left-hand side as a function of $\overset{2}{a^+}$ and $\overset{1}{a^-}$. Should a^+ and a^- be arbitrary operators, such a representation would be impossible. However, a^+ and a^- satisfy the commutation relation

$$[a^+, a^-] \overset{\text{def}}{=} a^+a^- - a^-a^+ = -1,$$

which enables us to compute the symbol of the product on the left-hand side of (I.9).

Compute first the product $W_- = a^-[\![G(\overset{2}{a^+}, \overset{1}{a^-})]\!]$. Here we apply a trick standard in "operator arithmetic". Clearly, we have

$$W_- = \overset{3}{a^-}G(\overset{2}{a^+}, \overset{1}{a^-}),$$

and the problem is to permute $\overset{2}{a^+}$ to the last place so as to identify the arguments $\overset{3}{a^-}$ and $\overset{1}{a^-}$. Note that if $G(\gamma, y)$ were a linear function, this problem could be solved immediately by applying the commutation relation, which can be rewritten in the form

$$\overset{2}{a^+}(\overset{1}{a^-} - \overset{3}{a^-}) = 1.$$

If $G(\xi, y)$ is a polynomial in ξ, one could apply induction on the order of $G(\xi, y)$. In fact, we can avoid this cumbersome procedure and obtain the result for general symbols as follows:

$$\begin{aligned} W_- &= \overset{3}{a^-}G(\overset{4}{a^+}, \overset{1}{a^-}) + \overset{3}{a^-}(G(\overset{2}{a^+}, \overset{1}{a^-}) - G(\overset{4}{a^+}, \overset{1}{a^-})) \\ &= \overset{3}{a^-}G(\overset{4}{a^+}, \overset{1}{a^-}) + \overset{3}{a^-}(\overset{2}{a^+} - \overset{4}{a^+})\frac{\delta G}{\delta \xi}(\overset{2}{a^+}, \overset{4}{a^+}, \overset{1}{a^-}), \end{aligned}$$

where

$$\frac{\delta G}{\delta \xi}(\xi, \eta, y) = \frac{G(\xi, y) - G(\eta, y)}{\xi - \eta}$$

is the difference derivative of $G(\xi, y)$ by ξ. The first term is already in the desired form, and we transform the second one by changing the Feynman indices:

$$\overset{3}{a^-}(\overset{2}{a^+} - \overset{4}{a^-})\frac{\delta H}{\delta \xi}(\overset{2}{a^+}, \overset{4}{a^+}, \overset{1}{a^-}) = \overset{3}{a^-}(\overset{2}{a^+} - \overset{4}{a^-})\frac{\delta H}{\delta \xi}(\overset{1}{a^+}, \overset{5}{a^+}, \overset{0}{a^-})$$

$$= \overset{3}{[\![}\overset{3}{a^-}(\overset{2}{a^+} - \overset{4}{a^+})]\!]\frac{\delta H}{\delta \xi}(\overset{1}{a^+}, \overset{5}{a^+}, \overset{0}{a^-})$$

(we can insert autonomous brackets since none of the operators outside them have a Feynman number in the interval $[2, 4]$). The commutation relation says that the operator in the autonomous brackets is equal to 1, and the last expression takes the form

$$\frac{\delta G}{\delta \xi}(\overset{1}{a^+}, \overset{5}{a^+}, \overset{0}{a^-}) = \frac{\delta G}{\delta \xi}(\overset{2}{a^+}, \overset{2}{a^+}, \overset{1}{a^-}) = \frac{\delta G}{\delta \xi}(\overset{2}{a^+}, \overset{1}{a^-})$$

(the difference derivative becomes the usual derivative on the diagonal $\xi = \eta$). Finally, we obtain

$$W_- = H(\overset{2}{a^+}, \overset{2}{a^-}),$$

where

$$H(\xi, y) = yG(\xi, y) + \frac{\partial G}{\partial \xi}(\xi, y).$$

We can now compute the product

$$W_+ = a^+ a^- [\![G(\overset{2}{a^+}, \overset{1}{a^1})]\!] = a^+ W_- = a^+ [\![H(\overset{2}{a^+}, \overset{1}{a^-}]\!].$$

In fact, this is trivial:

$$W_+ = a^+ [\![H(\overset{2}{a^+}, \overset{1}{a^-})]\!] = \overset{3}{a^+} H(\overset{2}{a^+}, \overset{1}{a^-}) = \overset{2}{a^+} H(\overset{2}{a^+}, \overset{1}{a^-}) = H_+(\overset{2}{a^+}, \overset{1}{a^-}),$$

where

$$H_+(\xi, y) = \xi H(\xi, y).$$

(The change of Feynman indices is valid, since their order over noncommuting operators was preserved.)

Let us substitute the result of our computation into the left-hand side of (I.9) and equate the symbols of operators on both sides of the resulting equation. For $G(\xi, y)$ we obtain the equation

$$\xi \left(y + \frac{\partial}{\partial \xi} \right) G(\xi, y) = \lambda G(\xi, y).$$

The remaining part of the solution is, in fact, purely technical. We have obtained an ordinary differential equation, whose general solution has the form

$$G(\xi, y) = e^{-\xi y} \xi^{\lambda} c(y),$$

where $c(y)$ is an arbitrary function. We substitute $\overset{2}{a^+}$ instead of ξ and $\overset{1}{a^-}$ instead of y into $G(\xi, y)$ and substitute the resulting operator in (I.8):

$$\Psi(x) = e^{-\overset{2}{a^+}\overset{1}{a^-}}(\overset{2}{a^+})^\lambda c(\overset{1}{a^-})v(x).$$

Extracting $(a^+)^\lambda$, we obtain

$$\Psi(x) = (a^+)^\lambda \varphi(x),$$

where

$$\varphi(x) = e^{-\overset{2}{a^+}\overset{1}{a^-}} c(\overset{1}{a^-})v(x).$$

Although $v(x)$ is an arbitrary function the same is not true of $\varphi(x)$. In fact, we intend to show that the operator $\exp(-\overset{2}{a^+}\overset{1}{a^-})c(\overset{1}{a^-})$ is one-dimensional[3]. Indeed, apply a^- to $\varphi(x)$:

$$a^-\varphi(x) = a^-[\![e^{-\overset{2}{a^+}\overset{1}{a^-}} c(\overset{1}{a^-})]\!]v(x).$$

We compute the operator on the right-hand side in the same way as W_- and obtain

$$a^-[\![e^{-\overset{2}{a^+}\overset{1}{a^-}} c(\overset{1}{a^-})]\!] = f(\overset{2}{a^+}, \overset{1}{a^-}),$$

where

$$f(\xi, y) = ye^{-\xi y}c(y) + \frac{\partial}{\partial \xi} e^{-\xi y}c(y) = 0.$$

Henceforth,

$$a^-\varphi(x) = -i(2\omega h)^{-1/2}\left(h\frac{\partial}{\partial x} + \omega x\right)\varphi(x) = 0,$$

so that

$$\varphi(x) = \text{const} \cdot e^{-x^2\omega/ih}.$$

This function decays together with all its derivatives more rapidly then any polynomial, so for each integer nonnegative λ the function $\Psi(x)$ lies in $L_2(\mathbb{R}^2)$.

Thus the function

$$\Psi_n(x) = (a^+)^n e^{-x^2\omega/ih} = \left(\frac{-i}{\sqrt{i\omega h}}\right)^n \left(\omega x - h\frac{\partial}{\partial x}\right)^n e^{-x^2\omega/ih}$$

is an eigenfunction of a^+a and hence of $\hat{H}$, and the corresponding eigenvalue is

$$E_n = \omega h(n + 1/2).$$

This is a classical result in quantum mechanics.

Remark I.1 How to prove that there are no other eigenvalues? The simplest way is to prove that the system of functions obtained is complete in $L_2(\mathbb{R}^2)$; however, one can also prove directly that the function $(a^+)^\lambda\varphi(x)$ does not lie in $L_2(\mathbb{R}^1)$ for noninteger λ.

[3]That is, its range is one-dimensional.

1.7 T-Exponentials, Trotter Formulas, and Path Integrals

Now that we are acquainted with autonomous brackets, let us return to the T-exponentials introduced in Subsection 1.1.

Consider the Schrödinger equation

$$-i\frac{\partial\psi}{\partial t} + H\left(\overset{2}{x}, -i\overset{1}{\frac{\partial}{\partial x}}, t\right)\psi = 0, \quad x \in \mathbb{R}^n, \quad t \in \mathbb{R}^1,$$

with time-dependent energy operator (we assume a system of units in which Planck's constant h equals 1). Its solutions can be expressed via the T-exponential,

$$\psi(x,t) = \exp\left(-i\int_0^t \overset{\tau}{[\![} H\left(\overset{2}{x}, -i\overset{1}{\frac{\partial}{\partial x}}, \tau\right)]\!] d\tau\right)\psi_0(x),$$

where $\psi_0(x)$ is the initial value.

The integrand in the exponent contains autonomous brackets. It would be interesting to remove them, so let us try it (as we shall see, the result is quite intriguing). To this end, consider first the T-exponential

$$F(t) = \exp\int_0^t \overset{\tau}{[\![} f(\overset{1}{A}, \overset{2}{B}, \tau)]\!] d\tau,$$

where A and B are bounded operators (e.g., matrices). By definition,

$$F(t) = \lim_{\substack{N\to\infty \\ \Delta t = \max_i \Delta t_i \to 0}} \exp([\![f(\overset{1}{A}, \overset{2}{B}, \tau_N)]\!]\Delta t_N)\dots\exp([\![f(\overset{1}{A}, \overset{2}{B}, \tau_1]\!]\Delta t_1),$$

where $0 = t_0 < t_1 < \dots < t_N = t$, $\Delta t_i = t_i - t_{i-1}$, and $\tau_i \in [t_{i-1}, t_i]$. However, we have

$$\begin{aligned}\exp([\![f(\overset{1}{A}, \overset{2}{B}, \tau_i)]\!]\Delta t_i) &= 1 + f(\overset{1}{A}, \overset{2}{B}, \tau_i)\Delta t_i + O(\Delta t^2)\\ &= \exp(f(\overset{1}{A}, \overset{2}{B}, \tau_i)\Delta t_i) + O(\Delta t^2).\end{aligned}$$

The remainder terms $O(\Delta t^2)$ result in $O(\Delta t)$, which vanishes as $\Delta t \to 0$ in the overall product. Hence the autonomous brackets can be removed,

$$\begin{aligned}&\exp\int_0^t \overset{\tau}{[\![} f(\overset{1}{A}, \overset{2}{B}, \tau)]\!] d\tau\\ &= \lim_{\substack{N\to\infty \\ \Delta t = \max_i \Delta t_i \to 0}} \exp(f(\overset{2N-1}{A}, \overset{2N}{B}, \tau_N)\Delta t_N)\dots\exp(f(\overset{1}{A}, \overset{2}{B}, \tau_1)\Delta t_1).\end{aligned}$$

It is natural to denote the limit on the right-hand side of the last equation by

$$\exp\left(\int_0^t f(\overset{\tau}{A}, \overset{\tau+0}{B}, \tau)\,d\tau\right),$$

that is, we have

$$\exp\int_0^t [\![f(\overset{1}{A}, \overset{2}{B}, \tau)]\!]\,d\tau = \exp\left(\int_0^t f(\overset{\tau}{A}, \overset{\tau+0}{B}, \tau)\,d\tau\right),$$

where the "+0" over B stands there to indicate that at each time τ the operator B acts *after* the operator A. Hence, the autonomous brackets can safely (and almost without trace) be removed in T-exponentials.

The $O(\Delta t^2)$ argument is no longer usable with functions of unbounded operators. However, under appropriate functional-analytic conditions (not to be discussed here) the conclusion remains the same.

Consider the simplest case in which

$$f(x, y, t) = x + y.$$

Then

$$\exp\left(\int_0^t [\![f(\overset{1}{A}, \overset{2}{B}, \tau)]\!]\,d\tau\right) = e^{[\![A+B]\!]t},$$

and, by setting $t_i = t/N$, $i = 0, \dots, N$, we obtain the *Trotter product formula* [180]

$$e^{[\![A+B]\!]t} = \lim_{N\to\infty} \underbrace{e^{(t/N)\,B} e^{(t/N)\,A} \dots e^{(t/N)\,B} e^{(t/N)\,A}}_{N \text{ pairs of factors}}.$$

We see that the rule of removing autonomous brackets in T-exponentials is merely a generalization of the Trotter formula.

Let us return to the T-exponential solution of the Schrödinger equation. According to our reasoning, we get

$$\psi(x, t) = \exp\left(-i\int_0^t H\left(\overset{\tau+0}{x}, -i\overset{\tau}{\frac{\partial}{\partial x}}, \tau\right) d\tau\right)\psi_0(x)$$

or

$$\psi(x, t) = \lim_{\substack{N\to\infty \\ \Delta t = \max_i \Delta t_i \to 0}} \left\{\prod_{k=1}^{N} \exp\left(-iH\left(\overset{2k}{x}, -i\overset{2k-1}{\frac{\partial}{\partial x}}, \tau_k\right)\Delta t_k\right)\psi_0(x)\right\}.$$

The operators included in the product can be expressed by the formula

$$\exp\left(-iH\left(\overset{2k}{x}, -i\overset{2k-1}{\frac{\partial}{\partial x}}, \tau_k\right)\Delta t_k\right)u(x)$$
$$= \frac{1}{(2\pi)^n}\int\limits_{\mathbb{R}^2} e^{ip(x-y)}e^{-iH(x,p,\tau_k)\Delta t_k}u(y)\,dy\,dp,$$

(this is one of the usual representations of a Fourier integral operator, see Subsection 1.3). Substituting this expression into the preceding equation, we obtain

$$\psi(x,t) = \lim_{\substack{N\to\infty \\ \Delta t=\max\limits_i \Delta t_i\to 0}} \int\limits_{\mathbb{R}^{2Nn}} \exp\left\{i\left(\sum_{k=0}^{N-1} p_k(x_{k+1}-x_k)\right.\right.$$
$$\left.\left.-H(x_{k+1}, p_k, \tau_k)\Delta t_k\right)\right\}\psi_0(x_0)\prod_{k=0}^{N-1}\frac{dx_k\,dp_k}{(2\pi)^n},$$

where x_k and p_k are n-vectors, the expression $p_k(x_{k+1}-x_k)$ is the inner product of p_k by $(x_{k+1}-x_k)$, and $x_N = x$.

Let us interpret (x_k, p_k) as the point at time $t = t_k$ of some trajectory $(q(t), p(t))$ in the phase space. Then, as $\Delta t_k \to 0$, we have

$$x_{k+1} - x_k = \dot{q}(\tau_k)\Delta t_k + O(\Delta t_k)^2,$$

and the expression in the exponent can be considered as a Riemann sum for the action integral

$$\int_0^t [p(\tau)\dot{q}(\tau) - H(q(\tau), p(\tau), \tau)]\,d\tau.$$

Hence, the limit on the right-hand side of the obtained equation is denoted by

$$\psi(x,t) = \iint [dq]\left[\frac{dp}{2\pi}\right]\exp\left\{i\int_0^t \left[p(\tau)\frac{dq(\tau)}{d\tau} - H(q(\tau), p(\tau), \tau)\right]d\tau\right\}\psi_0(x_0)\,dx_0;$$

the inner integral is known as the *Feynman path integral*; it is taken over all phase trajectories

$$(p(\tau), q(\tau))_{\tau\in[0,t]}$$

such that

$$q(0) = x_0 \quad\text{and}\quad q(t) = x.$$

Thus, at the formal level, the path integral expression for the solution of the Schrödinger equation can be obtained directly from the T-exponential expression by removing the autonomous brackets.

2 Functions of Noncommuting Operators: the Construction and Main Properties

In the preceding section we considered several examples of functions of noncommuting operators and described convenient notation (Feynman indices and autonomous brackets) permitting one to define the arrangement of operators in operator expressions. However, we were acting rather intuitively, having no definition for functions of noncommuting operators. The meaning of operator expressions is clear as long as the functions involved are reasonably simple (e.g., polynomials or exponents). However, the examples considered in Section 1 suggest that, in view of applications, noncommutative analysis cannot be limited to elementary functions. One is forced to deal with expressions of the form $f(\overset{1}{A_1}, \dots, \overset{n}{A_n})$, where $A_1, \dots, A_n$ are given operators and the symbol $f(x_1, \dots, x_n)$ is a function of quite general form. Hence, let us consider what meaning can be assigned to such expressions and study their basic properties.

2.1 Motivations

We intend to learn how to solve the following problem. Suppose that some operators $A_1, \dots, A_n$ are given (that is, elements of an operator algebra $\mathcal{A}$, noncommutative in general) and $\mathcal{F}_n$ is some class of functions of n arguments $x_1, \dots, x_n$. Let $f \in \mathcal{F}_n$. Then, what is $f(\overset{1}{A_1}, \dots, \overset{n}{A_n})$?

Let us first study the trivial case in which all the considered symbols are polynomials, that is, $\mathcal{F}_n = \mathbb{C}[x_1, \dots, x_n]$ is the algebra of polynomials in n variables. Here the definition is evident: if $f(x_1, \dots, x_n)$ is the polynomial

$$f(x_1, \dots, x_n) = \sum_{\alpha_1+\dots+\alpha_n \le m} C_{\alpha_1 \dots \alpha_n} x_1^{\alpha_1} \dots x_n^{\alpha_n},$$

then the corresponding operator has the form

$$f(\overset{1}{A_1}, \dots, \overset{n}{A_n}) = \sum_{\alpha_1+\dots+\alpha_n \le m} C_{\alpha_1 \dots \alpha_n} A_n^{\alpha_n} \dots A_1^{\alpha_1}$$

(the ordering of factors in each term on the right-hand side is determined by their Feynman indices: the larger the index, the further to the left the factor occurs).

Thus, a mapping

$$\mu_{\overset{1}{A_1}, \dots, \overset{n}{A_n}} : \mathbb{C}[x_1, \dots, x_n] \to \mathcal{A}$$

is determined that takes each symbol into the corresponding operator.

Note that this mapping has the following three properties and is uniquely defined by these properties:

1°. The mapping $\mu_{\substack{1 \\ A_1,\dots,A_n}}{}^{\!\!\!n}$ is linear.

2°. For $n = 1$ the product of symbols is taken into the product of corresponding operators, so that μ_A is a homomorphism of algebras with identity element. Moreover, if $f(x) = x$, then $\mu_A(f) = A$.

3°. If $f(x_1, \dots, x_n) = f_1(x_1)\, f_2(x_2)\, \dots\, f_n(x_n)$, then

$$\mu_{\substack{1 \quad n \\ A_1,\dots,A_n}}(f) = \mu_{A_n}(f_n) \dots \mu_{A_2}(f_2)\, \mu_{A_1}(f_1).$$

Now suppose that a larger symbol space $\mathcal{F}_n \supset \mathbb{C}[x_1, \dots, x_n]$ is considered. Then it is natural to try to extend the mapping $\mu_{\substack{1 \quad n \\ A_1,\dots,A_n}}$ from $\mathbb{C}[x_1, \dots, x_n]$ to $\mathcal{F}_n$ with properties 1°–3° being preserved and then, if it is possible, to set

$$f(\overset{1}{A_1}, \dots, \overset{n}{A_n}) \overset{\text{def}}{=} \mu_{\substack{1 \quad n \\ A_1,\dots,A_n}}(f).$$

This approach can lead to valuable results if the desired extension not only exists (its existence is merely a restriction on the class of admissible operators $A_1, \dots, A_n$) but is *unique*. Indeed, under lack of uniqueness the equation $A = B$ apparently would not imply that $f(A) = f(B)$, and the theory would appear to be stillborn.

The uniqueness problem resides in two different places.

a) Uniqueness for $n = 1$.

Condition 2° requires that μ_A is an algebra homomorphism[1]

$$\mu_A : \mathcal{F}_1 \to A.$$

Hence $\mathcal{F}_1$ should be an algebra (with respect to pointwise multiplication of functions) and we arrive at the following question:

Is the homomorphism μ_A uniquely determined by the condition

$$\mu_A(x) = A,$$

i.e., by the value it takes on the function $f(x) = x$? (with polynomial symbols, the answer is evident).

It turns out (see Subsection 2.2) that, under quite general conditions, the answer is "yes" if we consider only *continuous* homomorphisms. From now on we tacitly assume that the notion of convergence is defined in the symbol spaces $\mathcal{F}_n$ and the operator algebra $\mathcal{A}$ and all mappings in question are continuous[2].

We say that an element $A \in \mathcal{A}$ is a *generator* if some homomorphism μ_A satisfying the above condition is fixed.

---

[1]Throughout the exposition all algebras considered contain the identity element, and homomorphisms are assumed to take 1 into 1.

[2]The detailed and rigorous analysis of all questions pertaining to admissible symbol and operator classes, convergence, estimates in function spaces, etc., is postponed until Chapter IV.

b) Uniqueness for $n > 1$.

If the operators A_j, $j = 1, \dots, n$, are generators, that is, the homomorphisms μ_{A_j}, $j = 1, \dots, n$, are defined and fixed, then conditions 1° and 3° define the action of the mapping $\mu_{\overset{1}{A_1},\dots,\overset{n}{A_n}}$ uniquely on functions $f(x_1, \dots, x_n)$ representable as finite linear combinations

$$f(x_1, \dots, x_n) = \sum_{s=1}^{N} f_{s1}(x_1) \dots f_{sn}(x_n), \quad f_{sk} \in \mathcal{F}_1 \tag{I.10}$$

(such linear combinations comprise the so-called *tensor product*

$$\underbrace{\mathcal{F}_1 \otimes \cdots \otimes \mathcal{F}_1}_{n \text{ copies}} = \mathcal{F}_1^{\otimes n}$$

of n copies of the space $\mathcal{F}_1$). In contrast to the case of polynomial symbols, we cannot hope that the linear combinations (I.10) fill the entire space $\mathcal{F}_n$, i.e., that $\mathcal{F}_n = \mathcal{F}_1^{\otimes n}$. Instead, let us require that each symbol $g \in \mathcal{F}_n$ can be approximated by such linear combinations. Then the mapping $\mu_{\overset{1}{A_1},\dots,\overset{n}{A_n}}$, which is uniquely defined on $\mathcal{F}_1^{\otimes n}$, uniquely extends to the entire space $\mathcal{F}_n$ by continuity.

Prior to formal definitions, let us consider two simple examples.

Example I.1 (Entire functions of bounded operators) Let $\mathcal{A}$ be the algebra of bounded operators in a Hilbert space $\mathcal{H}$. For the space $\mathcal{F}_n$ of n-ary symbols we take the space $\mathcal{O}(\mathbb{C}^n)$ of entire analytic functions on $\mathbb{C}^n$ equipped with the topology of uniform convergence on compact subsets in $\mathbb{C}^n$.

Let $A \in \mathcal{A}$ be an arbitrary operator. If

$$f(x) = \sum_{k=0}^{\infty} f_k x^k \in \mathcal{F}_1$$

is an arbitrary function, then the series $\sum_{k=0}^{\infty} f_k A^k$ clearly converges in the operator norm and defines a continuous operator in $\mathcal{H}$, which will be denoted by $f(A)$. The mapping

$$\mu_A : f \mapsto f(A)$$

is obviously continuous. Furthermore, $f(A)$ can be obtained as the limit

$$f(A) = \lim_{N \to \infty} \sum_{k=0}^{N} f_k A^k$$

of polynomials in A, which implies that μ_A is an algebra homomorphism uniquely defined by the condition $\mu_A(x) = A$. Thus, each element of $\mathcal{A}$ is a generator.

The definition of functions of several operators is now evident. By passing to the limit we easily show that for each function

$$f(x_1, \dots, x_n) = \sum_{|\alpha|=0}^{\infty} f_k x_1^{\alpha_1}, \dots, x_n^{\alpha_n} \in \mathcal{F}_n$$

and any $A_1, \dots, A_n \in \mathcal{A}$ we have

$$f(\overset{1}{A_1}, \dots, \overset{n}{A_n}) = \sum_{|\alpha|=0}^{\infty} f_\alpha A_n^{\alpha_n} \dots A_1^{\alpha_1},$$

and the series on the right-hand side converges in the operator norm.

Example I.2 (Continuous functions of bounded self-adjoint operators) Let $\mathcal{F}_n = C(\mathbb{R}^n)$ be the algebra of bounded continuous functions on $\mathbb{R}^n$ with uniform convergence on compact subsets of $\mathbb{R}^n$, and let $\mathcal{A}$ be the same algebra as in Example I.1. If $A \in \mathcal{A}$ is a self-adjoint operator, then for each function $f(x) \in \mathcal{F}_1$ the operator

$$f(A) \overset{\text{def}}{=} \int_{-\infty}^{\infty} f(\lambda) dE_\lambda(A)$$

is well-defined (see [42]); here $E_\lambda(A)$ is the spectral function of the operator A, and the Stieltjes integral on the right-hand side is in fact taken over a finite interval.

The mapping

$$\mu_A : f \mapsto f(A)$$

is continuous and coincides with the "natural" one on polynomial symbols; since by the Weierstrass theorem a continuous function on a compactum can be approximated by polynomials, it is clear that μ_A is an algebra homomorphism uniquely determined by the condition $\mu_A(x) = A$. Hence A is a generator.

A function of several Feynman-ordered operators can be defined as follows:

$$f(\overset{1}{A_1}, \dots, \overset{n}{A_n}) = \int_{-\infty}^{\infty} \cdots \int_{-\infty}^{\infty} f(\lambda_1, \dots, \lambda_n) dE_{\lambda_n}(A_n) \dots dE_{\lambda_1}(A_1),$$

(the integral on the right-hand side is an iterated Stieltjes integral).

2.2 The Definition and the Uniqueness Theorem

Let us now make the construction of the preceding subsection into a precise definition.

Let $\mathcal{A}$ be some algebra, whose elements will be referred to as *operators* (in applications of noncommutative analysis, as a rule, $\mathcal{A}$ is an algebra of operators in some linear space; however, within the framework of the general theory it is convenient to pay no attention to this fact and assume that $\mathcal{A}$ is merely an associative algebra with 1). Furthermore, let a *space $\mathcal{F}_1$ of unary symbols* be given. The elements of $\mathcal{F}_1$ are functions of a real (or complex) variable x.[3] We assume that $\mathcal{F}_1$ is an algebra with respect to pointwise multiplication of functions and contains the function $f(x) \equiv x$ (and, consequently, all polynomials).

Definition I.1 An operator $A \in \mathcal{A}$ is said to be a *generator* if there exists a continuous homomorphism

$$\mu_A : \mathcal{F}_1 \to \mathcal{A}$$

such that $\mu_A(x) = A$. If A is a generator, then we set

$$f(A) = \mu_A(f)$$

for each $f \in \mathcal{F}_1$.

The property of being a generator obviously depends on the choice of the symbol class $\mathcal{F}_1$, so we mention $\mathcal{F}_1$ explicitly, by saying that A is an *$\mathcal{F}_1$-generator*, whenever ambiguity is possible.

Since the uniqueness of μ_A has not yet been proved (see Theorem I.1 below in this subsection), we temporarily assume where necessary that for all the operators involved the corresponding μ-homomorphism is chosen and fixed.

Let $\mathcal{F}_n$ be the space of n-ary symbols $f(x_1, \dots, x_n)$. According to our strategy, the elements of the tensor product $\mathcal{F}_1^{\otimes n}$ should be dense in $\mathcal{F}_n$. We assume that $\mathcal{F}_n$ is obtained from $\mathcal{F}_1^{\otimes n}$ by *completion*, that is to say, by adding the limits of Cauchy sequences of elements (I.10); we denote this by

$$\mathcal{F}_n = \underbrace{\mathcal{F}_1 \hat{\otimes} \cdots \hat{\otimes} \mathcal{F}_1}_{n \text{ copies}} = \mathcal{F}_1^{\hat{\otimes} n}$$

(the hat over $\otimes$ denotes completion).

Instead, we could impose a weaker condition that the embedding $\mathcal{F}_1^{\hat{\otimes} n} \hookrightarrow \mathcal{F}_n$ be continuous and dense. However, we would have to require additionally that the mapping (I.11) be continuous as a mapping from $\mathcal{F}_n$ to $\mathcal{A}$, which holds automatically in our case. This would lead to slight modifications in the subsequent statements.

Let $A_1, \dots, A_n$ be generators. The mapping

$$\mu_{\substack{1 \quad n \\ A_1, \dots, A_n}} : \mathcal{F}_1^{\otimes n} \to \mathcal{A}, \tag{I.11}$$

$$f_1(x_1)\, f_2(x_2) \dots f_n(x_n) \mapsto f_n(x_n) \dots f_2(x_2)\, f_1(x_1) \tag{I.12}$$

extends by continuity to the entire space $\mathcal{F}_n$.

[3]We do not assume that x ranges over the entire space $\mathbb{R}$ or $\mathbb{C}$.

Definition I.2 We set

$$f(\overset{1}{A_1}, \ldots, \overset{n}{A_n}) = \mu_{\overset{1}{A_1}, \ldots, \overset{n}{A_n}}(f);$$

a tuple $A = (\overset{1}{A_1}, \ldots, \overset{n}{A_n})$ of generators equipped with Feynman indices will be referred to as a *Feynman tuple.*

Functions of a Feynman tuple $(\overset{j_1}{A_1}, \ldots, \overset{j_n}{A_n})$ with arbitrary pairwise distinct Feynman indices $j_1, \ldots, j_n$ are defined in a similar way: the factors on the right-hand side of (I.12) should be arranged so that their Feynman indices form a descending sequence.

Remark I.2 In order to avoid clumsy notation, we assume that all the operators $A_1, \ldots, A_n$ are $\mathcal{F}_1$-generators with the same symbol class $\mathcal{F}_1$. In principle, one may well consider operator expressions involving generators with respect to various symbol classes. In that case $\mathcal{F}_n$ would be the tensor product of various spaces of unary symbols rather that $\mathcal{F}_1^{\hat{\otimes} n}$, each space associated with the corresponding operator argument. In the sequel we often make use of a particular case of this situation in which some of the operators occur only linearly or polynomially in the expression considered. If so, the corresponding symbol classes may be chosen to consist of polynomials, which eliminates the necessity of checking any additional conditions (with polynomial symbol, each operator is a generator).

Remark I.3 The presented construction of functions of several operators suggests the following general way of proving this or that identity, assertion, etc. in noncommutative analysis: first, they should be established for decomposable symbols of the form $f = f_1 \otimes \cdots \otimes f_n$ and then extended by linearity to $\mathcal{F}_1^{\otimes n}$ and by continuity to the entire $\mathcal{F}_n$, cf. the proof of Propositions I.1 and I.2 given below.

Our definitions do not allow coinciding Feynman indices over arguments of an operator expression (later on, when it will be shown that the ordering of Feynman indices over commuting operators does not affect the value of an operator expression, commuting operators will be allowed to bear the same Feynman indices). However, the formula $f(\overset{1}{A}, \overset{1}{A})$ is not erroneous. It means the following: one should restrict the function $f(x, y)$ to the diagonal $x = y$ by setting $g(x) = f(x, x)$ and substitute the operator $\overset{1}{A}$ into $g(x)$ instead of x. Thus,

$$f(\overset{1}{A}, \overset{1}{A}) = g(A).$$

The same interpretation is used if f has additional operator arguments, e.g.

$$f(\overset{1}{A}, \overset{1}{A}, \overset{2}{B}) \overset{\text{def}}{=} g(\overset{1}{A}, \overset{2}{B}),$$

where $g(x, z) = f(x, x, z)$.

It turns out that in this situation the Feynman indices over an operator A can be moved apart, i.e., set different.

Proposition I.1 (Moving indices apart) i) *The operator of restriction to the diagonal*

$$[\iota f](x) = f(x, x).$$

is a continuous operator from $\mathcal{F}_2$ to $\mathcal{F}_1$.

ii) *Let $A \in \mathcal{A}$ be a generator. For each symbol $f(x, y) \in \mathcal{F}_2$ we have*

$$f(\overset{1}{A}, \overset{2}{A}) = f(\overset{2}{A}, \overset{1}{A}) = f(\overset{1}{A}, \overset{1}{A})$$

(note that the right-hand side is nothing other than $[\iota f](A)$.)

Proof. Let $f(x, y) \in \mathcal{F}_2$ have the form $f = g \otimes h$, that is, $f(x, y) = g(x)\, h(y)$, so that

$$\iota(g \otimes h) = gh.$$

Since $\mathcal{F}_1$ is an algebra and multiplication in $\mathcal{F}_1$ is continuous, the operator ι is continuous from $\mathcal{F}_1 \hat{\otimes} \mathcal{F}_1 = \mathcal{F}_2$ to $\mathcal{F}_1$. Moreover, since μ_A is a homomorphism, we have

$$(g \otimes h)(\overset{1}{A}, \overset{2}{A}) = h(A)\, g(A) = hg(A) = \iota[g \otimes h](A),$$

and similarly for $(g \otimes h)(\overset{2}{A}, \overset{1}{A})$. Thus the identity in ii) holds on $\mathcal{F}_1 \otimes \mathcal{F}_1$ and, by continuity, on the entire $\mathcal{F}_2$. The proposition is proved. □

Evidently, the statement of Proposition I.1 remains valid also in the presence of additional operator arguments. The Feynman indices over A can be moved apart but only in such a way that the ordering relation between them and Feynman indices of other operators remains unchanged. More precisely,

$$f(\overset{j}{A}, \overset{j}{A}, \overset{s_1}{B_1}, \dots, \overset{s_m}{B_m}) = f(\overset{k}{A}, \overset{l}{A}, \overset{s_1}{B_1}, \dots, \overset{s_m}{B_m})$$

provided that j, k, and l all lie in some interval $[a, b]$ that contains none of the numbers $s_1, \dots, s_m$. Since our definitions make it evident that the value of an operator expression depends on the ordering relation between Feynman indices rather than on the indices themselves, we arrive at the following conclusion:

Feynman indices in an operator expression can be changed arbitrarily without affecting its value, provided that the ordering relation between Feynman indices over different operators is preserved.

For example,

$$f(\overset{1}{A}, \overset{2}{A}, \overset{3}{B}) = f(\overset{2}{A}, \overset{1}{A}, \overset{3}{B}) = f(\overset{1}{A}, \overset{1}{A}, \overset{5}{B}),$$

but

$$f(\overset{1}{A}, \overset{2}{A}, \overset{3}{B}) \neq f(\overset{4}{A}, \overset{2}{A}, \overset{3}{B})$$

in general, since the order of Feynman indices of the first and the third argument has been changed.

The next proposition is stated in its general form (i.e., with an arbitrary number of operator arguments).

Proposition I.2 (Extraction of a linear factor) *Let $f(x_1, \dots, x_n) \in \mathcal{F}_n$ be a symbol of the form*

$$f(x_1, \dots, x_n) = g(x_1, \dots, x_k)\, h(x_{k+1}, \dots, x_n),$$

and let $A = (\overset{j_1}{A_1}, \dots, \overset{j_n}{A_n})$ be a Feynman tuple satisfying the following condition: the Feynman indices $j_1, \dots, j_n$ lie in some interval (a, b) that contains none of the indices $j_{k+1}, \dots, j_n$. Then

$$f(\overset{j_1}{A_1}, \dots, \overset{j_n}{A_n}) = \overset{s}{[\![} g(\overset{j_1}{A_1}, \dots, \overset{j_k}{A_k})]\!]\, h(\overset{k+1}{A}_{k+1}, \dots, \overset{j_n}{A_n}),$$

where $s \in (a, b)$ can be chosen arbitrarily.

Remark I.4 Recall that the autonomous brackets $[\![\ ]\!]$ have the following meaning: one should evaluate the expression they enclose and use the result as a single new operator in the subsequent evaluation, no longer caring about the Feynman indices that occurred *within* the brackets. The index s over the left bracket is just the Feynman index to be assigned to this new operator. The operator $g(\overset{j_1}{A_1}, \dots, \overset{j_k}{A_k})$ occurs linearly in this identity, so, according to Remark I.2, there is no need to impose any additional requirements on this operator.

Proof. Assume that $f(x_1, \dots, x_n)$ is a decomposable symbol,

$$f(x_1, \dots, x_n) = f_1(x_1) \dots f_n(x_n).$$

Then

$$g(x_1, \dots, x_k) = f_1(x_1) \dots f_k(x_k)$$

and

$$h(x_{k+1}, \dots, x_n) = f_{k+1}(x_{k+1}) \dots f_n(x_n).$$

The factors $f_1(A_1), \dots, f_{k_{j_n}}(A_k)$ in the product $f_1(\overset{j_1}{A_1}) \dots f_n(\overset{j_n}{A_n})$ stand next to each other due to the condition imposed on the Feynman indices. Hence, we can group these factors together, thus obtaining the factor

$$f_1(\overset{j_1}{A_1}) \dots f_k(\overset{j_k}{A_k}) = g(\overset{j_1}{A_1}, \dots, \overset{j_k}{A_k})$$

in the considered product.

Hence, the proposition is proved for decomposable symbols. By linearity and continuity, it extends to arbitrary symbols $g \in \mathcal{F}_k$, $h \in \mathcal{F}_{n-k}$ (cf. Remark I.3). The proof is now complete. □

Let us consider two examples. We have

$$g(\overset{1}{A}, \overset{5}{B})\, h(\overset{2}{C}, \overset{3}{D}) = [\![h(\overset{2}{C}, \overset{3}{D}]\!]\, g(\overset{1}{A}, \overset{5}{B}) = \overset{2}{K}\, g(\overset{1}{A}, \overset{5}{B}),$$

where $K = h(\overset{2}{C}, \overset{3}{D})$. Similarly, we have

$$(\overset{2}{A} - \overset{4}{A}) \sin(\overset{1}{B} + \overset{5}{B}) = [\![\overset{3\,2}{A} - \overset{4}{A}]\!] \sin(\overset{1}{B} + \overset{5}{B}) = 0,$$

since

$$\overset{2}{A} - \overset{4}{A} = 0.$$

However,

$$(\overset{2}{A} - \overset{4}{A}) \sin(\overset{1}{B} + \overset{3}{B}) \neq 0$$

in general, since the operator $\overset{3}{B}$ acts between $\overset{2}{A}$ and $\overset{4}{A}$.

Proposition I.1 and I.2 enable us to prove a theorem stating the uniqueness of the homomorphism μ_A for any generator A.

Theorem I.1 *Suppose that the symbol class $\mathcal{F}_1$ satisfies the following condition: the difference derivative*

$$\frac{\delta}{\delta x} : f(x) \mapsto \frac{\delta f}{\delta x}(x, y) = \begin{cases} (f(x) - f(y))/(x - y), & x \neq y, \\ f'(x), & x = y \end{cases}$$

is a continuous operator form $\mathcal{F}_1$ to $\mathcal{F}_2$. Then for any generator $A \in \mathcal{A}$ the homomorphism μ_A described in Definition I.1 *is uniquely defined.*

Proof. Let A be a generator in $\mathcal{A}$, and let μ_1 and μ_2 be two (possibly, different) choices of the homomorphism μ_A. It suffices to prove that $\mu_1 = \mu_2$. For notational convenience, we set $C = A$, $\mu_A = \mu_1$, and $\mu_C = \mu_2$; in other words, we denote the operator argument differently, according to which of the two homomorphisms is used:

$$f(A) = \mu_A(f), \quad f(C) = \mu_C(f).$$

All that we need is to establish that $f(A) = f(C)$ for every $f \in \mathcal{F}_1$. By substituting $\overset{1}{A}$ for x and $\overset{2}{C}$ for y in the identity

$$f(x) - f(y) = (x - y)\frac{\delta f}{\delta x}(x, y),$$

we obtain

$$f(A) - f(C) = f(\overset{1}{A}) - f(\overset{2}{C}) = (\overset{1}{A} - \overset{2}{C})\frac{\delta f}{\delta x}(\overset{1}{A}, \overset{2}{C}).$$

By the conditions of the theorem, $\delta f/\delta x(x, y) \in \mathcal{F}_2$, and consequently,

$$(z - w)\delta f/\delta x(x, y) \in \mathcal{F}_4.$$

Hence we can apply Proposition I.1 so as to move the Feynman indices over A apart, and the same thing can be done for C:

$$f(A) - f(C) = (\overset{1}{A} - \overset{2}{C})\frac{\delta f}{\delta x}(\overset{0}{A}, \overset{3}{C}).$$

Next, we use Proposition I.2 to enclose the factor $(\overset{1}{A} - \overset{2}{C})$ in autonomous brackets:

$$(\overset{1}{A} - \overset{2}{C})\frac{\delta f}{\delta x}(\overset{0}{A}, \overset{3}{C}) = \overset{2}{[\![}\overset{1}{A} - \overset{2}{C}]\!]\frac{\delta f}{\delta x}(\overset{0}{A}, \overset{3}{C}),$$

and the last expression is equal to zero since

$$\overset{1}{A} - \overset{2}{C} = A - C = 0$$

(the reader should not be too suspicious about this conclusion, since the homomorphisms μ_A and μ_C coincide on the symbol $\varphi(x) = x$ by definition). Hence we have shown that $f(A) = f(C)$. The theorem is proved. □

In what follows we assume that the condition stated in Theorem I.1 is fulfilled.

Remark I.5 This proof is so simple that it seems to be a mere trick. However, it is perfectly rigorous. Its simplicity is just due to the wizardry of our notation, which displays itself in quite a few places thereafter.

Remark I.6 According to Proposition I.1, the operator ι of restriction to the diagonal acts continuously from $\mathcal{F}_2$ to $\mathcal{F}_1$. Since $\iota\,\delta/\delta x = \partial/\partial x$ (on the diagonal the difference derivative coincides with the usual one), we see that $\frac{d}{dx}$ is a continuous operator from $\mathcal{F}_1$ to $\mathcal{F}_1$ and consequently all symbols in $\mathcal{F}_1$ (and hence, in each $\mathcal{F}_n$) are infinitely differentiable.

We could choose another way and postulate that the operator

$$\frac{d}{dx} : \mathcal{F}_1 \to \mathcal{F}_1$$

is continuous. Then the formula

$$\frac{\delta f}{\delta x}(x, y) = \int_0^1 \frac{df}{dx}(\tau x + (1 - \tau)y)\, d\tau$$

would imply that the condition of Theorem I.1 is satisfied. However, in that case we would have to require that the domain where symbols are defined is arcwise connected and the mapping

$$(\tau, f(x)) \mapsto f(\tau x + (1 - \tau)y)$$

from $\mathbb{R} \times \mathcal{F}_1$ to $\mathcal{F}_2$ is continuous.

2.3 Basic Properties

Let $\mathcal{A}$ be an algebra of operators, and let

$$\mathcal{F}_1 = \mathcal{F}, \quad \mathcal{F}_2 = \mathcal{F}\hat{\otimes}\mathcal{F}, \ldots, \mathcal{F}_n = \mathcal{F}^{\hat{\otimes} n}, \ldots$$

be algebras of symbols. Given a tuple $A = (A_1, \ldots, A_n)$ of generators in $\mathcal{A}$, we intend to study the mapping

$$\mu_A = \mu_{\substack{1 \quad\; n\\ A_1,\ldots,A_n}} : \mathcal{F}_n \to \mathcal{A}.$$

In doing so it is convenient to use the *regular representations* of the algebra $\mathcal{A}$ on itself. Let $A \in \mathcal{A}$. The mappings

$$\begin{aligned} L_A &: \mathcal{A} \to \mathcal{A} \\ B &\mapsto L_A(B) = AB \end{aligned}$$

and

$$\begin{aligned} R_A &: \mathcal{A} \to \mathcal{A} \\ B &\mapsto R_A(B) = BA \end{aligned}$$

defined as left and right multiplication by A, respectively, are continuous linear operators on $\mathcal{A}$. The set of continuous linear operators on $\mathcal{A}$ will be denoted by $\mathcal{L}(\mathcal{A})$, so we can write $L_A \in \mathcal{L}(\mathcal{A})$ and $R_A \in \mathcal{L}(\mathcal{A})$. Consider the mapping

$$\begin{aligned} L &: \mathcal{A} \to \mathcal{L}(\mathcal{A}) \\ A &\mapsto L_A. \end{aligned}$$

It is a continuous linear mapping called the *left regular representation* of the algebra $\mathcal{A}$.

Similarly, the continuous linear mapping

$$\begin{aligned} R &: \mathcal{A} \to \mathcal{L}(\mathcal{A}) \\ A &\mapsto R_A \end{aligned}$$

is called the *right regular representation* of $\mathcal{A}$.

Note that L is a homomorphism of algebras

$$L_{AB} = L_A L_B,$$

whereas R is an antihomomorphism, that is, it inverses the order of factors

$$R_{AB} = R_B \, R_A.$$

Since $\mathcal{L}(\mathcal{A})$ is an algebra of continuous linear operators on a linear space, we may well consider the notion of a generator in $\mathcal{L}(\mathcal{A})$. It turns out that the supply of generators in $\mathcal{L}(\mathcal{A})$ is at least as rich as that in $\mathcal{A}$, as shown by the following theorem:

Theorem I.2 *The following conditions are equivalent:*

i) *A is a generator in $\mathcal{A}$.*
ii) *L_A is a generator in $\mathcal{L}(\mathcal{A})$.*
iii) *R_A is a generator in $\mathcal{L}(\mathcal{A})$.*

Proof. i) $\Leftrightarrow$ ii). Let A be a generator in $\mathcal{A}$ and $\mu_A : \mathcal{F} \to \mathcal{A}$ be the corresponding homomorphism. We define the mapping

$$\mu_{L_A} : \mathcal{F} \to \mathcal{L}(\mathcal{A})$$

by setting

$$\mu_{L_A}(f) = L_{\mu_A(f)}.$$

Clearly, this mapping is a continuous homomorphism as the composition of two continuous homomorphisms, namely of L and μ_A. We also have

$$\mu_{L_A}(x) = L_{\mu_A(x)} = L_A,$$

and so L_A matches the requirements of Definition I.1. Conversely, let $A \in \mathcal{A}$ be an operator such that L_A is a generator; in other words, for each symbol $f(x) \in \mathcal{F}$ the operator

$$f(L_A) = \mu_{L_A}(f)$$

is well-defined. To construct the corresponding mapping μ_A, note that for an arbitrary operator $B \in \mathcal{A}$ we have

$$[L_A, R_B] \stackrel{\text{def}}{=} L_A R_B - R_B L_A = 0,$$

that is, the operators L_A and R_B commute. Indeed, for any $C \in \mathcal{A}$ we have

$$L_A(R_B C) = L_A(CB) = A(CB) = (AC)B = R_B(AC) = R_B L_A C,$$

by the associativity of $\mathcal{A}$. We can now apply an argument similar to that used in the proof of Theorem I.1 to obtain

$$f(L_A)R_B = R_B \, f(L_A). \tag{I.13}$$

Indeed, we have

$$f(L_A)R_B - R_B f(L_A) = (f(\overset{3}{L_A}) - f(\overset{1}{L_A}))\overset{2}{R_B} = (\overset{3}{L_A} - \overset{1}{L_A})\overset{2}{R_B}\frac{\delta f}{\delta x}(\overset{3}{L_A}, \overset{1}{L_A}).$$

We can move apart the Feynman indices over L_A in this expression (Proposition I.1) and then extract a linear factor (Proposition I.2) thus obtaining

$$f(L_A R_B - R_B f(L_A) = [\![(\overset{1}{L_A} \overset{3}{-} \overset{1}{L_A})\overset{2}{R_B}]\!] \frac{\delta f}{\delta x}(\overset{3}{L_A}, \overset{1}{L_A}) = 0,$$

since the operator in the autonomous brackets is equal to zero.

Since B is arbitrary, equation (I.13) means exactly that $f(L_A)$ is the left regular representation of some element of $\mathcal{A}$, which will be denoted by $f(A)$. Specifically, we have

$$f(L_A)(X) = f(L_A)(1 \cdot X) = f(L_A)(R_X 1) = R_X f(L_A)(1) = (f(L_A)(1))X,$$

and consequently,

$$f(A) = f(L_A)(1).$$

The mapping $f \mapsto \mu_A(f) = f(A)$ is clearly continuous; moreover,

$$f(A)g(A) = f(L_A(1)g(L_A)(1) = f(L_A)g(L_A)(1) = (fg)(L_A(1) = fg(A),$$

so that it is a homomorphism.

i) $\Leftrightarrow$ iii). This can be proved similarly; the only point worth mentioning is that R is an *anti*homomorphism, and at first glance it may seem that the mapping

$$\mu_{R_A} = R\mu_A$$

fails to be a homomorphism. However, the image of μ_A is a *commutative* subalgebra of $\mathcal{A}$, and the restriction of R to it *is* a homomorphism.

The proof is complete. □

Now there is an evident formula:

$$f(\overset{1}{L_{A_1}}, \dots, \overset{n}{L_{A_n}}) = L_{f(\overset{1}{A_1}, \dots, \overset{n}{A_n})}$$

for any generators $A_1, \dots, A_n \in \mathcal{A}$ and symbol $f \in \mathcal{F}_n$. The proof goes by linearity and continuity: if $f(x) = f_1(x_1)\dots f_n(x_n)$ is a decomposable symbol, we have

$$\begin{aligned} f(\overset{1}{L_{A_1}}, \dots, \overset{n}{L_{A_n}}) &= f_n(L_{A_n})\dots f_1(L_{A_1}) = L_{f_n(A_n)}\dots L_{f_1(A_1)} \\ &= L_{f_n(A_n)\dots f_1(A_1)} = L_{f(\overset{1}{A_1}, \dots, \overset{n}{A_n})}, \end{aligned}$$

and this equation extends to the entire $\mathcal{F}_n$ by continuity.

Similarly,

$$f(\overset{1}{R_{A_1}}, \dots, \overset{n}{R_{A_n}}) = R_{f(\overset{n}{A_1}, \dots, \overset{1}{A_n})}$$

(the operators $A_1, \ldots, A_n$ on the right come in reversed Feynman ordering, since R is an antihomomorphism).

Let us consider a slightly more general situation in which a function contains both L and R arguments. Let $f(x, y) \in \mathcal{F}_2$ be an arbitrary binary symbol. What is $f(\overset{1}{L_A}, \overset{2}{R_B})$? Note that L_A and R_B commute, and so we can choose their Feynman indices arbitrarily; let us choose them in such a way that the index of L_A is greater than that of R_B; thus, we consider $f(\overset{3}{L_A}, \overset{1}{R_B})$. This operator is an element of $\mathcal{L}(\mathcal{A})$, and to describe it means to describe its action on an arbitrary element $C \in \mathcal{A}$. If f is decomposable, $f(x, y) = f_1(x)\, f_2(y)$, then we have

$$f(\overset{3}{L_A}, \overset{1}{R_B})(C) = f_1(L_A) f_2(R_B) C = f_1(A) C f_2(B) = f(\overset{3}{A}, \overset{2}{B})\overset{1}{C}.$$

By continuity, this identity extends to arbitrary symbols $f \in \mathcal{F}_2$. We now can state the result in a general form.

Theorem I.3 *Let*

$$A_1, \ldots, A_s, \quad B_1, \ldots, B_m \in \mathcal{A}$$

be generators, and let $j_1, \ldots, j_s$, $k_1, \ldots, k_m$ *be their (pairwise distinct) Feynman indices. Then*

$$f(\overset{j_1}{L_{A_1}}, \ldots, \overset{j_s}{L_{A_s}}, \overset{k_1}{R_{B_1}}, \ldots, \overset{k_m}{R_{B_m}})(C) = \overset{l}{C} f(\overset{j_1}{L_{A_1}}, \ldots, \overset{j_s}{L_{A_s}}, \overset{k_0-k_1}{B_1}, \ldots, \overset{k_0-k_m}{B_m}),$$

where k_0 *and* l *are chosen in such a way that* $k_0 - k_l < l < j_i$ *for any* $i \in \{1, \ldots, s\}$ *and* $l \in \{1, \ldots, m\}$.

The proof is modelled on the above argument, and we omit the trivial details.

Let us now state and prove the main properties of Feynman operator calculus. We fix some class $\mathcal{F}$ of unary symbols and some generator $A \in \mathcal{A}$.

Theorem I.4 *The Feynman operator calculus has the following naturality properties.*

1°. *Let* $\varphi : \mathcal{A} \to \mathcal{B}$ *be a continuous homomorphism of algebras. Then* $\varphi(A) \in \mathcal{B}$ *is a generator, and*

$$f(\varphi(A)) = \varphi(f(A))$$

for any $f \in \mathcal{F}$.

2°. *Let* $B \in \mathcal{A}$ *be an operator such that* $AB = BA$. *Then*

$$f(A)B = Bf(A)$$

for any $f \in \mathcal{F}$.

3°. *Suppose that*

$$AB = BC, \tag{I.14}$$

where $B, C \in \mathcal{A}$ and C is a generator. Then

$$f(A)B = Bf(C)$$

for any $f \in \mathcal{F}$.

4°. *Assume that $\mathcal{A}$ is an algebra of operators in a linear space V, $\mathcal{B}$ an algebra of operators in a linear space W, and*

$$\varkappa : W \to V$$

a continuous linear operator such that

$$A\varkappa = \varkappa C, \tag{I.15}$$

where $C \in \mathcal{B}$ is a generator. Then we have

$$f(A)\varkappa = \varkappa f(C)$$

for any $f \in \mathcal{F}$.

5°. *As in 4°, assume that $\mathcal{A}$ is an algebra of operators in a linear space V. Let $\xi \in V$ be an eigenvector of A with eigenvalue λ,*

$$A\xi = \lambda\xi.$$

Suppose also that $f(\lambda)$ is defined for any $f \in \mathcal{F}$. Then

$$f(A)\xi = f(\lambda)\xi$$

for any $f \in \mathcal{F}$.

Remark I.7 There are no Feynman indices in the statement of this theorem. However, the proof employs Feynman ordering heavily. See Theorem I.5 for generalization to multivariate symbols.

Proof. 1°. Let $\varphi : \mathcal{A} \to \mathcal{B}$ be a continuous homomorphism of algebras. Set

$$f(\varphi(A)) \stackrel{\text{def}}{=} \varphi(f(A)),$$

that is,

$$\mu_{\varphi(A)} = \varphi\, \mu_A.$$

Evidently, $\mu_{\varphi(A)}$ is a continuous homomorphism (as a composition of such mappings). Hence, $\varphi(A)$ is a generator in $\mathcal{B}$, and we obtain the desired result (recall that $\mu_{\varphi(A)}$ is unique by Theorem I.1).

2°. We proceed as in the proof of Theorem I.2. We can write

$$f(A)B - Bf(A) = \overset{2}{B}(f(\overset{3}{A}) - f(\overset{1}{A})) = \overset{2}{B}(\overset{3}{A} - \overset{1}{A})\frac{\delta f}{\delta x}(\overset{3}{A}, \overset{1}{A}).$$

Next, by moving indices apart and extracting a linear factor, we obtain

$$f(A)B - Bf(A) = [\![\overset{2\,2}{B}(\overset{3}{A} - \overset{1}{A})]\!]\frac{\delta f}{\delta x}(\overset{5}{A}, \overset{0}{A}) = 0$$

since the expression in autonomous brackets is zero:

$$\overset{2}{B}(\overset{3}{A} - \overset{1}{A}) = AB - BA = 0.$$

3°. In order to prove this, we repeat the above computation with $\overset{1}{A}$ replaced by $\overset{1}{C}$:

$$\begin{aligned} f(A)B - Bf(C) &= \overset{2}{B}(f(\overset{3}{A}) - f(\overset{1}{C})) \\ &= \overset{2}{B}(\overset{3}{A} - \overset{1}{C})\frac{\delta f}{\delta x}(\overset{3}{A}, \overset{1}{C}) \\ &= [\![\overset{2\,2}{B}(\overset{3}{A} - \overset{1}{C})]\!]\frac{\delta f}{\delta x}(\overset{5}{A}, \overset{0}{C}) = 0. \end{aligned}$$

Of course, 2° is a particular case of 3°.

4°. The condition of this item can be expressed by the commutative diagram

$$\begin{array}{ccc} V & \xrightarrow{A} & V \\ \uparrow{\scriptstyle \varkappa} & & \uparrow{\scriptstyle \varkappa} \\ W & \xrightarrow{C} & W \end{array}$$

We now make some preparations in order to reduce the proof of this item to that of item 3°.

Consider the direct sum $\boldsymbol{E} = V \oplus W$ of the linear spaces V and W. Continuous linear operators in $\boldsymbol{E}$ can be viewed as 2×2 matrices

$$\begin{pmatrix} E & F \\ G & H \end{pmatrix}$$

with continuous operator entries

$$\begin{aligned} E &: V \to V, \\ H &: W \to W, \\ F &: W \to V, \\ G &: V \to W. \end{aligned}$$

Relation (I.15) can be rewritten in the form

$$\begin{pmatrix} A & 0 \\ 0 & C \end{pmatrix}\begin{pmatrix} 0 & \varkappa \\ 0 & 0 \end{pmatrix}=\begin{pmatrix} 0 & \varkappa \\ 0 & 0 \end{pmatrix}\begin{pmatrix} A & 0 \\ 0 & C \end{pmatrix}.$$

Let $\tilde{\mathcal{A}}$ be some algebra of continuous linear operators in E containing the subalgebra $\mathcal{A} \oplus \mathcal{B}$ and the operator $\begin{pmatrix} 0 & \varkappa \\ 0 & 0 \end{pmatrix}$ (e.g., the minimal algebra generated by these objects). Then the last equation has the form (I.14) in the algebra $\tilde{\mathcal{A}}$. Furthermore, $\begin{pmatrix} A & 0 \\ 0 & C \end{pmatrix}$ is a generator in $\tilde{\mathcal{A}}$. Indeed, one can set

$$f\left(\begin{pmatrix} A & 0 \\ 0 & C \end{pmatrix}\right)=\begin{pmatrix} f(A) & 0 \\ 0 & f(C) \end{pmatrix}.$$

By applying item 3°, we obtain

$$f\left(\begin{pmatrix} A & 0 \\ 0 & C \end{pmatrix}\right)\begin{pmatrix} 0 & \varkappa \\ 0 & 0 \end{pmatrix}=\begin{pmatrix} 0 & \varkappa \\ 0 & 0 \end{pmatrix}f\left(\begin{pmatrix} A & 0 \\ 0 & C \end{pmatrix}\right),$$

or

$$\begin{pmatrix} f(A) & 0 \\ 0 & f(C) \end{pmatrix}\begin{pmatrix} 0 & \varkappa \\ 0 & 0 \end{pmatrix}=\begin{pmatrix} 0 & \varkappa \\ 0 & 0 \end{pmatrix}\begin{pmatrix} f(A) & 0 \\ 0 & f(C) \end{pmatrix}.$$

We perform the necessary multiplications and find that $f(A)\,\varkappa = \varkappa\, f(C)$, as desired.

5°. Strange as it may seem, but this is a particular case of 4°. Specifically, let $H \subset V$ be the one-dimensional subspace generated by the eigenvector ξ, and let

$$\varkappa : \mathbb{C} \to V$$

be the linear mapping such that $\varkappa(\alpha) = \alpha\xi$ for all $\alpha \in \mathbb{C}$. It is evident that $\varkappa(\mathbb{C}) = H$. The commutativity of the diagram

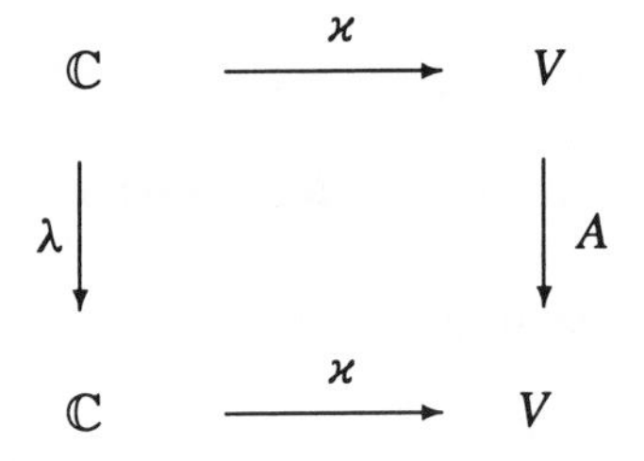

expresses the fact that $H = \varkappa(\mathbb{C})$ is an eigenspace of A with eigenvalue λ. An application of item 4° yields the desired result. The theorem is proved. □

We can now state the counterpart of Theorem I.4 for multivariate symbols. In the following we assume that the class $\mathcal{F}$ of unary symbols is fixed, $\mathcal{F}_n = \mathcal{F}^{\otimes n}$, and $A_1, \dots, A_n \in \mathcal{A}$ are arbitrary generators.

Theorem I.5 1°. *Let*

$$\varphi : \mathcal{A} \to \mathcal{B}$$

be a continuous homomorphism of algebras. Then we have

$$f(\varphi(\overset{1}{A_1}), \dots, \varphi(\overset{n}{A_n})) = \varphi(f(\overset{1}{A_1}, \dots, \overset{n}{A_n}))$$

for any symbol $f \in \mathcal{F}_n$.

2°. *Let* $B \in \mathcal{A}$ *be an operator commuting with all* A_j,

$$[A_j, B] \equiv A_j B - B A_j = 0.$$

Then B *commutes with functions of* $\overset{1}{A_1}, \dots, \overset{n}{A_n}$ *as well,*

$$[f(\overset{1}{A_1}, \dots, \overset{n}{A_n}), B] = [\![f(\overset{1}{A_1}, \dots, \overset{n}{A_n})]\!]B - Bf[\![(\overset{1}{A_1}, \dots, \overset{n}{A_n})]\!] = 0.$$

3°. *Let*

$$A_j B = B C_j, \quad j = 1, \dots, n,$$

where $B \in \mathcal{A}$, $C_j \in \mathcal{A}$, $j = 1, \dots, n$, *and each* C_j *is a generator. Then*

$$f(\overset{1}{A_1}, \dots, \overset{n}{A_n})B = Bf(\overset{1}{C_1}, \dots, \overset{n}{C_n})$$

for any symbol $f \in \mathcal{F}_n$.

4°. *Let* $A_1, \dots, A_n \in \mathcal{A}$ *be continuous linear operators in a linear space* V, $B_1, \dots, B_n \in \mathcal{B}$ *be continuous linear operators in a linear space* W, *and let*

$$\varkappa : W \to V$$

be a continuous linear operator such that

$$A_j \varkappa = \varkappa B_j, \quad j = 1, \dots, n$$

(in this case one says that $\varkappa$ *is an intertwining operator for the tuples* $(A_1, \dots, A_n)$ *and* $(B_1, \dots, B_n)$*).*

Assume that each B_j *is a generator in* $\mathcal{B}$. *Then*

$$f(\overset{1}{A_1}, \dots, \overset{n}{A_n}) \circ \varkappa = \varkappa\, f(\overset{1}{B_1}, \dots, \overset{n}{B_n}).$$

In other words, if $\varkappa$ *intertwines operator tuples, it also intertwines any function of these tuples.*

5°. *Let* $A_1, \dots, A_n \in \mathcal{A}$ *be linear operators in a linear space* V, *and let* $\xi \in V$ *be a common eigenvector of* $A_1, \dots, A_n$ *with eigenvalues* $\lambda = (\lambda_1, \dots, \lambda_n)$*:*

$$A_j \xi = \lambda_j \xi, \quad j = 1, \dots, n.$$

Then for each $f \in \mathcal{F}_n$ *the vector* ξ *is an eigenvector of* $f(\overset{1}{A_1}, \dots, \overset{n}{A_n})$ *with eigenvalue* $f(\lambda_1, \dots, \lambda_n)$,

$$f(\overset{1}{A_1}, \dots, \overset{n}{A_n})\xi = f(\lambda_1, \dots, \lambda_n)\xi.$$

Proof. The proof of this theorem can be carried out in the usual way. Namely, we first consider decomposable symbols; for these symbols the statement of each item follows directly from the corresponding item of Theorem I.4. Next we extend the result to arbitrary symbols by continuity. □

Let us make some remarks on Theorems I.3 and I.4. These theorems express naturality properties of the Feynman operator calculus: item 1° says that it behaves naturally under homomorphisms of algebras; item 4° says the same about its behaviour under homomorphisms of linear spaces in which the operators act; item 5° states that one may as well restrict consideration to common invariant subspaces where necessary, and finally, items 2° and 3° express quite natural behavior with respect to the commutation operation, in the spirit of von Neumann's definition of functions of operators via bicommutants.

It should be pointed out, however, that item 5° is less important as soon as functions of several operators are considered because, typically, a tuple of noncommuting operators has no common eigenvectors at all, and in any case, a system of such vectors cannot be complete (this is quite understandable since operators commute on each common eigenspace and hence on the sum of common eigenspaces.)

It is clear that the assertions of these theorems are far from being independent. In fact, we can draw the following diagram showing the dependence between some of our assertions:

$$\begin{array}{ccccccccc}
\text{The continuity} & & 2^o & = & 2^o & & & & \\
\text{of the mapping} & \Rightarrow & \Updownarrow & & \Updownarrow & \Rightarrow & \text{uniqueness} & \Rightarrow & 1^o. \\
\frac{\delta}{\delta x} : \mathcal{F} \to \mathcal{F} \hat{\otimes} \mathcal{F} & & 4^o & \Rightarrow & 3^o & & \text{of } \mu_A & &
\end{array}$$

We have already mentioned and used some of these implications in our proof. However, let us discuss those implications which remain not elucidated as yet. The implication $4^o \Rightarrow 3^o$ can be proved in a standard way: we consider the algebra $\mathcal{A} \oplus \mathcal{A}$ and rewrite the commutation relation (I.14) in the form

$$\begin{pmatrix} A & 0 \\ 0 & C \end{pmatrix} \begin{pmatrix} 0 & B \\ 0 & 0 \end{pmatrix} = \begin{pmatrix} 0 & B \\ 0 & 0 \end{pmatrix} \begin{pmatrix} A & 0 \\ 0 & C \end{pmatrix}.$$

It remains to follow the lines of proof of item 4°. The implication $4^o \Rightarrow 3^o$ can be proved as follows. We consider A, B, and C as linear operators in $\mathcal{A}$ acting by multiplication on the left, i.e., we pass to the left regular representation. We have

$$L_A L_B = L_B L_C,$$

and item 4° applies.

In conclusion let us state the following commutation theorem.

Theorem I.6 *Let $A, C \in \mathcal{A}$ be arbitrary generators; let $B, D \in \mathcal{A}$; and suppose that*

$$AB = BC + D.$$

Then for any symbol $f \in \mathcal{F}$ we have

$$f(A)B = Bf(C) + \overset{2}{D}\frac{\delta f}{\delta x}(\overset{3}{A}, \overset{1}{C}).$$

Proof. This can be proved by the following computation:

$$\begin{aligned} f(A)B - Bf(C) &= \overset{2}{B}(f(\overset{3}{A}) - f(\overset{1}{C})) = \overset{2}{B}(\overset{3}{A} - \overset{1}{C})\frac{\delta f}{\delta x}(\overset{3}{A}, \overset{1}{C}) \\ &= \overset{2}{B}(\overset{3}{A} - \overset{1}{C})\frac{\delta f}{\delta x}(\overset{4}{A}, \overset{0}{C}) \quad \text{(moving indices apart)} \\ &= \overset{2}{[\![}\overset{2}{B}(\overset{3}{A} - \overset{1}{C})]\!]\frac{\delta f}{\delta x}(\overset{4}{A}, \overset{0}{C}) \quad \text{(extraction of a linear factor)} \\ &= \overset{2}{D}\frac{\delta f}{\delta x}(\overset{3}{A}, \overset{1}{A}), \end{aligned}$$

as desired. □

2.4 Tempered Symbols and Generators of Tempered Groups

Though the discussion of functional-analytic questions has been postponed until Chapter IV, we consider here a particular realization of the construction presented in Subsection 2.2. This realization deals with symbols of tempered growth and the related class of generators and is particularly needed in applications to differential equations.

(a) The Symbol Classes.

We introduce the space $\mathcal{F}_n = S^\infty(\mathbb{R}^n)$ of n-ary symbols. By definition, the space $S^\infty(\mathbb{R}^n)$ consists of all functions $f(y_1, \dots, y_n)$, $(y_1, \dots, y_n) = y \in \mathbb{R}^n$, satisfying the estimate

$$\left|\left(\frac{\partial}{\partial y}\right)^\alpha f(y)\right| \le C_\alpha (1 + |y|)^r,$$

for any multi-index $\alpha = (\alpha_1, \dots, \alpha_n)$, with some constant r independent of α (but depending on f).

A (generalized) sequence $f_\beta \in S^\infty(\mathbb{R}^n)$ is said to be *convergent* to an $f \in S^\infty(\mathbb{R}^n)$ if there exists an r such that

$$\sup_{y \in \mathbb{R}^n} \left\{ \left\| \left(\frac{\partial}{\partial y}\right)^\alpha (f_\beta(y) - f(y)) \right\| (1 + |y|)^{-r} \right\} \to 0$$

for any multi-index α. The algebra $S^\infty(\mathbb{R}^n)$ is complete and the algebraic operations are continuous with respect to this convergence. Moreover, one has

$$S^\infty(\mathbb{R}^n) = S^\infty(\mathbb{R}^1)\hat{\otimes}\cdots\hat{\otimes}S^\infty(\mathbb{R}^1),$$

and the difference derivative is a continuous operator

$$\frac{\delta}{\delta y} : S^\infty(\mathbb{R}^1) \to S^\infty(\mathbb{R}^2).$$

These properties (whose proof we cannot carry out as yet since we do not have the material of Chapter IV at our disposal) show that the symbol spaces $S^\infty(\mathbb{R}^n)$ satisfy all the requirements imposed in Subsection 2.2.

(b) The Operator Classes.

We consider operators acting on a *Hilbert scale.* A Hilbert scale is a sequence of densely embedded Hilbert spaces

$$\cdots \subset H_{-2} \subset H_{-1} \subset H_0 \subset H_1 \subset H_2 \subset \cdots$$

indexed by $\mathbb{Z}$ or $\mathbb{R}$. To be definite, we assume that the indexing set is discrete and deal with $\mathbb{Z}$-indexed Hilbert scales. It is assumed that the set

$$H_{-\infty} = \bigcap_{j\in\mathbb{Z}} H_j$$

is dense in each H_s. A (generalized) sequence $h_\beta \in H_\infty = \bigcup_{j\in\mathbb{Z}} H_j$ is convergent to $h \in H_\infty$ if there exists a k such that for $\beta \geq \beta_0$ all $h_\beta \in H_k$ and $h_\beta \to h$ in H_k. A linear operator A on H_∞ is said to be bounded (of order r) in the scale $\{H_s\}$ if there exists an integer r such that for any s one has

$$AH_s \subset H_{s+r}$$

and the operator

$$A|_{H_s} : H_s \to H_{s+r}$$

is continuous. The minimal possible r is called the order of A and is denoted by ord A. If the set of possible r does not have a lower bound, we say that ord $A = -\infty$.

A bounded operator A in the scale $\{H_s\}$ is said to be the generator of a semigroup of tempered growth (or simply a *tempered generator*) if the Cauchy problem

$$-i\frac{du}{dt} = Au, \quad u|_{t=0} = u_0,$$

has a unique solution for any $u_0 \in H_\infty$, and there exists an l such that for any $s \in \mathbb{Z}$ the inclusion $u_0 \in H_s$ implies that $u(t) \in H_{s+l}$ for all t, is differentiable in H_{s+l}, and

$$||u(t)||_{s+l} \leq C(1+|t|)^N ||u_0||_s,$$

where $||h||_k$ is the norm of h in H_k and the constants C and N depend on s but are independent of $u_0 \in H_s$.

Thus, a tempered semigroup in a Hilbert scale grows at most polynomially as $t \to \infty$. We denote $u(t) = \exp(iAt)u_0$.

(c) The Functional Calculus.

Let $\mathcal{A}$ be the algebra of bounded operators in the scale $\{H_s\}$, and let $A \in \mathcal{A}$ be a tempered generator in $\{H_s\}$. Let us define the mapping

$$\mu_A : S^\infty(\mathbb{R}^1) \to \mathcal{A}$$

by setting

$$\mu_A(f) = \frac{(i+A)^m}{\sqrt{2\pi i}} \int_{\infty}^{\infty} \exp(iAt)\tilde{g}(t)dt,$$

where the integral is in the sense of strong convergence,

$$g(y) = \frac{f(y)}{(y+i)^m},$$

$$\tilde{g}(t) = \frac{1}{\sqrt{-2\pi i}} \int_{-\infty}^{\infty} e^{-ity} g(y)dy$$

is the Fourier transform of $g(y)$, and the number m is chosen large enough to ensure that $\tilde{g}(t)$ is continuous.

It is easy to prove, using our definitions and the properties of the Fourier transform, that any possible choice of m gives the same result and that the resulting operator is bounded in $\{H_s\}$. Moreover, μ_A takes y into A and is an algebra homomorphism; let us give a formal calculation proving the latter statement: if

$$\mu_A(h) = \frac{(i+A)^{m_1}}{\sqrt{-2\pi i}} \int_{-\infty}^{\infty} \exp(iAt)\tilde{k}(t)\,dt,$$

where $\tilde{k}(t)$ is the Fourier transform of $k(y) = (i+y)^{-m_1}h(y)$, then

$$\begin{aligned}\mu_A(f)\mu_A(h) &= \frac{(i+A)^{m+m_1}}{2\pi i} \int_{-\infty}^{\infty}\int_{-\infty}^{\infty} \exp(iA(t+\tau))\tilde{g}(t)\tilde{k}(\tau)\,d\tau \\ &= \frac{(i+A)^{m+m_1}}{\sqrt{2\pi i}} \int_{-\infty}^{\infty} \exp(i\eta A)\left\{\int_{-\infty}^{\infty} \frac{\tilde{g}(\eta-\tau)\tilde{k}(\tau)}{\sqrt{2\pi i}}\,d\tau\right\} d\eta.\end{aligned}$$

By the properties of the Fourier transform of the convolution, the expression in braces in the integrand on the right-hand side of this equation is just the Fourier transform of the product $g(y)\,k(y)$, and so we obtain

$$\mu_A(f)\mu_A(h) = \mu_A(fh), \quad \text{or} \quad f(A)h(A) = fh(A),$$

as desired.

We can now define $f(\overset{1}{A_1}, \dots, \overset{n}{A_n})$ for a function $f \in S^\infty(\mathbb{R}^n)$ and a Feynman tuple $(\overset{1}{A_1}, \dots, \overset{n}{A_n})$ of tempered generators in a usual way (see Remark I.3). Obviously, we get

$$f(\overset{1}{A_1}, \dots, \overset{n}{A_n}) = \left(\frac{1}{2\pi i}\right)^{n/2} \int \tilde{f}(t_1, \dots, t_n) \exp(iA_n t_n) \dots \exp(iA_1 t_1) dt_1 \dots dt_n,$$

where

$$\tilde{f}(t_1, \dots, t_n) = \left(\frac{i}{2\pi}\right)^{n/2} \int f(y_1, \dots, y_n) \exp(-it_1 y_1 - \dots - it_n y_n) dy_1 \dots dy_n$$

is the Fourier transform of f and the integral is understood in the sense of strong convergence.

We have omitted here several important details, which will be clarified in Chapter IV.

2.5 The Influence of the Symbol Classes on the Properties of Generators

The choice of the class $\mathcal{F}$ of unary symbols in fact determines the possible properties of generators in a rather restrictive manner; this is evident in itself from general considerations, and this was confirmed by the example considered in the preceding subsection. Here we present some more simple examples clarifying the subject.

Example I.3 We begin with a simple remark that the structure of $\mathcal{F}$ is closely related to the possible spectrum of a generator. For instance, if the symbol

$$f(x) = (x - \alpha)^{-1}$$

belongs to $\mathcal{F}$, then necessarily $\alpha \notin \sigma(A)$ for any generator A (here $\sigma(A)$ is the spectrum of A). Indeed, we have

$$(A - \alpha)f(A) = f(A)(A - \alpha) = 1,$$

and so $A - \alpha$ is right and left invertible. Going back to Section 2.4, we see that $(x - \alpha)^{-1} \in S^\infty(\mathbb{R}^1)$ for any nonreal α. Therefore, the spectrum of each generator lies completely on the real axis, so that the polynomial estimates for the growth of $\exp(iAt)$ become less surprising (in fact, the position of the spectrum of A on the real axis is not sufficient for the polynomial growth of $\exp(itA)$; some estimates of the resolvent of A are also needed; however, these estimates can also be derived from the fact that

$$\mu_A : f \mapsto f(A)$$

is defined and continuous on $S^\infty(\mathbb{R}^1)$.

Example I.4 This example is somewhat less trivial. Suppose that the symbol class $\mathcal{F}_3$ contains the function

$$f(y_1, y_2, y_3) = e^{i(y_3 - y_1)y_2}.$$

Suppose also that operators A and B satisfy

$$[A, B] = -i$$

(for example, $A = -i\frac{d}{dx}$ and $B = x$). Then at least one of the operators A and B is *not* an $\mathcal{F}$-generator.

Indeed, we have, by Theorem I.10,

$$\begin{aligned} f(\overset{1}{A}, \overset{2}{B}, \overset{3}{A}) &= f(\overset{1}{A}, \overset{3}{B}, \overset{2}{A}) + \overset{4}{[A, B]} \frac{\delta^2 f}{\delta y_2 \delta y_3}(\overset{1}{A}, \overset{3}{B}, \overset{5}{B}, \overset{2}{A}, \overset{6}{A}) \\ &= f(\overset{1}{A}, \overset{3}{B}, \overset{2}{A}) - i \frac{\delta}{\delta y_3} \frac{\partial f}{\partial y_2}(\overset{1}{A}, \overset{3}{B}, \overset{2}{A}, \overset{4}{A}), \end{aligned}$$

since the commutator $[A, B]$ commutes with B. Furthermore,

$$\frac{\delta}{\delta y_3} \frac{\partial f}{\partial y_2}(y_1, y_2, y_3, y_4) = \left\{ \frac{\delta}{\delta y_3} i(y_3 - y_1) e^{i(y_3 - y_1)y_2} \right\} (y_1, y_2, y_3, y_4).$$

We use the Leibniz rule for difference derivative

$$\frac{\delta}{\delta x}(f(x)g(x)) = \frac{\delta f}{\delta x}(x_1, x_2) g(x_1) + f(x_2) \frac{\delta g}{\delta x}(x_1, x_2)$$

and obtain

$$\frac{\delta}{\delta y_3} \frac{\partial}{\partial y_2}(y_1, y_2, y_3, y_4) = e^{i(y_4 - y_1)y_2} + i(y_3 - y_1)\varphi(y_1, y_2, y_3, y_4),$$

where

$$\varphi(y_1, y_2, y_3, y_4) = \frac{e^{i(y_3 - y_1)y_2} - e^{i(y_4 - y_1)y_2}}{y_2 - y_4}.$$

By substituting the last expression into the previously obtained expression for $f(\overset{1}{A}, \overset{2}{B}, \overset{3}{A})$, we find that

$$e^{i(\overset{3}{A} - \overset{1}{A})\overset{2}{B}} = e^{i(\overset{2}{A} - \overset{1}{A})\overset{3}{B}} + e^{i(\overset{4}{A} - \overset{1}{A})\overset{3}{B}} + (\overset{2}{A} - \overset{1}{A})\varphi(\overset{1}{A}, \overset{3}{A}, \overset{2}{A}, \overset{4}{A}).$$

However,

$$e^{i(\overset{2}{A}-\overset{1}{A})\overset{3}{B}} = e^{i(\overset{2}{A}-\overset{2}{A})\overset{3}{B}} = 1$$

by Proposition I.1,

$$e^{i(\overset{4}{A}-\overset{1}{A})\overset{3}{B}} = e^{i(\overset{3}{A}-\overset{1}{A})\overset{2}{B}},$$

and

$$(\overset{2}{A}-\overset{1}{A})\varphi(\overset{1}{A},\overset{3}{B},\overset{2}{A},\overset{4}{A}) = [\![\overset{12}{A}-\overset{1}{A}]\!]\varphi(\overset{0}{A},\overset{4}{B},\overset{3}{A},\overset{5}{B}) = 0.$$

Thus we obtain

$$e^{i(\overset{3}{A}-\overset{1}{A})\overset{2}{B}} = 1 + e^{i(\overset{3}{A}-\overset{1}{A})\overset{2}{B}},$$

which is a contradiction.

Thus, A and B cannot be generators simultaneously.

Example I.5 Let Γ be the Riemann surface of the function $\sqrt{z} = \sqrt{x+iy}$. It is a double covering of $\mathbb{C}\backslash\{0\}$. We will consider Γ as a manifold with local coordinates (x, y).

Consider the operators

$$A = -i\frac{\partial}{\partial x} \quad \text{and} \quad B = -i\frac{\partial}{\partial y}$$

on the Hilbert space $L_2(\Gamma)$ (the inner product on $L_2(\Gamma)$ is defined by the integral over the Lebesgue measure $dxdy$). These operators are essentially self-adjoint on the dense subspace $C_0^\infty(\Gamma) \subset L_2(\Gamma)$ of smooth compactly supported functions and commute on $C_0^\infty(\Gamma)$,

$$\frac{\partial^2 f}{\partial y \partial x} = \frac{\partial^2 f}{\partial x \partial y}$$

for any $f \in C_0^\infty(\Gamma)$. Consequently, they generate unitary groups on $L_2(\Gamma)$.

Thus, $[A, B] = 0$, and, according to the permutation formula, we should apparently have

$$g(\overset{1}{A}, \overset{2}{B}) = g(\overset{2}{A}, \overset{1}{B})$$

for any symbol $g \in S^\infty(\mathbb{R}^2)$. However, this is not the case. Indeed, consider the symbol

$$g(z_1, z_2) = e^{itz_1}e^{i\tau z_2}$$

(here t and τ are parameters). The corresponding operators

$$g(\overset{1}{A}, \overset{2}{B}) = e^{\tau\frac{\partial}{\partial y}}e^{t\frac{\partial}{\partial x}} \quad \text{and} \quad g(\overset{2}{A}, \overset{1}{B}) = e^{t\frac{\partial}{\partial x}}e^{\tau\frac{\partial}{\partial y}}$$

are simply the products of the corresponding unitary groups, which are the translations by t and τ along the x- and y-axis, respectively. But the translations along the two axes do not commute on Γ: starting from the point $(x, y) = (1, 1)$ and performing the translations by -2 along both axes, we move to different sheets of the Riemann surface depending on the order of the translations performed.

Thus A and B cannot be generators simultaneously in any Hilbert scale including $L_2(\Gamma)$. The reason for this phenomenon is that the product of the groups $e^{itA}e^{i\tau B}$ is not strongly differentiable on $C_0^\infty(\Gamma)$ and $C_0^\infty(\Gamma)$ is not invariant under these semigroups: as soon as in the course of a translation the support of a function collides with the branching point $(0, 0)$, the function loses its differentiability.

2.6 Weyl Quantization

We have given the definition of functions of Feynman tuples of noncommuting operators. The approach was to equip the operators with Feynman indices and arrange them in products according to the ordering of these indices. In the literature one can find another approach to the construction of the functional calculus of several noncommuting operators, known as the *Weyl* functional calculus or *Weyl quantization*. We neither study nor use the Weyl functional calculus in this book but only say a few words about it in this subsection.

In contrast to Feynman quantization, Weyl quantization is *symmetric*; this means that neither of the operators acts first or last. In a sense, they all act "simultaneously".

Let $A_1, \dots, A_n$ be a tuple of operators and f a polynomial of the form

$$f(z_1, \dots, z_n) = \varphi(\alpha_1 z_1 + \alpha_2 z_2 + \cdots + \alpha_n z_n),$$

where $\alpha_1, \dots, \alpha_n$ are numbers and $\varphi(y)$ is a univariate polynomial. Then the Weyl-quantized function of $A_1, \dots, A_n$ with symbol f is defined by

$$f_W(A_1, \dots, A_N) = \varphi([\![\alpha_1 A_1 + \cdots + \alpha_n A_n]\!])$$

(the autonomous brackets are not obligatory here but we let them stand, for clarity). The subscript W is an acronym for "Weyl". Now if $f(z_1, \dots, z_n)$ is an arbitrary polynomial, it can be represented in the form

$$f(z_1, \dots, z_n) = \sum_{\varphi, \alpha_1, \dots, \alpha_n} \varphi(\alpha_1 z_1 + \cdots + \alpha_n z_n),$$

where φ are univariate polynomials, $\alpha_1, \dots, \alpha_n$ are complex numbers, and the sum is finite. We set

$$f_W(A_1, \dots, A_n) = \sum_{\varphi, \alpha_1, \dots, \alpha_n} \varphi([\![\alpha_1 A_1 + \cdots + \alpha_n A_n]\!]);$$

this is well-defined.

The main property of Weyl quantization is its *affine covariance*. Let M be an $n \times n$ matrix and let

$$f(z) = g(Mz);$$

then

$$f(A) = g(MA),$$

where MA is the tuple of operators defined by

$$(MA)_j = \sum_{k=1}^{n} M_{jk} A_k.$$

To define Weyl quantization for general symbol classes, it is necessary to make some additional assumptions, namely:

1°. Any linear combination $\alpha_1 A_1 + \cdots + \alpha_n A_n$ is an $\mathcal{F}$-generator, where $\mathcal{F}$ is the class of univariate symbols.

2°. The set of functions $f(\alpha_1 z_1 + \cdots + \alpha_n z_n)$ for various $f \in \mathcal{F}$ and $\alpha_1, \ldots, \alpha_n$ is dense in the class $\mathcal{F}_n$ of n-ary symbols. (The admissible values of the scalars $\alpha_1, \ldots, \alpha_n$ can be taken either complex or real, depending on the specific case, but simultaneously in 1° and 2°).

Under these assumptions, the affine covariance property in conjunction with the requirement that $f(A)$ is defined as usual for unary symbols determines the Weyl quantization uniquely.

Let us consider two examples.

Example I.6 (Weyl functions of self-adjoint operators) Let $A_1, \ldots, A_n$ be self-adjoint operators on a Hilbert space $\mathcal{H}$. We assume that there exists a dense linear subset $D \subset \mathcal{H}$ such that for any $\alpha_1, \ldots, \alpha_n \in \mathbb{R}$ the linear combination $\alpha A = \alpha_1 A_1 + \cdots + \alpha_n A_n$ is essentially self-adjoint on D (that is, αA is closable and its closure ia a self-adjoint operator on $\mathcal{H}$. Then the exponential $U(\alpha) = \exp(\alpha A)$ is well-defined for any $\alpha \in \mathbb{R}^n$ and is strongly continuous with respect to α (see [179]).

We define $f_W(A_1, \ldots, A_n)$ for $f(z) \in C_0^\infty(\mathbb{R}^n)$ by the formula

$$f_W(A_1, \ldots, A_n) = \left(\frac{1}{2\pi i}\right)^{n/2} \int_{\mathbb{R}^n} \tilde{f}(\alpha) U(\alpha) d\alpha,$$

where $\tilde{f}(\alpha)$ is the Fourier transform of $f(z)$ and the integral is convergent in the strong sense, or else by the formula

$$f_W(A_1, \ldots, A_n) = \left(\frac{1}{2\pi i}\right)^{n/2} \int_{\mathbb{R}^n} f(z) \tilde{U}(z)\, dz,$$

where the operator-valued distribution $\tilde{U}(z)$ is the Fourier transform (in the sense of strong convergence) of $U(\alpha)$ (see [2] and [3]).

If D is invariant under $A_1, \ldots, A_n$, then the above definition can be extended to symbols of tempered growth (i.e., symbols growing polynomially as $|z| \to \infty$ with all derivatives).

Weyl quantization has a minor advantage from the quantum-mechanical point of view. Namely, if $A_1, \ldots, A_n$ are self-adjoint, then Weyl quantization takes real-valued functions (Hamiltonian functions in quantum mechanics) into self-adjoint operators. This is not at all essential in Cartesian coordinates since the Hamiltonian function usually has the form

$$H(x, p) = \frac{p^2}{2m} + V(x),$$

the position and momenta operators are not mixed and the Feynman and Weyl quantizations give the same result. However, in curvilinear generalized coordinates the Hamiltonian function becomes

$$H(q, p) = \sum a_{ij}(q) p_i p_j + V(q),$$

and the choice of ordering is essential. It is Weyl quantization that takes $H(q, p)$ into a self-adjoint operator.

Example I.7 (Analytic Weyl functions of bounded operators) Let $A_1, \ldots, A_n$ be bounded operators in a Banach space B such that for any $(\alpha_1, \ldots, \alpha_n) \in \mathbb{C}^n$ with $|\alpha_1|^2 + \cdots + |\alpha_n|^2 \leq 1$ the spectrum $\sigma(\alpha_1 A_1 + \cdots + \alpha_n A_n)$ lies inside the unit ball U in $\mathbb{C}^n$ centered at the origin. Let $f(z)$, $z \in \mathbb{C}^n$, be a holomorphic function in a neighbourhood of the unit ball. Then the Cauchy–Fantappiè formula

$$f(z) = \frac{(n-1)!}{(2\pi i)^n} \int_{\{|\zeta|=1\}} f(\zeta) \frac{\sum_{\nu=1}^{n} (-1)^{\nu-1} \bar{\zeta}_\nu \bar{\zeta}_1 \wedge \cdots \wedge d\bar{\zeta}_{\nu-1} \wedge d\bar{\zeta}_{\nu+1} \wedge \cdots \wedge d\bar{\zeta}_n \wedge d\zeta}{\left(1 - \sum_{\nu=1}^{n} \bar{\zeta}_\nu z_\nu\right)^n}$$

is valid for $z \in U$. For $\zeta \in \partial U$ we have $\sigma(\bar{\zeta} A) \Subset U$, and, consequently, the resolvent $(1 - \bar{\zeta} A)^{-1}$ is well-defined. We use the Cauchy–Fantappiè formula to define the Weyl-quantized function $f_W(A_1, \ldots, A_n)$,

$$\begin{aligned} f_W(A_1, \ldots, A_n) &= \frac{(n-1)!}{(2\pi i)^n} \int_{\{|\zeta|=1\}} f(\zeta)(1 - \bar{\zeta} A)^{-n} \\ &\quad \times \sum_{nu=1}^{n} (-1)^{\nu-1} \bar{\zeta}_\nu \bar{\zeta}_1 \wedge \cdots \wedge d\bar{\zeta}_{\nu-1} \wedge d\bar{\zeta}_{\nu+1} \wedge \cdots \wedge d\bar{\zeta}_n \wedge d\zeta. \end{aligned}$$

We considered here the simplest case in which the domain is the unit ball. Of course, more complicated spectrum structures can be considered as well by using the Cauchy–Fantappiè formula for more complicated domains; we leave this to the reader as an exercise.

3 Noncommutative Differential Calculus

The conventional differential calculus deals with the local behaviour of functions; given a function $f(x)$ and a point x_0, the main question is as follows: what can be said about $f(x_0 + \Delta x)$ as $\Delta x \to 0$? The ultimate answer is that, provided $f(x)$ is a "well-behaved" function, one has

$$f(x_0+\Delta x)=f(x_0)+f'(x_0)\Delta x+\frac{f''(x_0)}{2}(\Delta x)^2+\cdots+\frac{f^{(N)}(x_0)}{N!}(\Delta x)^N+O(\Delta x^{N+1}),$$

i.e., $f(x_0 + \Delta x)$ can be expanded into a Taylor series in powers of Δx. (Of course all of differential calculus is not reduced to this; but this is the essence.)

We will pose a similar question for functions of noncommuting operators, and the machinery developed to answer it will be called *noncommutative differential calculus*. We will see that the matter is rich in subtleties and try to explain them as clearly as possible.

Thus, we should consider $f(A + \Delta A)$, where ΔA is "small", $\Delta A \to 0$; but what does this requirement mean? We follow the usual practice of perturbation theory and take

$$\Delta A = \varepsilon B,$$

where ε is a small *numerical* parameter. This permits us to keep track of infinitesimals of different orders easily, since the orders are indicated by the powers of ε.

In Subsection 1.4 we have already considered two examples of computation for $f(A+\varepsilon B)$, namely,

$$f(x) = \frac{1}{\lambda - x}$$

and $f(x) = \exp(x)$. Here we deal with the general setting.

In the conventional differential calculus, the first-order term $f'(x_0)\Delta x$ of the Taylor expansion is of extreme importance; it bears the special name of the *first differential* of f, and teaching differential calculus usually begins with its intensive study. By analogy with that case, we are specifically interested in the term of order ε in the expression

$$f(A+\varepsilon B) = f(A) + \varepsilon C_1 + \varepsilon^2 C_2 + \cdots .$$

It would be wise to ask why such an expansion exists at all; however, as is the case for the conventional calculus, the computational method at the same time provides justification for the expansion.

The coefficient of ε can be obtained as

$$C_1 = \frac{d}{d\varepsilon}\,[f(A+\varepsilon B)]|_{\varepsilon=0}\,, \tag{I.16}$$

i.e., by differentiating with respect to ε followed by setting $\varepsilon = 0$. However, we will find an expression for a more general "derivative", of which this is a particular case.

3.1 The Derivation Formula

Let $\mathcal{A}$ be an operator algebra. A *derivation* of $\mathcal{A}$ is an arbitrary linear mapping

$$D : \mathcal{A} \to \mathcal{A} \tag{I.17}$$

satisfying the *Leibniz rule*

$$D(uv) = (Du)v + u(Dv), \quad u, v \in \mathcal{A}.$$

Let $A \in \mathcal{A}$ be a generator and $f \in \mathcal{F}$ a symbol. We claim that

$$D[f(A)] = \overset{2}{[DA]} \frac{\delta f}{\delta x}(\overset{1}{A}, \overset{3}{A}) \tag{I.18}$$

for an arbitrary derivation (I.17).

The proof will be divided into several stages. First, we consider a special class of derivations.

Definition I.3 Derivations of the form

$$D(A) = D_B(A) \equiv BA - AB,$$

where B is an arbitrary element of $\mathcal{A}$, are called *inner derivations* of $\mathcal{A}$.

It is easy to show that the commutator indeed defines a derivation; the corresponding computation is rather simple and is left to the reader.

Proposition I.3 *Formula* (I.18) *is valid for inner derivations.*

Proof. We need to compute the commutator

$$[B, f(A)] = Bf(A) - f(A)B.$$

So far, there are no Feynman indices in this expression; let us introduce some.

Note that they can be chosen independently for either summand on the right; we may prefer

$$\overset{2}{B} f(\overset{1}{A}) - f(\overset{2}{A})\overset{1}{B}, \quad \text{or} \quad \overset{10}{B} f(\overset{7}{A}) - f(\overset{5}{A})\overset{2}{B},$$

or whichever other choice provided that the indices are consistent with the order of factors. We would like to factor out B, and so we write

$$[B, f(A)] = \overset{2}{B} f(\overset{1}{A}) - f(\overset{3}{A})\overset{2}{B} = \overset{2}{B}(f(\overset{1}{A}) - f(\overset{3}{A})).$$

Of course, we cannot simply put $f(\overset{1}{A}) - f(\overset{3}{A}) = 0$, because the Feynman index of B lies just between those of the first and the second A.

The subsequent computations are quite simple: we obtain, by multiplying and dividing by $\overset{1}{A} - \overset{3}{A}$,

$$[B, f(A)] = \overset{2}{B}(\overset{1}{A} - \overset{3}{A})\frac{f(\overset{1}{A}) - f(\overset{3}{A})}{\overset{1}{A} - \overset{3}{A}} = \overset{2}{B}(\overset{1}{A} - \overset{3}{A})\frac{\delta f}{\delta x}(\overset{1}{A}, \overset{3}{A})$$

(the division by $\overset{1}{A} - \overset{3}{A}$ seems to be very dangerous, but in fact is not so; actually, we used the identity

$$f(x) - f(y) = (x - y)\frac{\delta f}{\delta x}(x, y),$$

which is valid everywhere including the diagonal $x = y$, where

$$\frac{\delta f}{\delta x} f(x, x) = f'(x).$$

We now move apart the Feynman indices over the A's, thus obtaining

$$\overset{2}{B}(\overset{1}{A} - \overset{3}{A})\frac{\delta f}{\delta x}(\overset{1}{A}, \overset{3}{A}) = \overset{2}{B}(\overset{1}{A} - \overset{3}{A})\frac{\delta f}{\delta x}(\overset{0}{A}, \overset{4}{A}) = [\![\overset{2}{\overset{2}{B}}(\overset{1}{A} - \overset{3}{A}]\!]\frac{\delta f}{\delta x}(\overset{1}{A}, \overset{3}{A})$$

(it is easy to see that the introduction of autonomous brackets is valid). It remains to notice that the operator in brackets is equal to

$$\overset{2}{B}(\overset{1}{A} - \overset{3}{A}) = BA - AB = [B, A],$$

and, consequently,

$$[B, f(A)] = \overset{2}{[B, A]}\frac{\delta f}{\delta x}(\overset{1}{A}, \overset{3}{A}).$$

The proposition is proved. □

Theorem I.7 *Formula* (I.18) *is valid for an arbitrary derivation of* $\mathcal{A}$.

In order to prove Theorem I.7, we need the following lemma.

Lemma I.1 *The left regular* L *representation takes each derivation* D *of* $\mathcal{A}$ *into the inner derivation* LDL^{-1} *of* $\mathcal{L}(\mathcal{A})$.

Proof. Let D be a derivation of $\mathcal{A}$ and $A \in \mathcal{A}$. For any $C \in \mathcal{A}$

$$L_{D(A)}(C) = D(A)C = D(AC) - AD(C)$$

by the Leibniz rule; we can thus write

$$L_{D(A)}(C) = D(L_A C) - L_A D(C) = [D, L_A](C),$$

i.e.,

$$L_{D(A)} = [D, L_A],$$

that is, we have the commutative diagram

$$\begin{array}{ccc} \mathcal{A} & \xrightarrow{D} & \mathcal{A} \\ L\downarrow & & \downarrow L \\ \mathcal{L}(\mathcal{A}) & \xrightarrow{[D,\,\cdot\,]} & \mathcal{L}(\mathcal{A}), \end{array}$$

which shows that any derivation D is transformed by L into the operator $[D, \cdot\,]$ of commutation with D in $\mathcal{L}(\mathcal{A})$. The lemma is proved. □

Proof of Theorem I.7. Let D be an arbitrary derivation of $\mathcal{A}$. By Proposition I.3,

$$[D, f(L_A)] = \overset{2}{[D, L_A]}\frac{\delta f}{\delta x}(\overset{1}{L_A}, \overset{3}{L_A}),$$

By Theorem I.2,

$$f(L_A) = L_{f(A)},$$

and so, by Lemma I.1, we obtain

$$L_{D(f(A))} = \overset{2}{L}_{D(A)}\frac{\delta f}{\delta x}(\overset{1}{L_A}, \overset{3}{L_A}) = L_{\overset{2}{D}(A)\frac{\delta f}{\delta x}(\overset{1}{A},\overset{3}{A})}$$

(we have applied Theorem I.2 one more time). Since L is a faithful representation, we obtain

$$D(f(A)) = \overset{2}{D}(A)\frac{\delta f}{\delta x}(\overset{1}{A}, \overset{3}{A}),$$

as desired. The proof is complete. □

3.2 The Daletskii–Krein Formula

Let us now return to our original problem of computing the coefficient C_1 defined in (I.16). To this end, consider the algebra $\mathcal{A}_{\{t\}}$ whose elements are (infinitely differentiable) families of elements of $\mathcal{A}$ depending on a numerical parameter t. Clearly, the mapping

$$\frac{d}{dt} : \mathcal{A}_{\{t\}} \to \mathcal{A}_{\{t\}}$$

taking each family $A(t) \in \mathcal{A}_{\{t\}}$ into its t-derivative is a continuous derivation of the algebra $\mathcal{A}_{\{t\}}$. By Theorem I.7 we obtain

$$\frac{d}{dt} f(A(t)) = \overset{2}{A'}(t)\frac{\delta f}{\delta x}(\overset{1}{A}(t), \overset{3}{A}(t)).$$

This is the famous *Daletskii–Krein formula*, obtained by these authors in [29] for the case of self-adjoint unbounded operators on a Hilbert space (their technique involved spectral families). From this we easily derive a formula for the coefficient (I.16). Namely, take $A(\varepsilon) = A + \varepsilon B$; then $A'(\varepsilon) \equiv B$, and we obtain

$$C_1 = \overset{2}{B}\frac{\delta f}{\delta x}(\overset{1}{A}, \overset{3}{A}),$$

i.e.,

$$f(A + \varepsilon B) = f(A) + \varepsilon \overset{2}{B}\frac{\delta f}{\delta x}(\overset{1}{A}, \overset{3}{A}) + O(\varepsilon^2).$$

Remark I.8 If $[B, A] = 0$, one can take the same Feynman index for both A's on the right-hand side of the last equation, thus obtaining the usual Taylor expression

$$f(A + \varepsilon B) = f(A) + \varepsilon B\frac{\partial f}{\partial x}(A) + O(\varepsilon^2).$$

3.3 Higher-Order Expansions

What we discussed in the preceding subsection was, in fact, the first-order infinitesimal calculus in a noncommutative setting. We have evaluated the *differential* of $f(A)$, that is, the linear part of the increment $f(C) - f(A)$ w.r.t. the difference $B = C - A$. It is given by the Daletskii–Krein formula, which can be rewritten as

$$df(A; B) \overset{\text{def}}{\equiv} \left[\frac{d}{d\varepsilon} f(A + \varepsilon B)\right]\bigg|_{\varepsilon=0} = \overset{2}{B}\frac{\delta f}{\delta x}(\overset{1}{A}, \overset{3}{A})$$

and employs at least the notation of noncommutative analysis, although there was no indication of it being useful on the left-hand side.

However, we should like to develop infinitesimal calculus in its full extent, which assumes deriving expansions of arbitrarily high order for the increment cited, with explicitly writing out the remainders.

Hence, let us consider the difference $f(C) - f(A)$ more comprehensively. Assume that

$$C - A = \varepsilon B,$$

where ε, as above, will be treated as a small parameter. Then it becomes much easier to keep track of the orders of various terms in our formulas and to explain convincingly the order of the accuracy of our expansions.

The simplest and clearest method to obtain the expansion of

$$f(C) - f(A) = f(A + \varepsilon B) - f(A)$$

in powers of ε is to apply the Daletskii–Krein formula successively. Namely, we can write down the usual MacLaurin expansion

$$f(A+\varepsilon B) - f(A) \cong \sum_{k=1}^{\infty} \frac{C_k \varepsilon^k}{k!},$$

where the coefficient C_k of ε^k is the kth ε-derivative of $f(A+\varepsilon B)$ at $\varepsilon = 0$. By the way, the Daletskii–Krein formula holds for $\varepsilon \neq 0$ as well:

$$\frac{d}{d\varepsilon} f(A+\varepsilon B) = \overset{2}{B} \frac{\delta f}{\delta x}(\overset{1}{[\![} A+\varepsilon B]\!], \overset{3}{[\![} A+\varepsilon B]\!]).$$

Thus we can differentiate it with respect to ε once more. There will clearly be two terms arising from the differentiation in the first and in the second argument of $\delta f/\delta x$; and the formula itself applies to each of these terms. Hence we obtain

$$\frac{d^2}{d\varepsilon^2} f(A+\varepsilon B) = 2\overset{2}{B}\overset{4}{B} \frac{\delta^2 f}{\delta x^2}(\overset{1}{[\![} A+\varepsilon B]\!], \overset{3}{[\![} A+\varepsilon B]\!], \overset{5}{[\![} A+\varepsilon B]\!])$$

(the factor 2 is due to the presence of two terms, which are equal to each other since $\delta f/\delta x(x, y)$ is symmetric,

$$\frac{\delta f}{\delta x}(x, y) = \frac{\delta f}{\delta x}(y, x),$$

and so it makes no difference with respect to which argument the subsequent derivatives are taken). At this stage, it is easy to predict the general result (and to prove it, which we leave to the reader):

$$\frac{d^k}{d\varepsilon^k} f(A+\varepsilon B) = k!\overset{2}{B}\overset{4}{B}\dots\overset{2k}{B} \frac{\delta^k f}{\delta x^k}(\overset{1}{[\![} A+\varepsilon B]\!], \overset{3}{[\![} A+\varepsilon B]\!], \dots, \overset{2k+1}{[\![} A+\varepsilon B]\!]).$$

Thus the factor $1/k!$ in the MacLaurin expansion cancels out, and we obtain the *Newton formula*

$$f(A+\varepsilon B) - f(A) \cong \sum_{k=1}^{\infty} \varepsilon^k \overset{2}{B}\overset{4}{B}\dots\overset{2k}{B} \frac{\delta^k f}{\delta x^k}(\overset{1}{A}, \overset{3}{A}, \dots, \overset{2k+1}{A})$$

or, forgetting about ε,

$$f(C) - f(A) \cong \sum_{k=1}^{\infty} \underbrace{\overset{2}{[\![} C-A]\!] \overset{4}{[\![} C-A]\!] \dots \overset{2k}{[\![} C-A]\!]}_{k \text{ copies}} \frac{\delta^k f}{\delta x^k} \underbrace{(\overset{1}{A}, \overset{3}{A}, \dots, \overset{2k+1}{A})}_{k+1 \text{ arguments}}.$$

However, these formulas are still of little practical importance, chiefly because of the mysterious sign "$\cong$" in the middle. What does it mean exactly? This is not so easy to explain without going into functional analytic peculiarities, but there is an alternative, stating that we will always be on the safe side if we have an explicit formula for the remainder. Fortunately, such a formula is at hand.

Theorem I.8 (Newton's formula with remainder) *Let A and C be generators. Then for any symbol $f(x)$ and for any positive integer N we have*

$$f(C) - f(A) = \sum_{k=1}^{N-1} \overset{2}{[\![} C - A]\!] \dots \overset{2k}{[\![} C - A]\!] \frac{\delta^k f}{\delta x^k}(\overset{1}{A}, \overset{3}{A}, \dots, \overset{2k+1}{A}) + R_N,$$

where the remainder R_N is given by the formula

$$R_N = \overset{2}{[\![} C - A]\!] \dots \overset{2N}{[\![} C - A]\!] \frac{\delta^N f}{\delta x^N}(\overset{1}{C}, \overset{3}{A}, \dots, \overset{2N+1}{A}).$$

Proof. We proceed by induction on N. Clearly,

$$f(C) - f(A) = f(\overset{1}{C}) - f(\overset{3}{A}) = (\overset{1}{C} - \overset{3}{A}) \frac{f(\overset{1}{C}) - f(\overset{3}{A})}{\overset{1}{C} - \overset{3}{A}}.$$

We can move the indices apart and then isolate the factor $C - A$,

$$(\overset{1}{C} - \overset{3}{A}) \frac{f(\overset{1}{C}) - f(\overset{3}{A})}{\overset{1}{C} - \overset{3}{A}} = (\overset{1}{C} - \overset{3}{A}) \frac{f(\overset{0}{C}) - f(\overset{4}{A})}{\overset{0}{C} - \overset{4}{A}} = \overset{2}{[\![} \overset{1}{C} - \overset{3}{A}]\!] \frac{f(\overset{0}{C}) - f(\overset{4}{A})}{\overset{0}{C} - \overset{4}{A}}.$$

Since $\overset{1}{C} - \overset{3}{A} = C - A$, it follows that

$$f(C) - f(A) = \overset{2}{[\![} C - A]\!] \frac{\delta f}{\delta x}(\overset{1}{C}, \overset{3}{A}), \tag{I.19}$$

which is none other than Newton's formula for $N = 1$. Let us carry out the induction step $N = 1 \Rightarrow N = 2$. To this end, we apply the last identity to the difference [4]

$$\overset{2}{[\![} C - A]\!] \frac{\delta f}{\delta x}(\overset{1}{C}, \overset{3}{A}) - \overset{2}{[\![} C - A]\!] \frac{\delta f}{\delta x}(\overset{1}{A}, \overset{3}{A}).$$

We obtain

$$\overset{2}{[\![} C - A]\!] \frac{\delta f}{\delta x}(\overset{1}{C}, \overset{3}{A}) = \overset{2}{[\![} C - A]\!] \frac{\delta f}{\delta x}(\overset{1}{A}, \overset{3}{A}) + \overset{0}{[\![} C - A]\!] \overset{2}{[\![} C - A]\!] \frac{\delta^2 f}{\delta x^2}(\overset{-1}{A}, \overset{1}{A}, \overset{3}{A}).$$

The second term, up to an inessential change in indices, is just the remainder R_2. Proceeding in a similar way, we accomplish the proof. □

---

[4]This means that we apply the identity (I.19) to the function $\overset{2}{[\![} C - A]\!] \delta f/\delta x(z, \overset{3}{A})$ substituting operators $\overset{1}{C}$ and $\overset{1}{A}$ instead of z.

The remainder is small in the following sense. If $C - A = \varepsilon B$, then ε^N factors out of R_N. Also, the distinguishing feature of Newton's formula is that all its terms, except for the remainder, are "multilinear forms in

$$\overset{2}{[\![}C - A]\!], \ldots, \overset{2k}{[\![}C - A]\!]$$

with various Feynman indices".

In the commuting case ($[A, C] = 0$) Newton's formula is reduced to the conventional Taylor formula with operator arguments. However, the commuting case is a distinguished one; it can be viewed as the "maximally degenerate case" of noncommutativity. Therefore, it is not surprising that there are a large variety of formulas generalizing the Taylor formula to the noncommutative case. Let us get acquainted with one of these formulas.

Theorem I.9 (Taylor's formula with remainder) *Under the hypotheses of Theorem* I.8 *one has*

$$f(C) - f(A) = \sum_{k=1}^{N} \frac{1}{k!} f^{(k)}(A)[\![(\overset{1}{C} - \overset{2}{A})^k]\!] + Q_N,$$

where the remainder Q_N is given by the formula

$$Q_N = \overset{2}{[\![}(\overset{1}{C} - \overset{2}{A})^N]\!] \frac{\delta^N f}{\delta x^N}(\overset{1}{C}, \overset{3}{A}, \ldots, \overset{3}{A}).$$

Proof. By the usual Taylor formula for functions of numerical arguments,

$$f(y) - f(x) = \sum_{k=1}^{N} \frac{1}{k!} f^{(k)}(x)\,(y - x)^k + Q_N(x, y),$$

where $Q_N(x, y)$ is easily proved to be equal to

$$Q_N(x, y) = \frac{\delta^N f}{\delta x^N}(y, x, , \ldots, x)(y - x)^N.$$

Let us insert $y \to \overset{1}{C}$, $x \to \overset{2}{A}$ into the formula for $f(y) - f(x)$. We obtain

$$f(C) - f(A) = \sum_{k=1}^{N} \frac{1}{k!} f^{(k)}(\overset{2}{A})(\overset{1}{C} - \overset{2}{A})^k + Q_N(\overset{2}{A}, \overset{1}{C}).$$

This yields, by moving indices apart and isolating appropriate factors, the desired formula. The proof is complete. □

Let us now make a comparative analysis of the Newton and Taylor formulas. Both formulas are quite easy to obtain; beyond this, there are more differences than similarities.

First of all, strange as it may seem, the Taylor formula *does not* generally provide an expansion in powers of ε if $C = A + \varepsilon B$. We claim that, unless A and B satisfy some algebraic relations, the remainder

$$Q_N = \overset{2}{[\![}(\overset{1}{C} - \overset{2}{A})^N]\!] \frac{\delta^N f}{\delta x^N}(\overset{1}{C}, \overset{3}{A}, \dots, \overset{3}{A})$$

is of order ε *regardless* of N. To understand this well, let us consider a few of the first terms of the expansion. The point is that although

$$\overset{1}{C} - \overset{2}{A} = C - A = \varepsilon B = O(\varepsilon),$$

this does not mean that

$$(\overset{1}{C} - \overset{2}{A})^N = O(\varepsilon^N).$$

Indeed, say,

$$(\overset{1}{C} - \overset{2}{A})^2 \neq (C - A)^2.$$

Instead, we have

$$(\overset{1}{C} - \overset{2}{A})^2 = C^2 + A^2 - 2AC,$$
$$(C - A)^2 = C^2 + A^2 - AC - CA,$$

and these two operators differ by $CA - AC = [C, A]$. However, $C = A + \varepsilon B$ and hence

$$[C, A] = \varepsilon[B, A],$$

because A commutes with itself. Eventually, we obtain

$$(\overset{1}{C} - \overset{2}{A})^2 = \varepsilon^2 B^2 + \varepsilon[B, A],$$

and this is not $O(\varepsilon^2)$ unless $[B, A] = 0$.

Next, let us compute $(\overset{1}{C} - \overset{2}{A})^3$. By similar computation, we are led to the relation

$$(\overset{1}{C} - \overset{2}{A})^3 = \varepsilon^2[A, B]B - \varepsilon^2[A, (B - A)^2] + \varepsilon[A, [A, B]].$$

We see that $(\overset{1}{C} - \overset{2}{A})^3 = O(\varepsilon)$ in general. However, if $[A, [A, B]] = 0$, then $(\overset{1}{C} - \overset{2}{A})^2) = O(\varepsilon^2)$. Similarly, if for some m_0

$$\underbrace{[A, [A, \dots [A,}_{m \text{ times}} B] \dots] = 0,$$

then the order in ε of summands in the Taylor formula increases with N, though less rapidly than in the Newton formula. This enables us to compute the derivatives $\left(\frac{d}{d\varepsilon}\right)^k f(A+\varepsilon B)\Big|_{\varepsilon=0}$ in terms of the derivatives of $f(x)$. Thus, if

$$[A,[A,B]]=0,$$

then we have

$$\begin{aligned}\left(\frac{d}{d\varepsilon}f(A+\varepsilon B)\right)\Big|_{\varepsilon=0} &= f'(A)\,B-\frac{1}{2}f''(A)[A,B],\\ \frac{d^2}{d\varepsilon^2}f(A+\varepsilon B)\Big|_{\varepsilon=0} &= f''(A)B^2-\frac{2}{3}f'''(A)[A,B]B-\frac{1}{3}f'''(A)B[A,B]\\ &\quad+\frac{1}{4}f''(A)[A,B]^2.\end{aligned}$$

We see that the derivative of order k of $f(A+\varepsilon B)$ w.r.t. ε can be expressed via *higher-order* derivatives of f taken at A.

Remark I.9 The Daletskii–Krein formula can also be considered as the usual formula for the differential. Namely, a symbol f can be viewed as a (partial) mapping from $\mathcal{A}$ to $\mathcal{A}$ that assigns $f(A)$ to each generator A. We may wish to consider the differential of this mapping. According to the general rule, the differential of a mapping $\varphi:\mathcal{A}\to\mathcal{A}$ is a linear mapping $\varphi_*:\mathcal{A}\to\mathcal{A}$, i.e., an element of $\mathcal{L}(A)$. Having this in mind, we can represent the Daletskii–Krein formula as follows:

$$f_*(A)=\frac{\delta f}{\delta x}(L_A,R_A).$$

This formula is understood in the sense that

$$df(A,B)=f_*(A)(B)=f_*(A)(B)=\frac{\delta f}{\delta x}(L_A,R_A)B=\overset{2}{B}\frac{\delta f}{\delta x}(\overset{1}{A},\overset{3}{A}).$$

We have considered various expansions for $f(A+\Delta A)$. These expansions are closely related to the conventional Taylor's formula in the analysis of functions of the numerical argument. However, there are also several topics specific to noncommutative differential calculus and having no counterpart in the usual calculus. We mean the index permutation formulas and the composite function formulas considered in the following two subsections.

3.4 Permutation of Feynman Indices

Given a binary symbol $f(x,y)$ and two generators A and B, one can define $f(A,B)$ in two different ways using Feynman's approach: one can take $f(\overset{1}{A},\overset{2}{B})$ or $f(\overset{2}{A},\overset{1}{B})$. How different will the results be? The answer is given by the following theorem.

Theorem I.10 (Index permutation formula)

$$\begin{aligned} f(\overset{1}{A}, \overset{2}{B}) &= f(\overset{2}{A}, \overset{1}{B}) + [\overset{3}{B, A}]\frac{\delta^2 f}{\delta x \delta y}(\overset{1}{A}, \overset{5}{A}, \overset{2}{B}, \overset{4}{B}) \\ &= f(\overset{2}{A}, \overset{1}{B}) + [\overset{3}{B, A}]\frac{\delta^2 f}{\delta x \delta y}(\overset{2}{A}, \overset{4}{A}, \overset{1}{B}, \overset{5}{B}). \end{aligned}$$

Proof. Let us consider the difference $[\![f(\overset{1}{A}, \overset{2}{B})]\!] - [\![f(\overset{2}{A}, \overset{1}{B})]\!]$. We can change the indices in this difference arbitrarily provided that

(i) the order of operators in each term remains unchanged;

(ii) noncommuting operators are assigned different indices.

This being done, we can omit the autonomous brackets. In particular, we can write

$$[\![f(\overset{1}{A}, \overset{2}{B})]\!] - [\![f(\overset{2}{A}, \overset{1}{B})]\!] = f(\overset{1}{A}, \overset{2}{B}) - f(\overset{3}{A}, \overset{2}{B}).$$

Let us now recall that once all operators in an expression are equipped with indices, we can transform the expression according to the rules of commutative algebra. Hence we have

$$f(\overset{1}{A}, \overset{2}{B}) - f(\overset{3}{A}, \overset{2}{B}) = (\overset{1}{A} - \overset{3}{A})\frac{f(\overset{1}{A}, \overset{2}{B}) - f(\overset{3}{A}, \overset{2}{B}))}{\overset{1}{A} - \overset{3}{A}} = (\overset{1}{A} - \overset{3}{A})\frac{\delta f}{\delta x} f(\overset{1}{A}, \overset{3}{A}, \overset{2}{B}).$$

Strange as it may seem, this computation is perfectly rigorous; in fact, it means nothing other than that we take the formula

$$f(x, y) - f(z, y) = (x - z)\frac{f(x, y) - f(z, y)}{x - z} = (x - z)\frac{\delta f}{\delta x} f(x, z; y)$$

and substitute the operators $\overset{1}{A}, \overset{2}{B}, \overset{3}{A}$ for the variables x, y, z. In the following, we always use the shortened form for computations like this.

The relations obtained give

$$f(\overset{1}{A}, \overset{2}{B}) - f(\overset{2}{A}, \overset{1}{B}) = (\overset{1}{A} - \overset{3}{A})\frac{\delta f}{\delta x}(\overset{1}{A}, \overset{3}{A}, \overset{2}{B}).$$

Note that we do not write autonomous brackets on the left-hand side of the last equation, although they must formally stand there; from now on we widely make such abuse of notation provided that this cannot lead to misunderstanding.

Changing the indices again according to (i) and (ii), we obtain

$$\begin{aligned} f(\overset{1}{A}, \overset{2}{B}) - f(\overset{2}{A}, \overset{1}{B}) &= \overset{1}{A}\frac{\delta f}{\delta x}(\overset{1}{A}, \overset{3}{A}, \overset{2}{B}) - \overset{3}{A}\frac{\delta f}{\delta x}(\overset{1}{A}, \overset{3}{A}; \overset{2}{B}) \\ &= \overset{3}{A}\frac{\delta f}{\delta x}(\overset{1}{A}, \overset{5}{A}, \overset{4}{B}) - \overset{3}{A}\frac{\delta f}{\delta x}(\overset{1}{A}, \overset{5}{A}; \overset{2}{B}) \\ &= \overset{3}{A}[\frac{\delta f}{\delta x}(\overset{1}{A}, \overset{5}{A}; \overset{4}{B}) - \frac{\delta f}{\delta x}(\overset{1}{A}, \overset{5}{A}; \overset{2}{B})]. \end{aligned}$$

We transform the right-hand side of this relation in a similar way, by introducing yet another difference derivative, and obtain

$$f(\overset{1}{A}, \overset{2}{B}) - f(\overset{2}{A}, \overset{1}{B}) = \overset{3}{A}(\overset{4}{B} - \overset{2}{B})\frac{\delta^2 f}{\delta x \delta y}(\overset{1}{A}, \overset{5}{A}; \overset{2}{B}, \overset{4}{B}),$$

or, by (i),

$$f(\overset{1}{A}, \overset{2}{B}) - f(\overset{2}{A}, \overset{1}{B}) = \overset{3}{A}(\overset{3,5}{B} - \overset{2,5}{B})\frac{\delta^2 f}{\delta x \delta y}(\overset{1}{A}, \overset{5}{A}; \overset{2}{B}, \overset{4}{B}).$$

We can now extract the linear factor $\overset{3}{A}(\overset{3,5}{B} - \overset{2,5}{B})$:

$$f(\overset{1}{A}, \overset{2}{B}) - f(\overset{2}{A}, \overset{1}{B}) = [\![\overset{3}{A}(\overset{3,5}{B} - \overset{2,5}{B})]\!]\frac{\delta^2 f}{\delta x \delta y}(\overset{1}{A}, \overset{5}{A}; \overset{2}{B}, \overset{4}{B}).$$

This is just the first variant of the index permutation formula since

$$\overset{3}{A}(\overset{3,5}{B} - \overset{2,5}{B}) = BA - AB = [B, A].$$

The proof of the second variant of the formula is left to the reader.
The theorem is proved. □

Remark I.10 Let us point out (though this is trivial) that expressions of the form

$$(\overset{i_1}{A} - \overset{i_2}{A})f(\overset{j_1}{B_1}, \dots, \overset{j_n}{B_n})$$

generally do not vanish if $[\overset{l}{B_i}, A] \neq 0$ for at least one l such that $i_1 < l < i_2$.

Remark I.11 By introducing additional operator arguments we find that

$$\begin{aligned} f(\overset{1}{A}, \overset{2}{B}, \overset{i_1}{C_1}, \dots, \overset{i_k}{C_k}) &= f(\overset{2}{A}, \overset{1}{B}, \overset{i_1}{C_1}, \dots, \overset{i_k}{C_k}) \\ &+ [\overset{1,5}{B}, A]\delta_{12}^2(\overset{1}{A}, \overset{2}{A}; \overset{1,25}{B}, \overset{1,75}{B}; \overset{i_1}{C_1}, \dots, \overset{i_k}{C_k}), \end{aligned}$$

where $i_s \notin [1, 2]$, $s = 1, \dots, k$.

Note that the commutation formula stated in Proposition I.3 is an easy consequence of Theorem I.10.

Corollary I.1 (Commutation formula) *One has*

$$\overset{1}{A}f(\overset{2}{B}) = \overset{2}{A}f(\overset{1}{B}) + [\overset{2}{B}, A]\frac{\delta f}{\delta y}(\overset{1}{B}, \overset{3}{B}).$$

or, in other form,

$$[A, f(B)] = [\overset{2}{A}, B]\frac{\delta f}{\delta y}(\overset{1}{B}, \overset{3}{B}).$$

Proof. This is just a particular case of the theorem with $f(x, y)$ replaced by $xf(y)$. Indeed, we have

$$\frac{\delta^2[xf(y)]}{\delta x \delta y} = \frac{\delta f}{\delta y}(y_1, y_2),$$

so that

$$\overset{1}{A} f(\overset{2}{B}) - \overset{2}{A} f(\overset{1}{B}) = \overset{2}{[B, A]} \frac{\delta^2[xf(y)]}{\delta x \delta y}(\overset{1}{A}, \overset{5}{A}; \overset{2}{B}, \overset{4}{B}) = \overset{2}{[B, A]} \frac{\delta f}{\delta y}(\overset{1}{B}, \overset{3}{B}).$$

□

We can also easily compute the commutator $[A, f(\overset{1}{B_1}, \dots, \overset{n}{B_n})]$. For this purpose, we consecutively permute A with $B_n, \dots, B_1$. After the first step we get

$$\begin{aligned}
[A, f(\overset{1}{B_1}, \dots, \overset{n}{B_n})] &= \overset{n+1}{A} f(\overset{1}{B_1}, \dots, \overset{n}{B_n}) - \overset{0}{A} f(\overset{1}{B_1}, \dots, \overset{n}{B_n}) \\
&= \overset{n}{A} f(\overset{1}{B_1}, \dots, \overset{n-1}{B}_{n-1}, \overset{n+1}{B}_n) + [\overset{n+1}{A}, B_n]\delta_n f(\overset{1}{B_1}; \dots; \overset{n-1}{B}_{n-1}; \overset{n}{B_n}, \overset{n+2}{B}_n) \\
&\quad - \overset{0}{A} f(\overset{1}{B_1}, \dots, \overset{n}{B_n}).
\end{aligned}$$

Next we proceed to B_{n-1}, etc. (Here δ_n stands for the difference derivative w.r.t. the n-th variable. Similar notations are used below).

Finally, the last term on the right-hand side cancels, and we obtain

$$[A, f(\overset{1}{B_1}, \dots, \overset{n}{B_n})] = \sum_{j=1}^{n} [\overset{j+1}{A}, B_j]\delta_j f(\overset{1}{B_1}; \overset{j-1}{B}_{j-1}; \overset{j}{B_j}, \overset{j+2}{B}_j; \overset{j+3}{B}_{j+1}; \dots; \overset{n+2}{B}_n).$$

Let us show how the formulas obtained can be used to derive some identities for pseudodifferential operators (that is, functions of x and $-i\partial/\partial x$). For convenience, we consider h-pseudodifferential operators, in which the operator $-i\partial/\partial x$ occurs with the factor h.

Example I.8 (Permutation of x and $-ih\partial/\partial x$) Let a symbol $P(x, p)$ be given. The usual correspondence

$$x \mapsto \overset{2}{x}, \quad p \mapsto -ih\overset{1}{\frac{\partial}{\partial x}}$$

gives the pseudodifferential operator $P\left(\overset{2}{x}, \overset{1}{-ih\partial/\partial x}\right)$. Let us rewrite this operator in the form $Q\left(\overset{1}{x}, \overset{2}{-ih\partial/\partial x}\right)$. By Theorem I.10, we have

$$P\left(\overset{2}{x}, -ih\overset{1}{\frac{\partial}{\partial x}}\right) = P\left(\overset{1}{x}, -ih\overset{2}{\frac{\partial}{\partial x}}\right) + \overset{3}{[x, -ih\frac{\partial}{\partial x}]} \frac{\delta^2 P}{\delta x \delta p}\left(\overset{2}{x}, \overset{4}{x}, -ih\overset{1}{\frac{\partial}{\partial x}}, -ih\overset{5}{\frac{\partial}{\partial x}}\right).$$

Since

$$\left[x, -ih\frac{\partial}{\partial x}\right] = ih,$$

we see that

$$\begin{aligned} P\left(\overset{2}{x}, -ih\overset{1}{\frac{\partial}{\partial x}}\right) &= P\left(\overset{1}{x}, -ih\overset{2}{\frac{\partial}{\partial x}}\right) + ih\frac{\delta^2 P}{\delta x \delta p}\left(\overset{2}{x}, \overset{2}{x}, -ih\overset{1}{\frac{\partial}{\partial x}}, -ih\overset{3}{\frac{\partial}{\partial x}}\right) \\ &= P\left(\overset{1}{x}, -ih\overset{2}{\frac{\partial}{\partial x}}\right) + ih\left[\frac{\partial}{\partial x}\frac{\delta P}{\delta p}\right]\left(\overset{2}{x}, -ih\overset{1}{\frac{\partial}{\partial x}}, -ih\overset{3}{\frac{\partial}{\partial x}}\right) \end{aligned}$$

(the last equality is due to the fact that $\delta f/\delta x(x, x) = f'(x)$).

Let us again permute the operators $-ih\overset{1}{\partial/\partial x}$ and $\overset{2}{x}$. We get

$$\begin{aligned} P\left(\overset{2}{x}, -ih\overset{1}{\frac{\partial}{\partial x}}\right) &= P\left(\overset{1}{x}, -ih\overset{2}{\frac{\partial}{\partial x}}\right) + ih\frac{\partial^2 P}{\partial p \partial x}\left(\overset{1}{x}, -ih\overset{2}{\frac{\partial}{\partial x}}\right) \\ &\quad + \left[\left(ih\frac{\partial}{\partial x} \circ \frac{\delta}{\delta p}\right)^2 P\right]\left(\overset{2}{x}, -ih\overset{1}{\frac{\partial}{\partial x}}, -ih\overset{3}{\frac{\partial}{\partial x}}, -ih\overset{3}{\frac{\partial}{\partial x}}\right). \end{aligned}$$

One can proceed with these manipulations and obtain the order changing formula modulo any power of h. Note that with this method we *obtain an exact formula for the remainder.*

Example I.9 (Product formula) Suppose that two h^{-1}-pseudodifferential operators $P\left(\overset{2}{x}, -ih\overset{1}{\partial/\partial x}\right)$ and $Q\left(\overset{2}{x}, -ih\overset{1}{\partial/\partial x}\right)$ are given. Let us compute their product. We have

$$[\![P\left(\overset{2}{x}, -ih\overset{1}{\frac{\partial}{\partial x}}\right)]\!][\![Q\left(\overset{2}{x}, -ih\overset{1}{\frac{\partial}{\partial x}}\right)]\!] = P\left(\overset{4}{x}, -ih\overset{3}{\frac{\partial}{\partial x}}\right) Q\left(\overset{2}{x}, -ih\overset{1}{\frac{\partial}{\partial x}}\right).$$

In order to reduce this expression to the standard ordering, we must permute the operators $-ih\overset{3}{\partial/\partial x}$ and $\overset{2}{x}$. By Theorem I.10, we have

$$\begin{aligned} [\![P\left(\overset{2}{x}, -ih\overset{1}{\frac{\partial}{\partial x}}\right)]\!][\![Q\left(\overset{2}{x}, -ih\overset{1}{\frac{\partial}{\partial x}}\right)]\!] &= P\left(\overset{2}{x}, -ih\overset{1}{\frac{\partial}{\partial x}}\right) Q\left(\overset{2}{x}, -ih\overset{1}{\frac{\partial}{\partial x}}\right) \\ &\quad + [\overset{4}{x}, -ih\frac{\partial}{\partial x}]\frac{\delta P}{\delta p}\left(\overset{7}{x}, -ih\overset{3}{\frac{\partial}{\partial x}}, -ih\overset{5}{\frac{\partial}{\partial x}}\right)\frac{\delta Q}{\delta x}\left(\overset{2}{x}, \overset{6}{x}, -ih\overset{1}{\frac{\partial}{\partial x}}\right). \end{aligned}$$

Since

$$\left[-ih\frac{\partial}{\partial x}, x\right] = -ih,$$

we obtain

$$[\![P\left(\overset{2}{x}, -ih\overset{1}{\frac{\partial}{\partial x}}\right)]\!] [\![Q\left(\overset{2}{x}, -ih\overset{1}{\frac{\partial}{\partial x}}\right)]\!] = P\left(\overset{2}{x}, -ih\overset{1}{\frac{\partial}{\partial x}}\right) Q\left(\overset{2}{x}, -ih\overset{1}{\frac{\partial}{\partial x}}\right)$$
$$-ih\frac{\delta P}{\delta p}\left(\overset{7}{x}, -ih\overset{3}{\frac{\partial}{\partial x}}, -ih\overset{5}{\frac{\partial}{\partial x}}\right) \frac{\delta Q}{\delta x}\left(\overset{2}{x}, \overset{6}{x}, -ih\overset{1}{\frac{\partial}{\partial x}}\right).$$

Taking into account that

$$\frac{\delta P}{\delta p}\left(\overset{7}{x}, -ih\overset{3}{\frac{\partial}{\partial x}}, -ih\overset{5}{\frac{\partial}{\partial x}}\right) = \frac{\delta P}{\delta p}\left(\overset{7}{x}, -ih\overset{3}{\frac{\partial}{\partial x}}, -ih\overset{3}{\frac{\partial}{\partial x}}\right) = \frac{\partial P}{\partial p}\left(\overset{7}{x}, -ih\overset{3}{\frac{\partial}{\partial x}}\right),$$

we get

$$[\![P\left(\overset{2}{x}, -ih\overset{1}{\frac{\partial}{\partial x}}\right)]\!] [\![Q\left(\overset{2}{x}, -ih\overset{1}{\frac{\partial}{\partial x}}\right)]\!] = P\left(\overset{2}{x}, -ih\overset{1}{\frac{\partial}{\partial x}}\right) Q\left(\overset{2}{x}, -ih\overset{1}{\frac{\partial}{\partial x}}\right)$$
$$-ih\frac{\partial P}{\partial p}\left(\overset{5}{x}, -ih\overset{3}{\frac{\partial}{\partial x}}\right) \frac{\delta Q}{\delta x}\left(\overset{2}{x}, \overset{4}{x}, -ih\overset{1}{\frac{\partial}{\partial x}}\right).$$

After permuting $\overset{2}{x}$ and $-ih\overset{3}{\partial/\partial x}$ in the last term, we obtain the following product formula

$$[\![P\left(\overset{2}{x}, -ih\overset{1}{\frac{\partial}{\partial x}}\right)]\!] [\![Q\left(\overset{2}{x}, -ih\overset{1}{\frac{\partial}{\partial x}}\right)]\!] = P\left(\overset{2}{x}, -ih\overset{1}{\frac{\partial}{\partial x}}\right) Q\left(\overset{2}{x}, -ih\overset{1}{\frac{\partial}{\partial x}}\right)$$
$$-ih\frac{\partial P}{\partial p}\left(\overset{5}{x}, -ih\overset{3}{\frac{\partial}{\partial x}}\right) \frac{\partial Q}{\partial x}\left(\overset{2}{x}, -ih\overset{1}{\frac{\partial}{\partial x}}\right)$$
$$+(-ih)^2\frac{\partial^2 P}{\partial p \partial q}\left(\overset{4}{x}, -ih\overset{3}{\frac{\partial}{\partial x}}\right) \delta_1^2\left(\overset{2}{x}, \overset{4}{x}, \overset{4}{x}; -ih\overset{1}{\frac{\partial}{\partial x}}\right).$$

Obviously, this process can be continued so as to obtain the subsequent terms of the expansion.

Remark I.12 In both preceding examples the more elegant way to get the answer would be to use the left ordered representation of the tuple $(x, -ih\partial/\partial x)$ (see Chapter II).

3.5 The Composite Function Formula

In this subsection we will consider the problem of rewriting the composite function $f([\![g(\overset{1}{A}, \overset{2}{B})]\!])$ via functions of A, B, and their commutators. Recall that $f(g(\overset{1}{A}, \overset{2}{B}))$ means $(f \circ g)(\overset{1}{A}, \overset{2}{B})$, as distinct from $f[\![g(\overset{1}{A}, \overset{2}{B})]\!])$, which means "$f(C)$ with $C = g(\overset{1}{A}, \overset{2}{B})$". We will derive a formula for $f([\![g(\overset{1}{A}, \overset{2}{B})]\!])$ whose leading term coincides with $f(g(\overset{1}{A}, \overset{2}{B}))$. First let us consider the relatively simple case in which $g(x, y) = x + y$.

Theorem I.11 *A function of the sum of two operators admits the following expansion:*

$$\begin{aligned} f([\![A+B]\!]) &= f(\overset{1}{A}+\overset{2}{B}) + \overset{3}{[A, B]}\frac{\overset{5}{\delta^2 f}}{\delta y^2}([\![\overset{1}{A}+B]\!], \overset{2}{A}+\overset{1}{B}, \overset{4}{A}+B) \\ &= f(\overset{1}{A}+\overset{2}{B}) + \overset{3}{[A, B]}\frac{\overset{4}{\delta^2 f}}{\delta y^2}([\![A+B]\!], \overset{2}{[\![A+B]\!]}, \overset{1}{A}+\overset{2}{B}). \end{aligned}$$

Proof. We have

$$\begin{aligned} f([\![A+B]\!]) - f(\overset{1}{A}+\overset{2}{B}) &= f(\overset{3}{[\![A+B]\!]}) - f(\overset{1}{A}+\overset{2}{B}) \\ &= (\overset{3}{[\![A+B]\!]} - \overset{1}{A} - \overset{2}{B})\frac{\overset{3}{\delta f}}{\delta y}([\![A+B]\!], \overset{1}{A}+\overset{2}{B}) \\ &= (\overset{3}{[\![A+B]\!]} - \overset{1}{A} - \overset{2}{B})\frac{\overset{4}{\delta f}}{\delta y}([\![A+B]\!], \overset{0}{A}+\overset{1,5}{B}). \end{aligned}$$

Should the operator $\overset{1,5}{B}$ not act between the operators $\overset{1}{A}$ and $\overset{2}{B}$ in the first factor, this expression would be zero. We now permute the operators $\overset{1}{A}$ and $\overset{1,5}{B}$ with the help of Theorem I.10 and get

$$\begin{aligned} f([\![A+B]\!]) - f(\overset{1}{A}+\overset{2}{B}) &= (\overset{3}{[\![A+B]\!]} - \overset{1,5}{A} - \overset{2}{B})\frac{\overset{4}{\delta f}}{\delta y}([\![A+B]\!], \overset{0}{A}+\overset{1}{B}) \\ &\quad - \overset{3}{[B, A]}\frac{\overset{4}{\delta^2 f}}{\delta y^2}([\![\overset{1}{A}+B]\!], \overset{2}{A}+\overset{1}{B}, \overset{4}{A}+B), \end{aligned}$$

since

$$\frac{\delta^2}{\delta x_2 \delta x_3}\{(x_5 - x_2 - x_4)\frac{\delta f}{\delta y}(x_6, x_1 + x_0)\} = -\frac{\delta^2 f}{\delta y^2}(x_6, x_1 + x_3', x_1 + x_3'').$$

The first term on the right-hand side of the penultimate formula vanishes, and we obtain the first version the desired formula. The proof of its second version is left to the reader as an exercise. □

Theorem I.11 can be used, in particular, for the case in which the commutator $[A, B]$ is in some sense "smaller" that the operators A and B (the examples are given below). In order to formalize the situation, we introduce the notion of a *degree* for arbitrary products of A, B, and their commutators. Our definition is recursive. We say that the degree of A and B is equal to zero; for any term C, the degree of the commutators $[A, C]$ and $[B, C]$ is equal to the degree of C plus one, and the degree of a product is the sum of the degrees of the factors. Thus, for example, the degree of $[A, [A, B]]$ is equal to 2 and the degree of $[A, B] \cdot [B, [A, B]]$ is equal to 3.

Note that the degree is *not* additive but only *semiadditive*: the degree of a sum may exceed the degree of the summands (for example, this is the case for the expression $AB - BA = [A, B]$).

Using Theorem I.11, we can expand $f([\![A + B]\!])$ up to terms of arbitrarily high degree. Let us consider an example of such an expansion.

Corollary I.2 *The expansion*

$$f([\![A+B]\!]) = f(\overset{1}{A}+\overset{2}{B}) + \frac{1}{2}\overset{2}{[A,B]}f''(\overset{1}{A}+\overset{3}{B})$$

$$+\overset{2}{[\![}[A,[A,B]] + \frac{1}{6}[[A,B],B]]\!]f^{(3)}\overset{1}{A}+\overset{3}{B}) + \frac{1}{8}\overset{2}{[\![}[A,B]^2]\!]f^{(4)}(\overset{1}{A}+\overset{3}{B})$$

is valid up to terms of degree ≥ 3.

Proof. We apply the first version of the equation in Theorem I.11 to the second term on its own right-hand side, thus obtaining

$$f([\![A+B]\!]) = f(\overset{1}{A}+\overset{2}{B}) + \overset{3}{[A,B]}\frac{\delta^2 f}{\delta y^2}(\overset{5}{A}+\overset{6}{B},\overset{1}{A}+\overset{2}{B},\overset{1}{A}+\overset{4}{B})$$

$$+\overset{3}{[A,B]}\frac{\delta^4 f}{\delta y^4}(\overset{9}{[\![}A+B]\!],\overset{1}{A}+\overset{2}{B},\overset{1}{A}+\overset{4}{B},\overset{5}{A}+\overset{6}{B},\overset{5}{A}+\overset{8}{B}).$$

In view of Theorem I.11, we can omit the autonomous brackets in the third term of the obtained expression and use any admissible indices over the operators, with the resultant remainder being of degree ≥ 3. Thus, we have

$$f([\![A+B]\!]) \cong f(\overset{1}{A}+\overset{2}{B}) + \overset{3}{[A,B]}\frac{\delta^2 f}{\delta y^2}(\overset{5}{A}+\overset{6}{B},\overset{1}{A}+\overset{2}{B},\overset{1}{A}+\overset{4}{B})$$

$$+\overset{2}{[\![}[A,B]^2]\!]\delta^4 f(\overset{1}{A}+\overset{3}{B},\overset{1}{A}+\overset{3}{B},\overset{1}{A}+\overset{3}{B},\overset{1}{A}+\overset{3}{B}),$$

where $\cong$ can be translated as "is equal modulo terms of degree ≥ 3".

In the second term on the right-hand side of the last equation we permute the operator $\overset{5}{A}$ consecutively with $\overset{4}{B}$ and $\overset{3}{[A,B]}$:

$$\overset{3}{[A,B]}\delta^2 f(\overset{5}{A}+\overset{6}{B},\overset{1}{A}+\overset{2}{B},\overset{1}{A}+\overset{4}{B}) \cong \overset{3}{[A,B]}\frac{\delta^2 f}{\delta y^2}(\overset{4}{A}+\overset{6}{B},\overset{1}{A}+\overset{2}{B},\overset{1}{A}+\overset{5}{B})$$

$$+\overset{3}{[A,B]}\overset{6}{[A,B]}\frac{\delta^4 f}{\delta y^4}(\overset{4}{A}+\overset{9}{B},\overset{8}{A}+\overset{9}{B},\overset{1}{A}+\overset{2}{B},\overset{1}{A}+\overset{5}{B},\overset{1}{A}+\overset{7}{B})$$

$$
\begin{aligned}
&= \overset{4}{[A, B]} \frac{\delta^2 f}{\delta y^2}(\overset{3}{A} + \overset{6}{B}, \overset{1}{A} + \overset{2}{B}, \overset{1}{A} + \overset{5}{B}) \\
&+ \overset{4}{[A, [A, B]]} \frac{\delta^3 f}{\delta y^3}(\overset{3}{A} + \overset{7}{B}, \overset{5}{A} + \overset{7}{B}, \overset{1}{A} + \overset{2}{B}, \overset{1}{A} + \overset{6}{B}) \\
&+ \overset{3}{[A, B]}\overset{6}{[A, B]} \frac{\delta^4 f}{\delta y^4}(\overset{4}{A} + \overset{9}{B}, \overset{8}{A} + \overset{9}{B}, \overset{1}{A} + \overset{2}{B}, \overset{1}{A} + \overset{5}{B}, \overset{1}{A} + \overset{7}{B}).
\end{aligned}
$$

In the second and third terms of the last expression we can use any admissible order of operators modulo terms of order ≥ 3. Hence,

$$
\begin{aligned}
f(\llbracket A + B \rrbracket) &= f(\overset{1}{A} + \overset{2}{B}) + \overset{4}{[A, B]} \frac{\delta^2 f}{\delta y^2}(\overset{3}{A} + \overset{6}{B}, \overset{1}{A} + \overset{2}{B}, \overset{1}{A} + \overset{5}{B}) \\
&+ \overset{2}{[A, [A, B]]} \frac{\delta^3 f}{\delta y^3}(\overset{1}{A} + \overset{3}{B}, \overset{1}{A} + \overset{3}{B}, \overset{1}{A} + \overset{3}{B}) \\
&+ 2\overset{2}{\llbracket [A + B]^2 \rrbracket} \frac{\delta^4 f}{\delta y^4}(\overset{1}{A} + \overset{3}{B}, \overset{1}{A} + \overset{3}{B}, \overset{1}{A} + \overset{3}{B}, \overset{1}{A} + \overset{3}{B}, \overset{1}{A} + \overset{3}{B}).
\end{aligned}
$$

In the second term on the right-hand side of this equation we permute $\overset{2}{B}$ with $\overset{3}{A}$ and $\overset{4}{[A, B]}$. Similar calculations yield

$$
\begin{aligned}
f(\llbracket A + B \rrbracket) &\cong f(\overset{1}{A} + \overset{2}{B}) + \overset{4}{[A, B]} \frac{\delta^2 f}{\delta y^2}(\overset{1}{A} + \overset{3}{B}, \overset{1}{A} + \overset{3}{B}, \overset{1}{A} + \overset{3}{B}) \\
&+ \overset{2}{\llbracket [A, [A, B]] + [[A, B], B] \rrbracket} \frac{\delta^3 f}{\delta y^3}(\overset{1}{A} + \overset{3}{B}, \overset{1}{A} + \overset{3}{B}, \overset{1}{A} + \overset{3}{B}, \overset{1}{A} + \overset{3}{B}) \\
&+ 3\overset{2}{\llbracket [A + B]^2 \rrbracket} \frac{\delta^4 f}{\delta y^4}(\overset{1}{A} + \overset{3}{B}, \overset{1}{A} + \overset{3}{B}, \overset{1}{A} + \overset{3}{B}, \overset{1}{A} + \overset{3}{B}, \overset{1}{A} + \overset{3}{B}).
\end{aligned}
$$

The last formula proves the corollary, since $\delta^k f/\delta x^k(x, \ldots, x) = \frac{f^{(k)}(x)}{k!}$. □

Let us now proceed to the general case.

Theorem I.12 *The following composite function formula holds:*

$$
\begin{aligned}
&f(\llbracket g(\overset{1}{A}, \overset{2}{B}) \rrbracket) = f(g(\overset{1}{A}, \overset{2}{B})) \\
&+ \overset{5}{[A, B]} \frac{\delta g}{\delta x_2}(\overset{3}{A}, \overset{4}{B}, \overset{6}{B}) \frac{\delta g}{\delta x_1}(\overset{2}{A}, \overset{7}{A}; \overset{8}{B}) \frac{\delta^2 f}{\delta y^2}(\llbracket g(\overset{1}{A}, \overset{2}{B}) \rrbracket, g(\overset{2}{A}, \overset{8}{B}), g(\overset{7}{A}, \overset{8}{B})).
\end{aligned}
$$

Proof. We have

$$
\begin{aligned}
f(\llbracket g(\overset{1}{A}, \overset{2}{B}) \rrbracket) - f(g(\overset{1}{A}, \overset{2}{B})) &= f(\overset{1}{\llbracket} g(\overset{1}{A}, \overset{2}{B}) \rrbracket) - f(g(\overset{1}{A}, \overset{2}{B})) \\
&= (\overset{1}{\llbracket} g(\overset{1}{A}, \overset{2}{B}) \rrbracket - g(\overset{2}{A}, \overset{3}{B})) \frac{\delta f}{\delta y}(\overset{1}{\llbracket} g(\overset{1}{A}, \overset{2}{B}) \rrbracket, g(\overset{2}{A}, \overset{3}{B})) \\
&= (\overset{2}{\llbracket} g(\overset{1}{A}, \overset{2}{B}) \rrbracket - g(\overset{4}{A}, \overset{5}{B})) \frac{\delta f}{\delta y}(\overset{1}{\llbracket} g(\overset{1}{A}, \overset{2}{B}) \rrbracket, g(\overset{3}{A}, \overset{6}{B})).
\end{aligned}
$$

By permuting the operators $\overset{3}{A}$ and $[\![g(\overset{2}{A}, \overset{1}{B})]\!]$ in the last expression, we obtain

$$f([\![g(\overset{1}{A}, \overset{2}{B})]\!]) - f(g(\overset{1}{A}, \overset{2}{B})) = ([\![g(\overset{3}{A}, \overset{1}{B})]\!] - g(\overset{4}{A}, \overset{5}{B}))\frac{\delta f}{\delta y}(\overset{1}{[\![g(\overset{1}{A}, \overset{2}{B})]\!]}, g(\overset{2}{A}, \overset{6}{B}))$$

$$+[\overset{3}{A}, [\![g(\overset{1}{A}, \overset{2}{B})]\!]]\frac{\delta^2 f}{\delta y^2}(\overset{1}{[\![g(\overset{1}{A}, \overset{2}{B})]\!]}, g(\overset{2}{A}, \overset{5}{B}), g(\overset{4}{A}, \overset{5}{B}))\frac{\delta g}{\delta x^1}(\overset{2}{A}, \overset{4}{A}, \overset{5}{B}).$$

Here the first term on the right-hand side vanishes, which can easily be proved by index manipulations (see Proposition I.1)

$$[A, [\![g(\overset{1}{A}, \overset{2}{B})]\!]] = [\overset{3}{A, B}]\frac{\delta g}{\delta x_2}(\overset{1}{A}; \overset{2}{B}, \overset{4}{B}).$$

The theorem is proved. □

Let us now apply the obtained formula to pseudodifferential operators.

Example I.10 Let $P\left(\overset{2}{x}, \overset{1}{-ih\partial/\partial x}\right)$ be an h^{-1}-pseudodifferential operator, and let a function $f(z)$ of a single variable z be given. We will rewrite the operator $f([\![P\left(\overset{2}{x}, \overset{1}{-ih\partial/\partial x}\right)]\!])$ in the form of an h^{-1}-pseudodifferential operator modulo $O(h^2)$. Due to Theorem I.12 we have

$$f([\![P\left(\overset{2}{x}, -ih\overset{1}{\frac{\partial}{\partial x}}\right)]\!]) = (f \circ P)\left(\overset{2}{x}, -ih\overset{1}{\frac{\partial}{\partial x}}\right)$$

$$+i\overset{5}{[}-ih\frac{\partial}{\partial x}, x]\frac{\delta P}{\delta x}\left(\overset{4}{x}, \overset{6}{x}, -ih\overset{3}{\frac{\partial}{\partial x}}\right)\frac{\delta P}{\delta p}\left(\overset{8}{x}, -ih\overset{2}{\frac{\partial}{\partial x}}, -ih\overset{7}{\frac{\partial}{\partial x}}\right)$$

$$\times\frac{\delta^2 f}{\delta z^2}(\overset{1}{[\![P}\left(x, -ih\frac{\partial}{\partial x}\right)]\!], P\left(\overset{8}{x}, -ih\overset{2}{\frac{\partial}{\partial x}}\right), P\left(\overset{8}{x}, -ih\overset{7}{\frac{\partial}{\partial x}}\right)).$$

Since $[-ih\partial/\partial x, x] = -ih$, we can omit the index 5. Consequently, the indices 4 and 6 can be set equal to each other. Taking into account that

$$\frac{\delta P}{\delta x}(x, x, p) = \frac{\partial P(x, p)}{\partial x},$$

we get

$$f([\![P\left(\overset{2}{x}, -ih\overset{1}{\frac{\partial}{\partial x}}\right)]\!]) = (f \circ P)\left(\overset{2}{x}, -ih\overset{1}{\frac{\partial}{\partial x}}\right)$$

$$-ih\frac{\partial P}{\partial x}\left(\overset{3}{x}, -ih\overset{2}{\frac{\partial}{\partial x}}\right)\frac{\delta P}{\delta p}\left(\overset{5}{x}, -ih\overset{2}{\frac{\partial}{\partial x}}, -ih\frac{\partial}{\partial x}\right)$$

$$\times \frac{\delta^2 f}{\delta z^2}\big(\overset{1}{[\![} P\big(\overset{2}{x}, -ih\overset{1}{\frac{\partial}{\partial x}}\big)]\!], P\big(\overset{5}{x}, -ih\overset{2}{\frac{\partial}{\partial x}}\big), P\big(\overset{5}{x}, -ih\overset{4}{\frac{\partial}{\partial x}}\big)\big).$$

Computing up to terms of order h^2, we can change arbitrarily the indices over operators in the second term on the right-hand side of the last formula. Hence,

$$f([\![P\big(\overset{2}{x}, -ih\overset{1}{\frac{\partial}{\partial x}}\big)]\!]) = (f \circ P)\big(\overset{2}{x}, -ih\overset{1}{\frac{\partial}{\partial x}}\big)$$
$$- \left[\frac{\partial P}{\partial x}\frac{\partial P}{\partial p}(f'' \circ P)\right]\big(\overset{2}{x}, -ih\overset{1}{\frac{\partial}{\partial x}}\big) + O(h^2).$$

Similarly, one can calculate the subsequent terms of this expansion.

4 The Campbell–Hausdorff Theorem and Dynkin's Formula

As a sample application of the techniques of noncommutative analysis, let us consider a famous old problem in the theory of Lie algebras and Lie groups. This problem was already mentioned in Subsection 1.5, but here we start from the very beginning and give a more detailed exposition. Those who are not familiar with Lie algebras and Lie groups at all may wish to consult the Appendix, where all the necessary information is provided; however, for better readability we reproduce here some of the definitions.

4.1 Statement of the Problem

Let L be a (finite-dimensional) Lie algebra. This means that L is a (finite-dimensional) linear space equipped with a bilinear operation $[\cdot, \cdot]$, referred to as *Lie bracket* and satisfying the following conditions:

i) $[a, b] = -[b, a]$ (antisymmetry);
ii) $[a, [b, c]] = [[a, b], c] + [b, [a, c]]$ (Jacobi identity).

For our aims it suffices to assume that L is realized by square matrices of size $m \times m$, with the usual matrix multiplication[1]. This assumption is, in fact, unnecessary, and we are only doing so in order to simplify the exposition by avoiding the consideration of unbounded operators.

[1] By Ado's theorem this is the general case.

Next, let G be a Lie group, i.e., a manifold equipped with smooth group operations. Again, we assume that, at least locally[2], G is a represented as a matrix group. Then the tangent space T_eG to G at the point e is naturally identified with a subspace of the space of matrices, and it can be shown to possess the structure of a Lie algebra. There arises a natural question: given a Lie algebra, can one reconstruct the multiplication law in the corresponding Lie group?

Let G be a Lie group and L the corresponding Lie algebra. There is a mapping

$$\exp : L \to G$$

taking each element $X \in L$ into the element $\exp(X) \in G$ defined as follows: let $\{g(t)\}$ be the one-parameter subgroup of G defined by the condition $\dot{g}(0) = X$. Then we set

$$\exp(X) = g(1).$$

Thus we obtain a coordinate system in the vicinity of the neutral element $e \in G$; this coordinate system is referred to as the *exponential coordinate system.*

Let us seek the multiplication law in the exponential coordinate system. If G is commutative, then the multiplication law has the form

$$A \cdot B = A + B.$$

If G is not commutative, there appear correction terms on the right-hand side of the last equation, namely,

$$A \cdot B = A + B + \frac{1}{2}[A, B] + \cdots.$$

The first correction is equal to $\frac{1}{2}[A, B]$ and this is completely determined by the commutation law in L. The Campbell–Hausdorff theorem [19] asserts that the same is true of all subsequent terms of this expansion, which implies that the multiplication law in G can be reconstructed given the commutation law in L. E. B. Dynkin [43] found an explicit exposition for all terms of the series. In this section, following Mosolova [139], we find a closed formula expressing $\ln(e^A e^B)$ in the form of an integral rather than a series. Then we use the conventional Fourier expansion so as to obtain the expression for $\ln(e^A e^B)$ via the commutators.

First of all, let us consider functions of commutation operators in more detail; this proves to be useful not only in the particular problem considered, but also in the general framework of noncommutative analysis.

4.2 The Commutation Operation

We have already seen how important a role commutators play in noncommutative analysis. It is clear that this operation is worth studying. In this subsection we obtain a simple expression for functions of the operator of commutation.

---

[2]That is, in a neighbourhood of the neutral element.

Let $\mathcal{A}$ be an algebra and $B \in \mathcal{A}$ some element. Denote by $\operatorname{ad}_B : \mathcal{A} \to \mathcal{A}$ the linear operator that takes each element $A \in \mathcal{A}$ into its commutator with the element B:

$$\begin{aligned} \operatorname{ad}_B : \quad & \mathcal{A} \to \mathcal{A} \\ & A \mapsto \operatorname{ad}_B(A) = [B, A]. \end{aligned}$$

Clearly,

$$\operatorname{ad}_B(A) : BA - AB = L_B(A) - R_B(A),$$

that is,

$$\operatorname{ad}_B = L_B - R_B,$$

where L_B and R_B are the operators of left and right regular representations introduced in Subsection 2.3.

The powers of ad_B give rise to multiple commutators with B:

$$(\operatorname{ad}_B)^k(A) = \underbrace{[B \ldots [}_{k \text{ times}} B, A] \ldots].$$

Let us find as expression for such commutators using Feynman indices. Note that

$$[B, A] = BA - AB = (\overset{3}{B} - \overset{1}{B})\overset{2}{A}.$$

Similarly,

$$(\operatorname{ad}_B)^k(A) = (L_B - R_B)^k(A) = (\overset{3}{B} - \overset{1}{B})^k \overset{2}{A},$$

according to the description of functions of the operators L_B and R_B given in Subsection 2.3.

Moreover, for any symbol $f(x)$ we have

$$f(\operatorname{ad}_B)(A) = f(\overset{3}{B} - \overset{1}{B})\overset{2}{A}. \tag{I.20}$$

Although this follows directly from the formulas given in Theorem I.3 let us present the proof for two particular cases in which the functions of operators are defined via the Fourier transform or the Cauchy integral formula.

A. The Cauchy integral formula. Since the mapping $f \mapsto f(A)$ is continuous, it suffices to consider the case $f(x) = (\lambda - x)^{-1}$.

Let

$$f(x) = (\lambda - x)^{-1},$$

where $\lambda \notin \sigma(B) - \sigma(B)$, that is, λ cannot be represented in the form $\lambda = \lambda_1 - \lambda_2$, where $\lambda_i \in \sigma(B)$, $i = 1, 2$. Then the function $(\lambda - x + y)^{-1}$ is analytic in the neighborhood of the product $\sigma(B) \times \sigma(B) \subset \mathbb{C}^2$, and hence the operator $(\lambda - \overset{3}{B} + \overset{1}{B})^{-1} \overset{2}{A}$ is well-defined. We have

$$(\lambda - \operatorname{ad}_B) \left(\frac{1}{\lambda - \overset{3}{B} + \overset{1}{B}} \overset{2}{A} \right) = (\lambda - \overset{3}{B} + \overset{1}{B}) [\![\overset{2}{\frac{\overset{2}{A}}{\lambda - \overset{3}{B} + \overset{1}{B}}}]\!]$$

$$= (\lambda - \overset{4}{B} + \overset{0}{B}) \frac{\overset{2}{A}}{\lambda - \overset{3}{B} + \overset{1}{B}} = (\lambda - \overset{3}{B} + \overset{1}{B}) \frac{\overset{2}{A}}{\lambda - \overset{3}{B} + \overset{1}{B}} = A$$

(we used extraction of a linear factor and moving indices apart). Hence the operator $\lambda - \operatorname{ad}_B$ is the left inverse of the operator

$$A \mapsto \frac{1}{\lambda - \overset{3}{B} + \overset{1}{B}} \overset{2}{A}.$$

Similarly, it can be proved that the operator $\lambda - \operatorname{ad}_B$ is the right inverse. Consequently,

$$R_\lambda(\operatorname{ad}_B)(A) = \frac{1}{\lambda - \overset{3}{B} + \overset{1}{B}} \overset{2}{A}.$$

□

B. The Fourier transform formula. Let

$$U(t)A = e^{it(\overset{3}{B} - \overset{1}{B})} \overset{2}{A}.$$

We intend to show that

$$U(t) = e^{it \operatorname{ad}_B}.$$

Indeed, let us differentiate $U(t)A$ by t. We obtain

$$\begin{aligned} \frac{d}{dt} U(t)A &= i(\overset{3}{B} - \overset{1}{B}) e^{it(\overset{3}{B} - \overset{1}{B})} \overset{2}{A} = i(\overset{4}{B} - \overset{0}{B}) e^{it(\overset{3}{B} - \overset{1}{B})} \overset{2}{A} \\ &= (\overset{4}{B} - \overset{0}{B}) [\![\overset{2}{e^{it(\overset{3}{B} - \overset{1}{B})} \overset{2}{A}}]\!] \\ &= i(\overset{3}{B} - \overset{1}{B}) \overset{2}{\overline{U(t)A}} = i \operatorname{ad}_B(U(t)A). \end{aligned}$$

Moreover, for $t = 0$ we have $U(0)A = A$, that is,

$$U(0) = \operatorname{id} = e^{it \operatorname{ad}_B} \Big|_{t=0}.$$

We see that $U(t)$ and $e^{it \operatorname{ad}_B}$ satisfy the same Cauchy problem and hence are equal, which implies the desired result for the case in which functions of operators are defined via the Fourier transform.

Let us now prove a technical lemma, which will be used in the next section.

Lemma I.2 *Suppose that $A, B, C \in \mathcal{A}$ are elements such that*

$$e^C = e^B e^A.$$

Then

$$e^{\mathrm{ad}_C} = e^{\mathrm{ad}_B} e^{\mathrm{ad}_A}.$$

Proof. We have

$$e^{\mathrm{ad}_B} e^{\mathrm{ad}_A}(D) = e^{\mathrm{ad}_B}\left(e^{\overset{3}{A}-\overset{1}{A}}\overset{2}{D}\right) = e^{\overset{4}{B}-\overset{0}{B}} e^{\overset{3}{A}-\overset{1}{A}}\overset{2}{D} = e^{\overset{3}{A}+\overset{4}{B}}\overset{2}{D}e^{-\overset{1}{A}-\overset{0}{B}}.$$

But

$$e^{\overset{3}{A}+\overset{4}{B}} = e^C, \quad e^{-\overset{1}{A}-\overset{0}{B}} = e^{-C}.$$

Thus,

$$e^{\mathrm{ad}_B} e^{\mathrm{ad}_A}(D) = e^{\overset{3}{C}}\overset{2}{D}e^{-\overset{1}{C}} = e^{\mathrm{ad}_C}(D),$$

as desired.

4.3 A Closed Formula for ln $(e^B e^A)$

For simplicity, assume that A and B are elements of a Banach algebra $\mathcal{A}$ with identity; this means that $\mathcal{A}$ is a complete normed linear space, and the multiplication defined on $\mathcal{A}$ has an identity element $1 \in \mathcal{A}$ and satisfies the inequality

$$||XY|| \leq ||X|| \cdot ||Y||$$

for any X and $Y \in \mathcal{A}$. Throughout the following exposition we only use analytic symbols and define functions of operators via Taylor series or Cauchy integrals.

Our intention is to find an explicit expression for the element $C = \ln(e^B e^A)$. Let us consider the family of elements

$$D(t) = \ln(e^{tB} e^A)$$

with parameter t. Clearly, we have $D(0) = A$ and $D(1) = C$. Moreover,

$$e^{D(t)} = e^{tB} e^A$$

by the definition of the logarithm.

We differentiate the last equation by t and multiply by $e^{-D(t)}$ on the right. This gives the following result:

$$B = \left(\frac{d}{dt}e^{D(t)}\right)e^{-D(t)}.$$

The latter equation can be transformed with the help of the Daletskii–Krein formula (see Subsection 3.2), which asserts that

$$\frac{d}{dt}e^{D(t)} = \overset{2}{D'(t)}\,\frac{e^{\overset{3}{D}(t)} - e^{\overset{1}{D}(t)}}{\overset{3}{D}(t) - \overset{1}{D}(t)}.$$

Thus,

$$B = \overset{2}{D'(t)}\,\frac{e^{\overset{3}{D}(t)} - e^{\overset{1}{D}(t)}}{\overset{3}{D}(t) - \overset{1}{D}(t)}\,\overset{0}{e^{-D(t)}}.$$

There are no operators with Feynman indices between 0 and 1 on the right- hand side. Therefore, the "moving indices apart" rule applies to the operators $\overset{0}{D}(t)$ and $\overset{1}{D}(t)$, and their indices can be set equal to each other:

$$B = \overset{2}{D'(t)}\,\frac{e^{\overset{3}{D}(t)-\overset{1}{D}(t)} - 1}{\overset{3}{D}(t) - \overset{1}{D}(t)}.$$

According to the results of Subsection 4.2,

$$f(\overset{3}{D} - \overset{1}{D})\overset{2}{E} = f(\mathrm{ad}_D)(E),$$

where ad_D is the operator of commutation with D. By taking $D = D(t)$, $E = D'(t)$, and

$$f(x) = \frac{e^x - 1}{x},$$

we obtain

$$B = f(\mathrm{ad}_{D(t)})(D'(t)).$$

We apply the operator $f(\mathrm{ad}_{D(t)})^{-1}$ to both sides of this equation and obtain

$$D'(t) = \frac{1}{f(\mathrm{ad}_{D(t)})}(B)$$

However, by Lemma I.2,

$$\mathrm{ad}_{D(t)} = \ln(e^{t\,\mathrm{ad}_B}e^{\mathrm{ad}_A}),$$

and so

$$D'(t) = \frac{1}{f(\ln(e^{t\,\mathrm{ad}_B} e^{\mathrm{ad}_A}))}(B) = \varphi(e^{t\,\mathrm{ad}_B} e^{\mathrm{ad}_A})(B),$$

where

$$\varphi(z) = \frac{1}{f(\ln z)} = \frac{\ln z}{z-1}.$$

Finally, we use the Newton–Leibniz formula and obtain

$$C = A + \int_0^1 D'(t)\,dt = A + \left[\int_0^t \varphi(e^{t\,\mathrm{ad}_B} e^{\mathrm{ad}_A})\,dt\right](B).$$

We see that C is expressed as a linear combination of A, B, and their commutators. Indeed, we can expand $\varphi(z)$ into a Taylor series in powers of $z-1$:

$$\varphi(z) = \frac{\ln z}{z-1} = \sum_{k=0}^{\infty} (-1)^k \frac{(z-1)^k}{k+1};$$

thus,

$$\varphi(e^{t\,\mathrm{ad}_B} e^{\mathrm{ad}_A}) = \sum_{k=0}^{\infty} (-1)^k \frac{(e^{t\,\mathrm{ad}_B} e^{\mathrm{ad}_A} - 1)^k}{k+1}.$$

Let us represent the exponentials in the form of Taylor series. We obtain

$$\varphi(e^{t\,\mathrm{ad}_B} e^{\mathrm{ad}_A}) = \sum_{k=0}^{\infty} (-1)^k \frac{\left(\sum\limits_{l+m>0} \frac{t^l (\mathrm{ad}_B)^l (\mathrm{ad}_A)^m}{l!m!}\right)^k}{k+1}$$

$$= \sum_{k=0}^{\infty} \frac{(-1)^k}{k+1} \sum_{l_i+m_i>0} \frac{t^{l_1+\dots+l_k} (\mathrm{ad}_B)^{l_1} (\mathrm{ad}_A)^{m_1} \dots (\mathrm{ad}_B)^{l_k} (\mathrm{ad}_A)^{m_k}}{l_1!m_1!\dots l_k!m_k!}.$$

We replace $\varphi(e^{t\,\mathrm{ad}_B} e^{\mathrm{ad}_A})$ by the last expression and perform term-by-term integration:

$$\begin{aligned} C &= A + B + \sum_{k=1}^{\infty} \frac{(-1)^k}{k+1} \sum_{l_i+m_i>0} \frac{(\mathrm{ad}_B)^{l_1} (\mathrm{ad}_A)^{m_1} \dots (\mathrm{ad}_B)^{l_k} (\mathrm{ad}_A)^{m_k}(B)}{(l_1+\dots+l_k+1) l_1!m_1!\dots l_k!m_k!} \\ &= A + B + \sum_{k=1}^{\infty} \frac{(-1)^k}{k+1} \sum_{l_i+m_i>0, l_i\ge 0, m_i\ge 0} \frac{[B^{l_1} A^{m_1} \dots B^{l_k} A^{m_k} B]}{l_1!m_1!\dots l_k!m_k!}, \end{aligned}$$

where

$$\begin{aligned} &[B^{l_1} A^{m_1} \dots B^{l_k} A^{m_k} B] \\ &= (l_1+\dots+l_k+1)^{-1} \underbrace{B[B\dots[B}_{l_1}\underbrace{[A\dots[A}_{m_1}\dots\underbrace{[B\dots[B}_{l_k}\underbrace{[A\dots[A}_{m_k}, B]\dots]. \end{aligned}$$

This expansion is slightly different from that obtained by Dynkin. In fact, Dynkin's expansion is more "symmetric" and can easily be obtained by our method if we consider the family

$$D(t) = e^{tB} e^{tA}$$

as was originally done by Mosolova [139]. The only reason for considering our family is that this leads to even simpler calculations.

Thus, we have proved that $\ln(e^B e^A)$ can be expressed as a linear combination of operators A, B and their commutators. There arises a natural question: where is the obtained formula valid? We do not go into a thorough investigation of this question here; let us only mention that $\varphi(z)$ is an analytic function in the disk $|z-1| < 1$; therefore, $\varphi(e^{t\,\mathrm{ad}_A} e^{\mathrm{ad}_B})$ is well-defined at least so long as

$$\|e^{t\,\mathrm{ad}_B} e^{\mathrm{ad}_A} - 1\| < 1.$$

The latter inequality is valid at least if $\|\mathrm{ad}_B\|$ and $\|\mathrm{ad}_A\|$ are small enough (one can easily write down the bounds for $\|\mathrm{ad}_B\|$ and $\|\mathrm{ad}_A\|$).

Next, if this condition is violated, we can proceed by analytic continuation. The function $\varphi(P(t))$, where

$$P(t) = e^{t\,\mathrm{ad}_B} e^{\mathrm{ad}_A}$$

can be defined via the Cauchy integral

$$\varphi(P(t)) = \frac{1}{2\pi i} \oint\limits_C \frac{\varphi(\lambda)}{\lambda - P(t)} d\lambda,$$

where C is a contour surrounding the spectrum of $P(t)$ and such that $\varphi(\lambda)$ is holomorphic in the closed domain bounded by C. The function $\varphi(z)$ has the unique ramification point $z = 0$; it is easy to see that a sufficient condition for $\varphi(P(t))$ can be defined is that the spectrum of $\varphi(P(t))$ does not surround the point $z = 0$ and does not contain this point. If this is true, we can find a simply connected domain D_t such that $\sigma(P(t)) \subset D_t$ and $0 \notin D_t$. We set $C = \partial D_t$.

Note that one should define $\varphi(P(t))$ for each $t \in [0, 1]$ in such a way that $\varphi(P(t))$ depends continuously on t. This imposes an additional condition on the behaviour of the domain D_t as t moves along the interval $[0, 1]$. For instance, if both A and B are positive definite self-adjoint operators, then the spectrum of $P(t)$ never intersects the negative part of the real axis and everything is all right. We refer the reader to M. V. Mosolova [139] for a detailed discussion of conditions under which the analytic continuation is possible. □

4.4 A Closed Formula for the Logarithm of a T-Exponential

The reasoning in the preceding subsection can easily be modified so as to solve the following problem.

Problem. Let a T-exponential

$$U = \exp\left(\int_0^1 \overset{\theta}{A}(\theta)d\theta\right)$$

be given, where $A(\theta)$ is a family of elements[3] of an operator algebra $\mathcal{A}$. Find an element $C \in \mathcal{A}$ such that

$$U = e^C.$$

This problem is an obvious continuous generalization of the discrete problem of finding $\log(e^B e^A)$.

Again, the method of solving this problem is to use noncommutative differential calculus. To this end, we introduce the family of operators

$$U(t) = \exp\left(\int_0^t \overset{\theta}{A}(\theta)d\theta\right), \quad t \in [0, 1],$$

and seek for an operator $C(t)$ such that

$$U(t) = e^{D(t)}.$$

Clearly, $U(1) = U$, and $D(1) = C$ is the desired operator. Let us differentiate both sides of the last equation with respect to t. By the definition of T-exponentials, we have

$$\frac{d}{dt}U(t) = A(t)U(t) = A(t)e^{D(t)};$$

the differentiation on the right-hand side gives

$$\frac{d}{dt}e^{D(t)} = \overset{2}{D'(t)} \frac{e^{\overset{3}{D}(t)} - e^{\overset{1}{D}(t)}}{\overset{3}{D}(t) - \overset{1}{D}(t)}$$

by the Daletskii–Krein formula. Combining the last two equations, we obtain

$$A(t) = \overset{2}{D'(t)} \frac{e^{\overset{3}{D}(t)-\overset{1}{D}(t)} - 1}{\overset{3}{D}(t) - \overset{1}{D}(t)},$$

or

$$D'(t) = \frac{\overset{3}{D}(t) - \overset{1}{D}(t)}{e^{\overset{3}{D}(t)-\overset{1}{D}(t)} - 1} \overset{2}{A}(t) = f(\mathrm{ad}_{D(t)})(A(t)),$$

[3] Our argument is a bit formal in this subsection since we do not dwell upon the conditions that would guarantee the existence of the T-exponential, which is a rather complicated topic in its own right.

where

$$f(z) = \frac{z}{e^z - 1}.$$

We will make use of the following lemma.

Lemma I.3 *The formula*

$$e^{\mathrm{ad}_{D(t)}} = \exp\left(\int_0^t \overset{\theta}{\mathrm{ad}_{A(\theta)}} d\theta \right)$$

is valid.

Proof. We have

$$e^{\mathrm{ad}_{D(t)}}(B) = e^{\overset{3}{D(t)} - \overset{1}{D(t)}} \overset{2}{B} = e^{D(t)} B e^{-D(t)} = U(t) B U(t)^{-1}$$

for any $B \in \mathcal{A}$ (cf. Section 4); in view of the definition of $U(t)$,

$$\frac{d}{dt} e^{\mathrm{ad}_{D(t)}}(B) = A(t) \circ U(t) B U(t)^{-1} - U(t) B U(t)^{-1} A = \mathrm{ad}_{A(t)}(e^{\mathrm{ad}_{D(t)}}(B)).$$

On the other hand,

$$\frac{d}{dt} \exp\left(\int_0^t \overset{\theta}{\mathrm{ad}_{A(\theta)}} d\theta \right)(B) = \mathrm{ad}_{A(t)} \left(\exp\left(\int_0^t \overset{\theta}{\mathrm{ad}_{A(\theta)}} d\theta \right) \right)(B)$$

by the definition of the T-exponential. We see that both sides of the equation in question satisfy the same first-order equation

$$\frac{dX}{dt} = \mathrm{ad}_{A(t)} X$$

and the same initial condition $X(0) = \mathrm{id}$ (identity operator). Thus, they coincide for all t. The lemma is proved. □

It follows from Lemma I.3 that

$$\mathrm{ad}_{D(t)} = \log\left([\![\exp\left(\int_0^t \overset{\theta}{\mathrm{ad}_{A(\theta)}} d\theta \right)]\!] \right).$$

Hence we obtain

$$D'(t) = \varphi\left([\![\exp\left(\int_0^t \overset{\theta}{\mathrm{ad}_{A(\theta)}} d\theta \right)]\!] \right)(A(t)),$$

where

$$\varphi(z) = f(\log z) = \frac{\log z}{z-1} = \sum_{k=0}^{\infty} \frac{(1-z)^k}{k+1}.$$

Since $D(0) = 0$, we obtain

$$D(t) = \int_0^t \varphi\left([\![\exp\left(\int_0^{\tau} \operatorname{ad}_{\overset{\theta}{A(\theta)}} d\theta \right)]\!] \right) A(\tau) d\tau$$

and, in particular

$$C = D(1) = \int_0^1 \varphi\left([\![\exp\left(\int_0^{t} \operatorname{ad}_{\overset{\theta}{A(\theta)}} d\theta \right)]\!] \right) A(t) dt.$$

This is a closed expression for

$$C = \log\left([\![\exp\left(\int_0^1 \overset{\theta}{A}(\theta) d\theta \right)]\!] \right)$$

via the operators $A(t)$ and their commutators. Proceeding by analogy with the preceding subsection, we can obtain an expansion of C whose Nth term contains Nth-order commutators for any N. To this end, we expand φ in a Taylor series, thus obtaining

$$C = \int_0^1 \sum_{k=0}^{\infty} \frac{\left(1 - [\![\exp\left(\int_0^t \operatorname{ad}_{\overset{\theta}{A(\theta)}} d\theta \right)]\!] \right)^k}{k+1} A(t) dt.$$

Next, we use the Taylor series expansion of the T-exponential:

$$\begin{aligned}
\exp\left(\int_0^t \operatorname{ad}_{\overset{\theta}{A(\theta)}} d\theta \right) &= 1 + \int_0^t \operatorname{ad}_{\overset{\theta}{A(\theta)}} d\theta + \frac{1}{2} \left(\int_0^t \operatorname{ad}_{\overset{\theta}{A(\theta)}} d\theta \right)^2 + \cdots \\
&= 1 + \int_0^t \operatorname{ad}_{A(\theta)} d\theta + \int_0^t \int_0^{\theta_1} \operatorname{ad}_{A(\theta_1)} \operatorname{ad}_{A(\theta_2)} d\theta_2 d\theta_1 + \cdots
\end{aligned}$$

(in the Nth-order term we divide the integration domain into $N!$ equal simplices on each of which the factors $\operatorname{ad}_{A(\theta_j)}$ appear in a fixed order; it is easy to see that all these simplices give the same contribution, and the factor $\frac{1}{N!}$ cancels out). On substituting this expansion of the T-exponential into the preceding formula we obtain

$$C = \sum_{s=0}^{\infty} \sum_{k=0}^{s} \frac{(-1)^k}{k+1}$$

$$\times \sum_{j_1+\cdots+j_k=s,\, j_i \geq 1} \int_0^1 \left(\int_0^t \cdots \int_0^{\theta_{j_1-1}} \mathrm{ad}_{A(\theta_1)}\, \mathrm{ad}_{A(\theta_2)} \ldots \mathrm{ad}_{A(\theta_{j_1})}\, d\theta_{j_1} \ldots d\theta_1 \right) \ldots$$

$$\times \left(\int_0^t \cdots \int_0^{\theta_{j_k}-1} \mathrm{ad}_{A(\theta_1)}\, \mathrm{ad}_{A(\theta_2)} \ldots \mathrm{ad}_{A(\theta_{jk})}\, d\theta_{j_k} \ldots d\theta_1 \right) A(t)dt.$$

The sth term of the outer sum is a finite sum of integrals; this sum involves all sth-order commutators involved in the expansion of C. Let us write out a few first terms of the expansion. We have

$$\begin{aligned} C \;=\; & \int_0^1 A(t)dt - \frac{1}{2} \int_0^1 \int_0^t \mathrm{ad}_{A(\theta)}(A(t))d\theta dt \\ & -\frac{1}{2} \int_0^1 \int_0^t \int_0^{\theta_1} \mathrm{ad}_{A(\theta_1)}\, \mathrm{ad}_{A(\theta_2)}(A(t))d\theta_2 d\theta_1 dt \\ & +\frac{1}{3} \int_0^1 \int_0^t \int_0^t \mathrm{ad}_{A(\theta_1)}\, \mathrm{ad}_{A(\theta_2)}(A(t))d\theta_2 d\theta_1 dt + \cdots \end{aligned}$$

or, in terms of commutators,

$$\begin{aligned} C \;=\; & \int_0^1 A(t)dt - \frac{1}{2} \int_0^1 \int_0^t [A(\theta), A(t)]d\theta dt \\ & -\frac{1}{2} \int_0^1 \int_0^t \int_0^{\theta_1} [A(\theta_1), [A(\theta_2), A(t)]]d\theta_2 d\theta_1 dt \\ & +\frac{1}{3} \int_0^1 \int_0^t \int_0^t [A(\theta_1), [A(\theta_2), A(t)]]d\theta_2 d\theta_1 dt + \cdots, \end{aligned}$$

where the dots stand for a sum of integrals of commutators whose order is greater than or equal to three.

Let us consider a simple illustrative example.

Example I.11 Consider the first-order differential equation

$$\frac{\partial u}{\partial t} = a(t)\frac{\partial u}{\partial x} + b(t)xu$$

with the initial condition

$$u(x,t)|_{t=0} = u_0(x).$$

The solution of this Cauchy problem can easily be found explicitly by straightforward computation. We consider the equation of characteristics

$$\dot{x} = a(t), \quad x(0) = x_0;$$

its solution is

$$x(x_0,t) = x_0 + \int_0^t a(\tau)d\tau.$$

Set

$$U(x_0,t) = u(x(x_0,t),t);$$

then for U we obtain the equation

$$\frac{dU}{dt} = b(t)\left(x_0 + \int_0^t a(\tau)d\tau\right)U,$$

whence

$$U(x_0,t) = u_0(x_0)\exp\left(\int_0^t b(\theta)\left(x_0 + \int_0^\theta a(\tau)d\tau\right)d\theta\right).$$

Making the inverse substitution

$$x_0 = x - \int_0^t a(\tau)d\tau,$$

we obtain

$$u(x,t) = u_0\left(x - \int_0^t a(\tau)d\tau\right)\exp\left(\int_0^t b(\theta)\left(x - \int_\theta^t a(\tau)d\tau\right)d\theta\right).$$

Now, let us compute the same solution by means of the T-exponential.

Let $u(t)$ denote the solution $u(x,t)$ considered as the function of t with values in functions of x. We have

$$u(t) = \exp\left(\int_0^t \overset{\tau}{A}(\tau)d\tau\right)u(0),$$

where

$$A(\tau) = a(\tau)\frac{\partial}{\partial x} + b(\tau)x.$$

Let us compute the T-exponential for $t = t_0$ (t_0 is arbitrary) using the formula we have just obtained. We shall see that this leads to the same result.

We have

$$[A(\theta), A(t)] = \left[a(\theta)\frac{\partial}{\partial x} + b(\theta)x, a(t)\frac{\partial}{\partial x} + b(t)x\right] = a(\theta)b(t) - a(t)b(\theta),$$

and the commutators of second and higher orders are all zero. Therefore, only two terms are retained in the expansion of the logarithm of the T-exponential. Namely, we have

$$\log\left(\exp\left(\int_0^{t_0} \overset{\tau}{A}(\tau)d\tau\right)\right) = \int_0^{t_0} A(t)dt - \frac{1}{2}\int_0^{t_0}\int_0^{t} [A(\theta), A(t)]d\theta dt$$
$$= \left(\int_0^{t_0} a(t)dt\right)\frac{\partial}{\partial x} + \left(\int_0^{t_0} b(t)dt\right)x - \frac{1}{2}\int_0^{t_0}\left(b(t)\int_0^{t} a(\theta)d\theta\right)dt$$
$$+\frac{1}{2}\int_0^{t_0}\left(a(t)\int_0^{t} b(\tau)d\tau\right)dt \overset{\text{def}}{=} C_1\frac{\partial}{\partial x} + C_2x + \lambda.$$

The value of the solution $u(x, t)$ at $t = t_0$, say, coincides with the solution at $t = 1$ of the Cauchy problem

$$\begin{cases} \dfrac{\partial v}{\partial t} = C_1\dfrac{\partial v}{\partial x} + C_2xv + \lambda v, \\ v|_{t=0} = u_0(x) \end{cases}$$

with coefficients independent of t.

We have

$$v(x, 1) = u_0(x - C_1)\exp\left(\int_0^1 C_2(x - (1 - t)C_1)dt + \lambda\right)$$
$$= u_0\left(x - \int_0^{t_0} a(t)dt\right)\exp\left\{x\int_0^{t_0} b(t)dt - \frac{1}{2}\int_0^{t_0} a(t)dt\int_0^{t_0} b(\tau)d\tau\right.$$
$$\left.+\frac{1}{2}\int_0^{t_0} a(t)\int_0^{t} b(\tau)d\tau\, dt - \frac{1}{2}\int_0^{t_0} a(t)\int_0^{t_0} b(\tau)d\tau\, dt\right\}$$

$$= u_0\left(x - \int_0^{t_0} a(t)dt\right)\exp\left\{\int_0^{t_0} b(t)\left(x - \int_t^{t_0} a(\tau)d\tau\right)dt\right\},$$

which coincides with the value obtained by the straightforward method.

The method shown in this example is quite useful if for each t the operator $A(t)$ is a linear combination of generators of a nilpotent finite-dimensional Lie algebra,

$$A(t) = \sum_{j=1}^{m} a_j(t)\, A_j,$$

where A_j are the generators. In this case the computation of the T-exponential $\exp\left(\int_0^t \overset{\theta}{A(\theta)}d\theta\right)$ is reduced to the computation of $\exp\left(\sum \alpha_j(t)A_j(t)\right)$, where the coefficients α_j can be obtained by appropriate integrations from the above expansion.

Even if the Lie algebra in question is not nilpotent, the problem is greatly simplified since we reduce the computation of $\exp\left(\int_0^t \overset{\theta}{A(\theta)}d\theta\right)$ (operators in an infinite-dimensional space) to the computation of

$$\int_0^1 \varphi\left(\left[\!\left[\exp\left(\int_0^t \overset{\theta}{\mathrm{ad}_{A(\theta)}}d\theta\right]\!\right]\right)\right) A(t)dt,$$

which in fact is carried out in the m-dimensional space of the Lie algebra.

5 Summary: Rules of "Operator Arithmetic" and Some Standard Techniques

In the preceding sections we have considered a number of notions and statements pertaining to noncommutative analysis. We hope that it is now clear to the reader why noncommutative analysis proves so useful in applications. Here we summarize the main notions and notation introduced and also give a few common techniques, or even "tricks", that are usually employed in noncommutative analysis.

So far we have been studying the "operator arithmetic", which comprises the rules used in treating operator expressions and the corresponding notation.

5.1 Notation

The subject of noncommutative analysis is functions of operators, more precisely, numerical functions of numerical arguments substituted by linear operators (or elements of some associative algebra). The definition of such a substitution encounters certain difficulties. First, even for functions of a single argument it is not clear a priori what the substitution means if the function is not a polynomial; second, if there is more than one argument, then an additional ambiguity arises, caused by the fact that a function of commuting numerical arguments cannot in principle carry any information about the arrangement of the arguments.

The first difficulty is overcome by the traditional method following from the familiar functional calculus of a single operator. The requirement that the mapping *symbol* $\to$ *operator* be a continuous homomorphism taking x into A uniquely determines the operator $f(A)$ for each symbol $f(x)$ for an appropriate class of symbol spaces (see Theorem I.1). The second difficulty can be removed by introducing new notation having no counterpart in classical analysis. This new notation includes Feynman indices and autonomous brackets.

Feynman Indices. In order to determine the order of action for operators occurring in an operator expression, we write an index (a real number) over each of the operators. The factors in all products are arranged so that these indices decrease from left to right (that is, the leftmost operator has the highest Feynman index, and the rightmost, the lowest). For example,

$$\overset{1}{A}\overset{3}{A}\overset{2}{B}\overset{0}{C} = ABAC.$$

If two operators commute, their indices are allowed to coincide; this does not lead to any ambiguity.

Autonomous Brackets. These are the special brackets $[\![\ ]\!]$ used to determine the order of computations in an operator expression. The expression within the brackets is computed first and then is used in the subsequent computations as a single operator (equipped where necessary with a Feynman index, which is written in this case over the left autonomous bracket). In particular, the indices outside and inside the brackets do not affect the action of each other. For example,

$$\overset{1}{A}\overset{1}{A}\overset{2}{B} = B\overset{2}{A}, \quad \overset{1}{A}\overset{0}{[\![}\overset{1}{A}\overset{2}{B}]\!] = ABA;$$

$$(\overset{1}{A} + \overset{2}{B})^2 = A^2 + B^2 + 2BA,$$

$$[\![\overset{1}{A} + \overset{2}{B}]\!]^2 = (A + B)^2 = A^2 + B^2 + AB + BA.$$

Nested autonomous brackets are allowed. For example, the expression

$$\overset{1}{A}\overset{4}{A}\overset{3}{[\![}(\overset{1}{A} + [\![\overset{2}{A} + B]\!])^2]\!]$$

should be computed as follows. First we compute the inner bracket

$$C = [\![A + B]\!] = A + B,$$

and then the outer one:

$$\begin{aligned} D &= [\![(\overset{1}{A} + \overset{2}{C})^2]\!] = A^2 + C^2 + 2CA \\ &= A^2 + (A + B)^2 + 2(A + B)A = 4A^2 + B^2 + 3BA + AB. \end{aligned}$$

Finally, the entire expression is computed:

$$\overset{1}{A}\overset{4}{A}\overset{3}{D} = A(4A^2 + B^2 + 3BA + AB)A = 4A^4 + AB^2A + 3ABA^2 + A^2BA.$$

5.2 Rules

It follows from the definition of Feynman ordering that it is the mutual order of the Feynman indices, not the indices themselves, that counts. Hence the indices can be changed without affecting the value of an operator expression. Let us now state a rule describing admissible changes of indices.

A pair of arguments A, B of an operator expression is said to be *critical* if the following conditions are satisfied:

(a) A and B do not commute, $[A, B] \neq 0$.

(b) The operator expression cannot be represented in the form $E_1 + E_2$, where $\overset{j}{A}$ occurs only in E_1 and $\overset{l}{B}$ in E_2.

(c) The expression does not contain autonomous brackets such that $\overset{j}{A}$ is inside the brackets and $\overset{l}{B}$ outside the brackets, or vice versa.

The Rule of Changing Feynman Indices. *Feynman indices in an expression can be changed arbitrarily provided that for any critical pair the ordering relation between the Feynman indices of the operators in the pair remains valid.*

Examples.

1) $\overset{1}{A} + \overset{2}{B} = \overset{2}{A} + \overset{1}{B}$ (the pair $\overset{1}{A}$, $\overset{2}{B}$ is not critical since condition (b) is violated).

2) $f(\overset{1}{A}, \overset{2}{A}, \overset{3}{B}) = f(\overset{1}{A}, \overset{1}{A}, \overset{2}{B}) \equiv g(\overset{1}{A}, \overset{1}{B})$, where $g(x, y) = f(x, x, y)$. Thus, the changing indices rule includes changing the number of arguments of a symbol by restriction to the diagonal ("moving indices apart", Proposition I.1).

3) $\overset{3}{B}\overset{2}{\llbracket}(\overset{1}{A} + \overset{5}{B})^2\rrbracket = \overset{3}{B}\overset{2}{\llbracket}(\overset{4}{A} + \overset{5}{B})^2\rrbracket$ (the pair $\overset{3}{B}$, $\overset{1}{A}$ is not critical since B and A are separated by an autonomous bracket).

The Rule of Deleting Autonomous Brackets. *Autonomous brackets can be removed if for any operator argument within the brackets its Feynman index is in the same relation with the Feynman indices of operators outside the brackets as the index of the brackets themselves.*

Naturally, this rule also tells us whether we can introduce the autonomous brackets.

Examples.

1) $\overset{2}{\llbracket}\overset{3}{A}(\overset{2}{B} - \overset{4}{B})\rrbracket\delta f/\delta x(\overset{1}{B}, \overset{5}{B}) = \overset{3}{A}(\overset{2}{B} - \overset{4}{B})\delta f/\delta x(\overset{1}{B}, \overset{5}{B})$.

2) $\overset{3}{\llbracket}(\overset{1}{A} + \overset{2}{B})^2\rrbracket(\overset{1}{A} - \overset{2}{B})^2 \neq (\overset{1}{A} + \overset{2}{B})^2(\overset{1}{A} - \overset{2}{B})^2$ in general, since the Feynman index of the $\overset{1}{A}$ inside the brackets, which have the index 3, is less than that of $\overset{2}{B}$ outside the brackets.

Surprisingly, this is almost all the new notation and rules to be introduced in order to permit us to deal easily with operator expressions.

5.3 Standard Techniques

When someone applies differential and integral calculus to solve a particular problem, he or she generally uses a lot of standard techniques such as integration by parts, L'Hôpital's rule, etc., being however scarcely aware of their presence, because they are used often enough to become fairly commonplace. Similarly, noncommutative analysis includes quite a few techniques of its own, which should always be at hand in noncommutative computations. Some of them were already presented in this chapter; and the others will be given in the sequel. Here we recall briefly what we have already seen above. The methods can be divided into a few groups.

First, there are some simple transformations following directly from definitions and from the rules of changing Feynman indices and inserting (deleting) autonomous brackets, given in the preceding subsection. It is probably convenient to present them as a number of axioms, as was done in [129].

(μ_1) (Homogeneity axiom). For any symbol f and number α we have

$$\llbracket\alpha f(\overset{n_1}{A_1}, \dots, \overset{n_k}{A_k})\rrbracket = \alpha\llbracket f(\overset{n_1}{A_1}, \dots, \overset{n_k}{A_k})\rrbracket.$$

(μ_2) (Index changing axiom). Let $n_1, \dots, n_k$ and $m_1, \dots, m_k$ be Feynman indices such that $n_i < n_j$ if and only if $m_i < m_j$. Then

$$f(\overset{n_1}{A_1}, \dots, \overset{n_k}{A_k}) = f(\overset{m_1}{A_1}, \dots, \overset{m_k}{A_k}).$$

If $n_i = n_j$ and $A_i = A_j = A$, then

$$f(\overset{n_1}{A_1}, \dots, \overset{n_i}{A_i}, \dots, \overset{n_j}{A_j}, \dots, \overset{n_k}{A_k}) = g(\overset{n_1}{A_1}, \dots, \overset{n_i}{A_i}, \dots, \overset{n_k}{A_k}),$$

where

$$g(x_1, \dots, x_i, \dots, x_i, \dots, x_k) = f(x_1, \dots, x_k)\big|_{x_j = x_i}.$$

(μ_3) (Correspondence axiom). If $f(x_1, \dots, x_k) = 1$, then $f(\overset{n_1}{A_1}, \dots, \overset{n_k}{A_k}) = 1$ (identity operator); if $f(x_1, \dots, x_k) = x_j$, then $f(\overset{n_1}{A_1}, \dots, \overset{n_k}{A_k}) = A_j$.

(μ_4) (The sum axiom).

$$f(\overset{n_1}{A_1}, \dots, \overset{n_k}{A_k}) + g(\overset{m_1}{B_1}, \dots, \overset{m_l}{B_l}) = [\![f(\overset{n_1}{A_1}, \dots, \overset{n_k}{A_k})]\!] + [\![g(\overset{m_1}{B_1}, \dots, \overset{m_l}{B_l})]\!].$$

(μ_5) (Product axiom). If $m_i < n_j$ for any i and j, then

$$f(\overset{n_1}{A_1}, \dots, \overset{n_k}{A_k}) g(\overset{m_1}{M_1}, \dots, \overset{m_l}{B_l}) = [\![f(\overset{n_1}{A_1}, \dots, \overset{n_k}{A_k})]\!][\![g(\overset{m_1}{B_1}, \dots, \overset{m_l}{B_l})]\!].$$

(μ_6) (Zero axiom). If $f(\overset{n_1}{A_1}, \dots, \overset{n_k}{A_k}) = 0$ and the indices

$$p_1, \dots, p_l, \quad r_1, \dots, r_m$$

satisfy $p_i < n_j$ and $r_i > n_j$ for any i and j, then

$$f(\overset{n_1}{A_1}, \dots, \overset{n_k}{A_k}) g(\overset{p_1}{B_1}, \dots, \overset{p_l}{B_l}, \overset{r_1}{C_1}, \dots, \overset{r_m}{C_m}) = 0$$

for any generators $B_1, \dots, B_l$ and $C_1, \dots, C_m$.

(T_1) (Continuity axiom I). The mapping

$$f(x_1, \dots, x_k) \mapsto f(\overset{n_1}{A_1}, \dots, \overset{n_k}{A_k})$$

is continuous.

(T_2) (Continuity axiom II). If $f_\alpha(\overset{n_1}{A_1}, \dots, \overset{n_k}{A_k}) \to 0$ and the indices $p_1, \dots, p_l$, $r_1, \dots, r_m$ satisfy the conditions of axiom (μ_6), then

$$f_\alpha(\overset{n_1}{A_1}, \dots, \overset{n_k}{A_k}) g(\overset{p_1}{B_1}, \dots, \overset{p_l}{B_l}, \overset{r_1}{C_1}, \dots, \overset{r_m}{C_m}) \to 0$$

for any generators $B_1, \dots, B_l$ and $C_1, \dots, C_m$.

Second, there is a group of techniques based on the naturality properties of the mapping symbol $\mapsto$ operator, as stated in Theorems I.4 and I.5. Actually, the main naturality property is that expressed by item 1° in the theorems cited, namely, that if

$$\varphi : \mathcal{A} \to \mathcal{B}$$

is a homomorphism of operator algebras, then φ takes generators into generators and preserves symbols when applied to functions of generators (i.e., $\varphi(f(A))$ is the function of $B = \varphi(A)$ with symbol f, and a similar statement is true of multivariate symbols). Let us recall some of these techniques.

(a) Passage to the left (or right) regular representation. The mapping

$$\mathcal{A} \xrightarrow{L} \mathcal{L}(\mathcal{A}),$$
$$A \mapsto L_A,$$

where

$$L_A(B) = AB, \quad B \in \mathcal{A}$$

is an algebra homomorphism (a representation of $\mathcal{A}$ in the linear space $\mathcal{A}$) and we have

$$f(L_A) = L_{f(A)},$$
$$f(\overset{1}{L}_{A_1}, \dots, \overset{n}{L}_{A_n}) = L_{f(\overset{1}{A}_1, \dots, \overset{n}{A}_n)}$$

(this was proved in Subsection 2.3 independently of Theorems I.4 and I.5). Since the representation is *faithful* (i.e., its kernel is zero), if suffices to reason in terms of the left regular representation operators, which is often much simpler.

For example, in proving the derivation formula

$$D[f(A)] = \overset{2}{\overline{DA}} \frac{\delta f}{\delta x}(\overset{1}{A}, \overset{3}{A})$$

the passage to L_A permits us to assume that D is an inner derivation (i.e., given by the commutator with some element), since

$$L_{DA} = [D, L_A].$$

(b) The conjugation transformation.

If U is an invertible element of an operator algebra $\mathcal{A}$, then the mapping

$$A \mapsto UAU^{-1} \overset{\text{def}}{=} \operatorname{Ad}_U(A)$$

is an automorphism of $\mathcal{A}$. Consequently, for any generator $A \in \mathcal{A}$ the operator UAU^{-1} is a generator as well, and we have

$$U[\![f(\overset{1}{A}_1, \dots, \overset{n}{A}_n)]\!]U^{-1} = f(\overset{1}{\overline{UA_1U^{-1}}}, \dots, \overset{n}{\overline{UA_nU^{-1}}})$$

for any symbol $f(y_1, \dots, y_n)$ and any Feynman tuple $(\overset{1}{A}_1, \dots, \overset{n}{A}_n)$ of generators.

Further examples of using the naturality property can be found in Chapter II.

Along with L_A, R_A, and Ad_U, it is often helpful to use two other classes of transformations on $\mathcal{A}$; namely, we speak of

$$\mathrm{ad}_A = L_A - R_A$$

(Commutation operator) and

$$\mathrm{an}_A = L_A + R_A$$

(Anticommutation operator). Suppose that A is a generator; then ad_A is a generator provided that $f(x-y) \in \mathcal{F}_2$ for any $f \in \mathcal{F}$, and the same is true of an_A if $f(x+y) \in \mathcal{F}_2$ for any $f \in \mathcal{F}$.

These operators satisfy numerous remarkable formulas. Here are some of them:

$$\begin{aligned}
&e^{\mathrm{ad}_A} = \mathrm{Ad}_{(e^{L_A})};\\
&[\mathrm{ad}_A, \mathrm{ad}_B] = [\mathrm{an}_A, \mathrm{an}_B] = \mathrm{ad}_{[A,B]};\\
&[\mathrm{ad}_A, \mathrm{an}_B] = \mathrm{an}_{[a,b]};\\
&[L_A, L_B] = [L_A, \mathrm{ad}_B] = [L_A, \mathrm{an}_B] = L_{[a,b]}.
\end{aligned}$$

Third, there is a very useful trick of considering *matrix* operators. It was widely used in the proofs of theorems in Section 2.

Without loss of generality it can be assumed that all operators considered act in linear spaces (this can always be achieved by passing to the left regular representation). Let $\mathcal{A}$ be an algebra of operators on V, $\mathcal{B}$ an algebra of operators on W, $\mathcal{G}$ an $(\mathcal{A}, \mathcal{B})$-bimodule of operators from W into V, and $\mathcal{D}$ a $(\mathcal{B}, \mathcal{A})$-bimodule of operators from V into W (we do not exclude the case in which $V = W$ and $\mathcal{A} = \mathcal{B} = \mathcal{G} = \mathcal{D}$). We may consider the algebra $\mathcal{U}$ of matrix operators

$$\begin{pmatrix} A & C \\ D & B \end{pmatrix}, \quad A \in \mathcal{A}, \quad B \in \mathcal{B}, \quad C \in \mathcal{G}, \quad \text{and} \quad D \in \mathcal{D},$$

acting on $V \oplus W$. Computations in such an algebra are sometimes much simpler than in the original algebra.

This is chiefly due to the following property.

Lemma I.4 *Let $A \in \mathcal{A}$ and $B \in \mathcal{B}$ be generators. Then for any $C \in \mathcal{G}$ the operator $\begin{pmatrix} A & C \\ 0 & B \end{pmatrix}$ is a generator in $\mathcal{U}$, and*

$$f\left(\begin{pmatrix} A & C \\ 0 & B \end{pmatrix}\right) = \begin{pmatrix} f(A) & \overset{2}{C}\dfrac{\delta f}{\delta y}(\overset{3}{A}, \overset{1}{B}) \\ 0 & f(B) \end{pmatrix}$$

for any symbol $f \in \mathcal{F}$.

Proof. The above formula clearly defines a continuous mapping $\mathcal{F} \to \mathcal{U}$. Next, this mapping takes $1 \mapsto 1$ and $y \mapsto \begin{pmatrix} A & C \\ 0 & B \end{pmatrix}$ since $\delta y/\delta y \equiv 1$. It remains to check that the mapping is a homomorphism. We have

$$\begin{aligned} & f\left(\begin{pmatrix} A & C \\ 0 & B \end{pmatrix}\right) g\left(\begin{pmatrix} A & C \\ 0 & B \end{pmatrix}\right) \\ &= \begin{pmatrix} f(A) & \overset{2}{C}\frac{\delta f}{\delta y}(\overset{3}{A}, \overset{1}{B}) \\ 0 & f(B) \end{pmatrix} \begin{pmatrix} g(A) & \overset{2}{C}\frac{\delta g}{\delta y}(\overset{3}{A}, \overset{1}{B}) \\ 0 & g(B) \end{pmatrix} \\ &= \begin{pmatrix} f(A)g(A) & f(A)[\![\overset{2}{C}\frac{\delta g}{\delta y}(\overset{3}{A}, \overset{1}{B})]\!] + [\![\overset{2}{C}\frac{\delta f}{\delta y}(\overset{3}{A}, \overset{1}{B})]\!] g(B) \\ 0 & f(B)g(B) \end{pmatrix}. \end{aligned}$$

The diagonal entries have the desired form; the superdiagonal entry can be transformed as follows:

$$f(A)[\![\overset{2}{C}\frac{\delta g}{\delta y}(\overset{3}{A}, \overset{1}{B})]\!] + [\![\overset{2}{C}\frac{\delta f}{\delta y}(\overset{3}{A}, \overset{1}{B})]\!] g(B) = \overset{2}{C} F(\overset{3}{A}, \overset{1}{B}),$$

where

$$\begin{aligned} F(x, y) &= f(x)\frac{\delta g}{\delta y}(x, y) + g(y)\frac{\delta f}{\delta y}(x, y) \\ &= \frac{f(x)g(x) - f(x)g(y) + g(y)f(x) - g(y)f(y)}{x - y} = \frac{\delta(fg)}{\delta y}(x, y). \end{aligned}$$

The lemma is proved. □

Let us consider two examples.

(a) Suppose that the identity

$$ac = cb + d$$

is valid for some elements $a, b, c, d \in \mathcal{A}$ where a and b are generators. In matrix form, we can write out this identity as

$$\begin{pmatrix} a & d \\ 0 & b \end{pmatrix} \begin{pmatrix} 1 & c \\ 0 & 0 \end{pmatrix} = \begin{pmatrix} 1 & c \\ 0 & 0 \end{pmatrix} \begin{pmatrix} a & d \\ 0 & b \end{pmatrix},$$

or

$$AC = CA,$$

where

$$A = \begin{pmatrix} a & d \\ 0 & b \end{pmatrix} \quad \text{and} \quad C = \begin{pmatrix} 1 & c \\ 0 & 0 \end{pmatrix}.$$

By Theorem I.4, 2°, we obtain

$$f(A)C = Cf(A)$$

for any symbol f, or, according to Lemma I.4,

$$\begin{pmatrix} f(a) & \overset{2}{d}\frac{\delta f}{\delta y}(\overset{3}{a}, \overset{1}{b}) \\ 0 & f(b) \end{pmatrix} \begin{pmatrix} 1 & c \\ 0 & 0 \end{pmatrix} = \begin{pmatrix} 1 & c \\ 0 & 0 \end{pmatrix} \begin{pmatrix} f(a) & \overset{2}{d}\frac{\delta f}{\delta y}(\overset{3}{a}, \overset{1}{b}) \\ 0 & f(b) \end{pmatrix}.$$

Multiplying the matrices on both sides of the last equation, we get

$$f(a)c = cf(b) + \overset{2}{d}\frac{\delta f}{\delta y}(\overset{3}{a}, \overset{1}{b}),$$

that is, we have obtained yet another proof of Theorem I.6.

(b) Our second example pertains to Lie algebras and employs $n \times n$ rather than 2×2 matrices. Operators $A_1, \dots, A_n \in \mathcal{A}$ are said to form a representation of a Lie algebra if

$$[A_j, A_k] = \sum_{l=1}^{n} \lambda^l_{jk} A_l,$$

where λ^l_{jk} are numbers, called the *structural constants* of the Lie algebra. These commutation relations can be rewritten in a very simple form if we introduce the matrices

$$A = \begin{pmatrix} A_1 & & & 0 \\ & A_3 & & \\ & & \ddots & \\ 0 & & & A_n \end{pmatrix}, \quad \mathcal{A}_j = \begin{pmatrix} A_j & & & 0 \\ & A_j & & \\ & & \ddots & \\ 0 & & & A_j \end{pmatrix},$$

and $\Lambda_j = ||\lambda^l_{kj}||^n_{k,l=1}$.

With this notation, the Lie commutation relations take the form

$$A\mathcal{A}_j = (\mathcal{A}_j + \Lambda_j)A,$$

which permits one to obtain various permutation formulas easily; for example, we have, by Theorem I.4, 3°

$$Af(\mathcal{A}_j) = f(\mathcal{A}_j + \Lambda_j)A.$$

Finally, let us note that the techniques of noncommutative analysis widely use the difference derivative in conjunction with moving indices apart; this can be seen throughout this chapter.

Of course, there are also numerous formulas such as those derived in Section 3. However, it would be redundant to list all these formulas here, and we refrain from doing so, with the exception of the following ones:

$$\frac{d}{dt} f(A(t)) = \overset{2}{\overline{A'(t)}} \frac{\delta f}{\delta y}(\overset{1}{A}(t), \overset{3}{A}(t))$$

(Daletskii–Krein formula);

$$f(C) = f(A) + \sum_{k=1}^{N-1} \overset{2}{[\![} C - A]\!] \dots \overset{2k}{[\![} C - A]\!] \frac{\delta^k f}{\delta y^k}(\overset{1}{A}, \dots, \overset{2k+1}{A})$$
$$+ \overset{2}{[\![} C - A]\!] \dots \overset{2N}{[\![} C - A]\!] \frac{\delta^N f}{\delta y^N}(\overset{1}{C}, \overset{3}{A}, \dots, \overset{2N+1}{A})$$

(Newton formula);

$$f(C) = f(A) + \sum_{N-1}^{k=1} \frac{f^{(k)}(A)}{A!} [\![(\overset{1}{C} - \overset{2}{A})^k]\!] + [\![(\overset{2}{C} - \overset{1}{A})^N]\!] \overset{2}{} \frac{\delta^N f}{\delta y^N}(\overset{1}{C}, \overset{3}{A}, \dots, \overset{3}{A})$$

(Taylor formula);

$$f(\overset{1}{A}, \overset{2}{B}) = f(\overset{2}{A}, \overset{1}{B}) + \overset{3}{[B, A]} \frac{\delta^2 f}{\delta y_1 \delta y_2}(\overset{1}{A}, \overset{5}{A}; \overset{2}{B}, \overset{4}{B})$$

(Index permutation formula);

$$\begin{aligned}
f([\![A + B]\!]) &= f(\overset{1}{A} + \overset{2}{B}) + \overset{3}{[A, B]} \frac{\delta^2 f}{\delta y^2}(\overset{4}{[\![}A + B]\!], \overset{2}{[\![}A + B]\!], \overset{1}{A} + \overset{2}{B}) \\
&= f(\overset{1}{A} + \overset{2}{B}) + \overset{3}{[A, B]} \frac{\delta^2 f}{\delta y^2}(\overset{5}{[\![}A + B]\!], \overset{1}{A} + \overset{2}{B}, \overset{1}{A} + \overset{4}{B}), \\
f([\![g(\overset{1}{A}, \overset{2}{B})]\!]) &= f(g(\overset{1}{A}, \overset{2}{B})) \\
&= \overset{5}{[A, B]} \frac{\delta g}{\delta y_1}(\overset{2}{A}, \overset{7}{A}, \overset{8}{B}) \frac{\delta g}{\delta y_2}(\overset{3}{A}, \overset{4}{B}, \overset{6}{B}) \\
&\quad \times \frac{\delta^2 f}{\delta y^2}(\overset{1}{[\![}g(\overset{1}{A}, \overset{2}{B})]\!], g(\overset{2}{A}, \overset{8}{B}), g(\overset{7}{A}, \overset{8}{B}))
\end{aligned}$$

(Composite function formulas);

$$f(\mathrm{ad}_B)(A) = f(\overset{3}{B} - \overset{1}{B})\overset{2}{A}$$

(functions of ad_B).

Chapter II
Method of Ordered Representation

1 Ordered Representation: Definition and Main Property

1.1 Wick Normal Form

We begin with an example.

In Subsection 1.2 we have posed the problem of calculating the Wick normal form for operators acting on a Hilbert state space, and in fact the answer was given for a rather particular case in Subsection 1.6. Let us recall the matter and clarify the subject.

For simplicity, suppose that we deal with the case in which there is only one basic state. So, we are given a Hilbert space $\mathcal{H}$ and two operators, a^+ and a^-, on $\mathcal{H}$ satisfying the commutation relation

$$[a^+, a^-] \equiv a^+a^- - a^-a^+ = -I,$$

where I is the identity operator on $\mathcal{H}$; the operators a^+ and a^- will be referred to as the *creation* and the *annihilation operator*, respectively.

For example, we can take $\mathcal{H} = L_2(\mathbb{R}^1)$ and

$$a^{\pm} = (2h\omega)^{-1/2}\left(-ih\frac{\partial}{\partial x} \pm i\omega x\right),$$

where h and ω are positive parameters, as was done in Subsection 1.6 in considering the eigenvalue problem for the quantum-mechanical oscillator.

The Wick normal form of an operator A acting on $\mathcal{H}$ is a representation of A in the form of a function of the Feynman tuple $(\overset{2}{a^+}, \overset{1}{a^-})$, namely,

$$A = f(\overset{2}{a^+}, \overset{1}{a^-})$$

where

$$f(z, w) = \sum_{\alpha,\beta\geq 0} c_{\alpha\beta} z^\alpha w^\beta$$

is a function of two arguments, which we call the *Wick symbol* of A (we avoid discussing exact requirements on the symbols and assume that $f(\overset{2}{a^+}, \overset{1}{a^-})$ is defined as

$$f(\overset{2}{a^+}, \overset{1}{a^-}) = \sum_{\alpha,\beta\geq 0} c_{\alpha\beta}(a^+)^\alpha (a^-)^\beta,$$

where the series is assumed to converge, say in the weak sense, on some dense subset of $\mathcal{H}$).

Given an operator A on $\mathcal{H}$, how can we reduce it to the Wick normal form? If A is arbitrary, this question is rather difficult.

Therefore, we will not consider this question in full generality, but will rather consider a particular class of operators A, namely, the operators which can be represented as functions of several appropriately numbered occurrences of a^+ and a^-.

First of all, consider the problem of reducing to the Wick normal form the product $A = A_1 A_2$, where A_1 and A_2 are already in normal form,

$$A_1 = f_1(\overset{2}{a^+}, \overset{1}{a^-}), \quad A_2 = f_2(\overset{2}{a^+}, \overset{1}{a^-}).$$

In a particular case (A_2 is arbitrary and $A_1 = a^+ a^-$) this problem has already been solved in Subsection 1.6. However, the general case is not much more complicated. Indeed, suppose for the moment that $f_1(z, w) = z$ or $f_1(z, w) = w$. For the former, we have

$$\overset{2}{a^+}[\![f_2(\overset{2}{a^+}, \overset{1}{a^-})]\!] = \overset{3}{a^+} f_2(\overset{2}{a^+}, \overset{1}{a^-}) = \overset{2}{a^+} f_2(\overset{2}{a^+}, \overset{1}{a^-}) \equiv g(\overset{2}{a^+}, \overset{1}{a^-}),$$

where

$$g(z, w) = z f_2(z, w).$$

For the latter, the calculation is a little longer, since we have to move the operator $\overset{3}{a^+}$ to the first position. In doing so, we use the permutation formula:

$$\overset{2}{a^-}[\![f_2(\overset{2}{a^+}, \overset{1}{a^-})]\!] = \overset{3}{a^-} f_2(\overset{2}{a^+}, \overset{1}{a^-}) = \overset{1}{a^-} f_2(\overset{2}{a^+}, \overset{1}{a^-}) + a^-, \overset{3}{a^+}] \frac{\delta f_2}{\delta z}(\overset{4}{a^+}, \overset{2}{a^+}, \overset{1}{a^-}).$$

However,

$$[a^-, a^+] = I,$$

and so we can continue this chain of equations by identifying the first and the third arguments in $\delta f_2 / \delta z$:

$$\overset{2}{a^-}[\![f_2(\overset{2}{a^+}, \overset{1}{a^-})]\!] = \overset{1}{a^-} f_2(\overset{2}{a^+}, \overset{1}{a^-}) + \frac{\partial f_2}{\partial z}(\overset{2}{a^+}, \overset{1}{a^-}) = h(\overset{2}{a^+}, \overset{1}{a^-}),$$

where

$$h(z, w) = wf_2(z, w) - \frac{\partial f_2}{\partial z}(z, w).$$

Denote by l^+ and l^- the operators taking f_2 into g and h, respectively, i.e.,

$$l^+ = z, \quad l^- = w - \frac{\partial}{\partial z}.$$

Thus, we have

$$a^+ [\![f(\overset{2}{a^+}, \overset{1}{a^-})]\!] = (l^+ f)(\overset{2}{a^+}, \overset{1}{a^-}),$$
$$a^- [\![f(\overset{2}{a^+}, \overset{1}{a^-})]\!] = (l^- f)(\overset{2}{a^+}, \overset{1}{a^-}).$$

The operators l^+ and l^- characterized by this property are called the *left ordered representation* operators for the Feynman tuple $(\overset{2}{a^+}, \overset{1}{a^-})$. As soon as these operators are evaluated, the problem of reducing the product of operators to the Wick normal form becomes trivial. Indeed, let

$$f_1(z, w) = \sum_{\alpha,\beta} b_{\alpha,\beta} z^\alpha w^\beta.$$

Then

$$[\![f_1(\overset{2}{a^+}, \overset{1}{a^-})]\!] [\![f_2(\overset{2}{a^+}, \overset{1}{a^-})]\!] = \sum_{\alpha,\beta} b_{\alpha,\beta} (a^+)^\alpha (a^-)^\beta [\![f_2(\overset{2}{a^+}, \overset{1}{a^-})]\!].$$

We can split off the operators a^- and a^+ in the products $(a^+)^\alpha (a^-)^\beta$ one by one. As a result, each of them will be replaced by the corresponding operator l^+ or l^- acting on f_2, and we obtain

$$\begin{aligned} [\![f_1(\overset{2}{a^+}, \overset{1}{a^-})]\!] [\![f_2(\overset{2}{a^+}, \overset{1}{a^-})]\!] &= \sum_{\alpha,\beta} b_{\alpha,\beta} ((l^+)^\alpha (l^-)^\beta f_2)(\overset{2}{a^+}, \overset{1}{a^-}) \\ &= [f_1(\overset{2}{l^+}, \overset{1}{l^-}) f_2](\overset{2}{a^+}, \overset{1}{a^-}) \end{aligned} \tag{II.1}$$

(we do not discuss the convergence of the series).

Thus, the symbol of the product

$$[\![f_1(\overset{2}{a^+}, \overset{1}{a^-})]\!] [\![f_2(\overset{2}{a^+}, \overset{1}{a^-})]\!]$$

can be calculated according to the following recipe:

Take the left ordered representation operators $\overset{2}{l^+}$ *and* $\overset{1}{l^-}$ *and substitute them into the symbol* f_1. *Then apply the obtained operator to the symbol* f_2.

We see that the introduction of left ordered representation operators in this example enables us to avoid direct calculations with the operators a^+ and a^- and with functions of these operators. Instead, we consider symbols and operators acting on symbols.

In fact, this is the basic idea of the method of ordered representations. Let us now discuss this idea and then proceed to general definitions.

1.2 Ordered Representation and Theorem on Products

Let

$$A = (\overset{1}{A_1}, \dots, \overset{n}{A_n})$$

be a Feynman tuple in some operator algebra $\mathcal{A}$.

We are interested in the problem of reducing an operator $A \in \mathcal{A}$ to a "normal form". What is a normal form? By this we mean a representation

$$B = f(\overset{1}{A_1}, \dots, \overset{n}{A_n}),$$

i.e., an operator is in normal form if it is represented as a function of the Feynman tuple A.

Let us point out that we make no attempts to reduce arbitrary operators to normal form. Our task is far less general. We start from some operators already in normal form, perform some algebraic operations and try to express the result in normal form. Clearly, the main difficulty is to represent, in the form desired, the product of two operators and the superposition $f([\![g(A)]\!])$. Once this is done, we will actually pass from analysis in the algebra $\mathcal{A}$ to analysis in symbol spaces.

Let us now give the precise definition. We assume that a class $\mathcal{F}$ of unary symbols is fixed and $\mathcal{F}_n = \mathcal{F}^{\hat{\otimes} n}$.

Let $A = (\overset{1}{A_1}, \dots, \overset{n}{A_n})$ be a Feynman tuple in an algebra $\mathcal{A}$ and suppose that each A_i is an $\mathcal{F}$-generator in $\mathcal{A}$.

Definition II.1 A Feynman tuple $l = (\overset{1}{l_1}, \dots, \overset{n}{l_n})$ of operators

$$l_j : \mathcal{F}_n \to \mathcal{F}_n, \quad j = 1, \dots, n,$$

is called the *left ordered representation* of the tuple $A = (\overset{1}{A_1}, \dots, \overset{n}{A_n})$ if the following two conditions hold:

i) for any $f \in \mathcal{F}_n$ and $j = 1, \dots, n$ we have

$$A_j[\![f(\overset{1}{A_1}, \dots, \overset{n}{A_n})]\!] = (l_j f)(\overset{1}{A_1}, \dots, \overset{n}{A_n});$$

ii) if a symbol $f \in \mathcal{F}_n$ does not depend on $y_{j+1}, \dots, y_n$, $f = f(y_1, \dots, y_j)$, then

$$l_j f = y_j f \tag{II.2}$$

Condition ii) will be referred to as the *regularity condition*. In view of i), this condition is quite natural; indeed,

$$A_j [\![f(\overset{1}{A_1}, \dots, \overset{j}{A_j})]\!] = \overset{j+1}{A}_j f(\overset{1}{A_1}, \dots, \overset{j}{A_j}) = \overset{j}{A_j} f(\overset{1}{A_1}, \dots, \overset{j}{A_j}),$$

so that the symbol of this product can be chosen equal to $y_j f(y_1, \dots, y_j)$, and condition ii) simply says that the operators of left regular representation are consistent with this natural choice[1].

We intend to establish a theorem generalizing (II.1) to the "abstract" situation of Definition II.1. Since we have no reason to assume that it is possible to insert the operators L_j into arbitrary symbols $f \in \mathcal{F}_n$, we need to be extremely careful in our statement.

Let $\tilde{\mathcal{F}} \subset \mathcal{F}$ be a continuously embedded subalgebra (we do not exclude the case $\tilde{\mathcal{F}} = \mathcal{F}$). Assume that the operators L_j are $\tilde{\mathcal{F}}$-generators in $\mathcal{L}(\mathcal{F}_n)$. Such a subalgebra $\tilde{\mathcal{F}}$ can always be found; in the worst case, it still contains all polynomials. Clearly, $\tilde{\mathcal{F}}_n = \tilde{\mathcal{F}}^{\hat{\otimes} n}$ is a continuously embedded subspace in the symbol space $\mathcal{F}_n$.

Theorem II.1 *Under the above assumptions any product*

$$[\![f(\overset{1}{A_1}, \dots, \overset{n}{A_n})]\!] \, [\![g(\overset{1}{A_1}, \dots, \overset{n}{A_n})]\!],$$

where $f \in \tilde{\mathcal{F}}_n$ and $g \in \mathcal{F}_n$, can be reduced to normal form. Namely,

$$[\![f(\overset{1}{A_1}, \dots, \overset{n}{A_n})]\!] \, [\![g(\overset{1}{A_1}, \dots, \overset{n}{A_n})]\!] = [f(\overset{1}{l_1}, \dots, \overset{n}{l_n}) g] \, (\overset{1}{A_1}, \dots, \overset{n}{A_n}).$$

Proof. In the following we will widely use the more convenient notation $f(A)$ for $f(\overset{1}{A_1}, \dots, \overset{n}{A_n})$; with this notation, the preceding equation reads

$$[\![f(A)]\!] \, [\![g(A)]\!] = [f(l) g](A),$$

or even

$$f(A) \circ g(A) = f(l) \, g(A)$$

(the small circle on the left-hand side serves as a substitute for autonomic brackets, which look somewhat clumsy in combination with the abridged notation).

Condition i) can be rewritten as the commutative diagram

---

[1] It is essential to impose this requirement since the uniqueness of the symbol of a given operator is not assumed anywhere.

$$\begin{array}{ccc} \mathcal{F}_n & \xrightarrow{\mu} & \mathcal{A} \\ l_j \downarrow & & \downarrow L_{A_j} \\ \mathcal{F}_n & \xrightarrow{\mu} & \mathcal{A}, \end{array}$$

where $\mu = \mu_A$ is the linear mapping sending each symbol $f(\overset{1}{x_1}, \dots, \overset{n}{x_n})$ into the corresponding operator $f(\overset{1}{A_1}, \dots, \overset{n}{A_n})$ and L_{A_j} is the operator of left regular representation, $L_{A_j} B = A_j B$ for any $B \in \mathcal{A}$.

The operators l_j and L_{A_j} are $\tilde{\mathcal{F}}$-generators in $\mathcal{L}(\mathcal{F}_n)$ and $\mathcal{L}(\mathcal{A})$, respectively, and the continuous mapping μ intertwines the tuples $l = (\overset{1}{l_1}, \dots, \overset{n}{l_n})$ and $L_A = (\overset{1}{L_{A_1}}, \dots, \overset{n}{L_{A_n}})$, as expressed by the diagram. By Theorem I.4, we have

$$f(L_A)\,\mu = \mu\, f(l)$$

for any symbol $f \in \tilde{\mathcal{F}}_n$. Let us apply both sides of the last equation to an arbitrary symbol $g \in \mathcal{F}_n$. We obtain

$$f(L_A)\,(g(A)) = (f(l)g)\,(A),$$

whence the statement of the theorem follows immediately, since

$$f(L_A)\,(g(A)) = f(A)g(A)$$

(see Subsection 2.3). The proof is complete. □

1.3 Reduction to Normal Form

We now consider the case $\tilde{\mathcal{F}} = \mathcal{F}$. In this case the statement of Theorem II.1 can be given the following interpretation. We introduce a bilinear operation $*$ on $\mathcal{F}_n$ by setting

$$f * g = f(l)\, g.$$

This operation will be referred to as the "twisted product". It makes $\mathcal{F}_n$ into an algebra (possibly nonassociative; the associativity conditions will be discussed later in this chapter). Then Theorem II.1 states that the mapping

$$\mu : (\mathcal{F}_n, *) \to \mathcal{A}$$

(where $\mathcal{A}$ is equipped with the usual operator multiplication) is an algebra homomorphism.

Next, assuming that $\tilde{\mathcal{F}} = \mathcal{F}$, we can give a somewhat more elegant statement of condition ii) in Definition II.1.

Lemma II.1 *If all l_j are $\mathcal{F}$-generators in $\mathcal{F}_n$, then condition* ii) *in Definition* II.1 *is equivalent to the following condition:*

ii′) *for any $g \in \mathcal{F}_n$ one has*

$$g(\overset{1}{l_1}, \ldots, \overset{n}{l_n})\, 1 = g(y_1, \ldots, y_n),$$

where the operator on the left is applied to the symbol identically equal to 1.

Proof. ii′) ⇒ ii). Let $f(y_1, \ldots, y_j) \in \mathcal{F}_n$; set

$$g(y_1, \ldots, y_j) = y_j f(y_1, \ldots, y_j).$$

By condition ii′), we have

$$\begin{aligned} y_j f(y_1, \ldots, y_j) &= \overset{j}{l_j} f(\overset{1}{l_1}, \ldots, \overset{j}{l_j})1 = \overset{j+1}{l_j} f(\overset{1}{l_1}, \ldots, \overset{j}{l_j})\, 1 \\ &= l_j(f(\overset{1}{l_1}, \ldots, \overset{j}{l_j})\, 1) = l_j f(y_1, \ldots, y_j), \end{aligned}$$

the last equality also due to ii′).

ii) ⇒ ii′). It suffices to prove ii′) for decomposable symbols of the form

$$g(y_1, \ldots, y_n) = g_1(y_1) \ldots g_n(y_n).$$

Condition ii) implies that the subspace $\mathcal{F}_j \subset \mathcal{F}_n$ consisting of symbols independent of $y_{j+1}, \ldots, y_n$ is an invariant subspace of the operator l_j, and the restriction $l_j|_{\mathcal{F}_j}$ coincides with multiplication by y_j; that is, there is a commutative diagram

$$\begin{array}{ccc} \mathcal{F}_j & \longrightarrow & \mathcal{F}_n \\ {\scriptstyle y_j}\downarrow & & \downarrow{\scriptstyle l_j} \\ \mathcal{F}_j & \longrightarrow & \mathcal{F}_n, \end{array}$$

where the horizontal arrows are the embeddings. Furthermore, l_j is an $\mathcal{F}$-generator in $\mathcal{F}_n$ by assumption and y_j is an $\mathcal{F}$-generator in $\mathcal{F}_j$ (for any $f \in \mathcal{F}$ the corresponding function of y_j is merely the operator of multiplication by $f(y_j)$). Hence the embedding $\mathcal{F}_j \to \mathcal{F}_n$ intertwines the $\mathcal{F}$-generators y_j and l_j, and by Theorem I.4 we have

$$f(l_j)|_{\mathcal{F}_j} = f(y_j)$$

for any $f \in \mathcal{F}$.

We now apply this identity successively for $j = 1, 2, \ldots, n$, and obtain

$$\begin{aligned} g(\overset{1}{l_1}, \ldots, \overset{n}{l_n})\, 1 &= g_n(l_n) \ldots g_2(l_2)\, g_1(l_1)\, 1 \\ &= g_n(l_n) \ldots g_2(l_2) g_1(y_1) \\ &= g_n(l_n) \ldots g_3(l_3) g_2(y_2) g_1(y_1) \\ &\ldots \\ &= g(y_1, \ldots, y_n). \end{aligned}$$

This equation extends by continuity to the entire symbol space $\mathcal{F}_n$. The lemma is proved. □

We have considered the problem of reduction to normal form for the product $C_1 C_2$ of two operators in normal form. Let us now consider the general case. Let $i_1, \ldots, i_s$ and $j_1, \ldots, j_s$ be two sequences of indices such that $j_l \neq j_r$ whenever $[A_{j_l}, A_{j_r}] \neq 0$. Let $\varphi(y_1, \ldots, y_s)$ be a given symbol. We consider the operator

$$C = \varphi(\overset{j_1}{A_{i_1}}, \ldots, \overset{j_s}{A_{i_s}})$$

and try to reduce it to normal form.

Theorem II.2 *Suppose that the left regular representation* $l = (\overset{1}{l_1}, \ldots, \overset{n}{l_n})$ *of the Feynman tuple* $A = (\overset{1}{A_1}, \ldots, \overset{n}{A_n})$ *exists and that each* l_j *is an* $\tilde{\mathcal{F}}$*-generator in* $\mathcal{F}_n$. *Assume that* $\varphi(y_1, \ldots, y_s) \in \tilde{\mathcal{F}}_s$. *Then the operator* C *can be reduced to a normal form. Namely,*

$$\varphi(\overset{j_1}{A_{i_1}}, \ldots, \overset{j_s}{A_{i_s}}) = \left\{ \varphi(\overset{j_1}{l}_{i_1}, \ldots, \overset{j_s}{l}_{i_s})\, 1 \right\} (A).$$

Proof. By analogy with the proof of Theorem II.1, we see that the mapping

$$\begin{aligned} \mu_A : \quad & \mathcal{F}_n \to \mathcal{A}, \\ & f \mapsto f(\overset{1}{A_1}, \ldots, \overset{n}{A_n}) \end{aligned}$$

intertwines the Feynman tuples $(\overset{j_1}{L}_{A_{i_1}}, \ldots, \overset{j_s}{L}_{A_{i_s}})$ and $(\overset{j_1}{l}_{i_1}, \ldots, \overset{j_s}{l}_{i_s})$. Therefore, we obtain

$$\begin{aligned} \varphi(\overset{j_1}{A_{i_1}}, \ldots, \overset{j_s}{A_{i_s}}) &= \varphi(\overset{j_1}{L}_{A_{i_1}}, \ldots, \overset{j_s}{L}_{A_{i_s}})\, (1) \\ &= \varphi(\overset{j_1}{L}_{A_{i_1}}, \ldots, \overset{j_s}{L}_{A_{i_s}})\, \mu_A(1) \\ &= \mu_A \varphi(\overset{j_1}{l}_{i_1}, \ldots, \overset{j_s}{l}_{i_s})\, (1) = f(\overset{1}{A_1}, \ldots, \overset{n}{A_n}), \end{aligned}$$

where

$$f(y_1, \dots, y_n) = \varphi(\overset{j_1}{l}_{i_1}, \dots, \overset{j_s}{l}_{i_s})(1).$$

The theorem is proved. □

Remark II.1 We cannot guarantee that the symbol $f(y_1, \dots, y_n)$ lies in $\tilde{\mathcal{F}}_n$ since there is no reason for $\tilde{\mathcal{F}}_n$ to be invariant under functions of L. However, the paradox is that we can still compute products of the form $f(A)\, g(A)$ with such a symbol f; we need only bear in mind that $f(A)$ can be represented as $\varphi(\overset{j_1}{A}_{i_1}, \dots, \overset{j_s}{A}_{i_s})$, and so

$$f(A) \circ g(A) = \left\{ \varphi(\overset{j_1}{l}_{i_1}, \dots, \overset{j_s}{l}_{i_s})\, g \right\} (A)$$

Let us now consider how to perform reduction to normal form for composite functions. Let $f(z)$ be a function of a single variable z, let $g \in \mathcal{F}_n$ be an n-ary symbol, and suppose that the tuple $A = (\overset{1}{A}_1, \dots, \overset{n}{A}_n)$ of $\mathcal{F}$-generators has a left ordered representation $l = (\overset{1}{l}_1, \dots, \overset{n}{l}_n)$ consisting of $\mathcal{F}$-generators.

Theorem II.3 *Let $f \in \tilde{\mathcal{F}}$, where $\tilde{\mathcal{F}}$ is some symbol class, and suppose that the operators $g(A)$ and $g(l)$ are $\tilde{\mathcal{F}}$-generators in $\mathcal{A}$ and $\mathcal{L}(\mathcal{F}_n)$, respectively. Then*

$$f([\![g(A)]\!]) = (f([\![g(l)]\!])\, 1)\,(A).$$

In other words, $f([\![g(A)]\!])$ can be reduced to normal form, $f([\![g(A)]\!]) = \varphi(A)$, where the symbol $\varphi(y) \in \mathcal{F}_n$ is given by the action of the operator $f([\![g(l)]\!])$ on the constant function $1 \in \mathcal{F}_n$.

Proof. This is quite standard stuff. Indeed, the mapping μ_A clearly intertwines $g(l)$ and $g(L_A)$, that is, the diagram

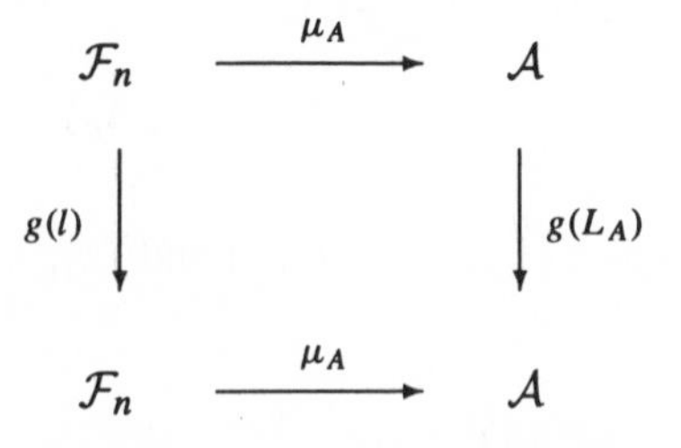

$$\begin{array}{ccc} \mathcal{F}_n & \xrightarrow{\mu_A} & \mathcal{A} \\ \downarrow{\scriptstyle g(l)} & & \downarrow{\scriptstyle g(L_A)} \\ \mathcal{F}_n & \xrightarrow{\mu_A} & \mathcal{A} \end{array}$$

commutes. Next, $g(l)$ and $g(L_A)$ are $\tilde{\mathcal{F}}$-generators in $\mathcal{L}(\mathcal{F}_n)$ and $\mathcal{L}(\mathcal{A})$, respectively, so that we have, by Theorem I.4,

$$\mu_A\, f([\![g(l)]\!]) = f([\![g(L_A)]\!])\, \mu_A,$$

and consequently,

$$\begin{aligned} f([\![g(A)]\!]) &= L_{f([\![g(A)]\!])}(1) = f(L_{g(A)})\,(1) = f(L_{g(A)})\mu_A(1) \\ &= f([\![g(L_A)]\!])\mu_A(1) = \mu_A f([\![g(l)]\!])\,(1), \end{aligned}$$

whence the assertion of the theorem follows immediately.

The proof is complete. □

We have defined the *left* ordered representation of a Feynman tuple

$$(\overset{1}{A_1}, \dots, \overset{n}{A_n}).$$

For symmetry reasons, it is clear that one can define the right ordered representation in a similar way.

Definition II.2 A Feynman tuple $r = (\overset{-1}{r}_1, \dots, \overset{-n}{r}_n)$ of operators

$$r_j : \mathcal{F}_n \to \mathcal{F}_n, \quad j = 1, \dots, n$$

is called the *right ordered representation* of the tuple $A = (\overset{1}{A_1}, \dots, \overset{n}{A_n})$ if the following two conditions hold:

i) for any $f \in \mathcal{F}_n$ and $j = 1, \dots, n$ we have

$$[\![f(\overset{1}{A_1}, \dots, \overset{n}{A_n})]\!]A_j = (r_j f)\,(\overset{1}{A_1}, \dots, \overset{n}{A_n});$$

ii) if a symbol $f = f(y_j, \dots, y_n) \in \mathcal{F}$ does not depend on $y_1, \dots, y_{j-1}$, then

$$r_j f = y_j f.$$

Remark II.2 Note that the Feynman indices of the tuple r have the minus sign. This is because the natural ordering of these operators is the opposite of that of $A_1, \dots, A_n$.

Let us state, without proof (which follows the lines of that for the left ordered representation) the main properties of the right ordered representation.

Theorem II.4 *Let the operators $r_1, \dots, r_n$ of the right ordered representation be $\tilde{\mathcal{F}}$-generators for some continuously embedded subalgebra $\tilde{\mathcal{F}} \subset \mathcal{F}$. Then*

(i) *for any $f \in \mathcal{F}_n$ and $g \in \tilde{\mathcal{F}}_n$ we have*

$$[\![f(\overset{1}{A_1}, \dots, \overset{n}{A_n})]\!]\,[\![g(\overset{1}{A_1}, \dots, \overset{n}{A_n})]\!] = [g(\overset{-1}{r_1}, \dots, \overset{-n}{r_n})f]\,(\overset{1}{A_1}, \dots, \overset{n}{A_n})$$

or, in the shorthand notation,

$$f(A)\,g(A) = [g(r)\,f]\,(A);$$

(ii) *for any* $f \in \tilde{\mathcal{F}}_s$ *and any sequences of indices* $i_1, \dots, i_s$ *and* $j_1, \dots, j_s$ *such that* $j_k \neq j_l$ *whenever* $[A_{i_k}, A_{i_l}] \neq 0$ *we have*

$$f(\overset{j_1}{A_{i_1}}, \dots, \overset{j_s}{A_{i_s}}) = [f(\overset{-j_1}{r_{i_1}}, \dots, \overset{-j_s}{r_{i_s}})\, 1]\,(A);$$

(iii) *If* $\tilde{\mathcal{F}} = \mathcal{F}$, *then condition* ii) *of Definition* II.2 *is equivalent to the condition*

$$ii') \quad f(r)\, 1 = f(y) \quad \text{for any} \quad f \in \mathcal{F}_n;$$

(iv) *let* $r_1, \dots, r_n$ *be* $\mathcal{F}$*-generators in* $\mathcal{L}(\mathcal{F}_n)$, *and suppose that* $g \in \mathcal{F}_n$ *and* $f \in \tilde{\mathcal{F}}$, *where* $\tilde{\mathcal{F}}$ *is some symbol class, and moreover,* $g(L_A)$ *and* $g(r)$ *are* $\tilde{\mathcal{F}}$*-generators in* $\mathcal{A}$ *and* $\mathcal{F}_n$, *respectively. Then*

$$f([\![g(A)]\!]) = (f([\![g(r)]\!])\, 1)\,(A).$$

We remark that the existence of the right ordered representation is equivalent to the existence of the left ordered representation. More precisely, the following statement is valid.

Theorem II.5 *Suppose that the tuple* $A = (\overset{1}{A_1}, \dots, \overset{n}{A_n})$ *has a left ordered representation* $l = (\overset{1}{l_1}, \dots, \overset{n}{l_n})$ *such that all the operators* l_j *are* $\mathcal{F}$*-generators. Then there exists a right regular representation* $r = (\overset{-1}{r_1}, \dots, \overset{-n}{r_n})$ *of this tuple. It is given by*

$$(r_j f)\,(y) = f(\overset{1}{l_1}, \dots, \overset{n}{l_n})\,(y_j).$$

Proof. The proof goes by straightforward verification. Indeed we have

$$(r_j f)(A) = (f(l) y_j)(A) = f(A)\, A_j$$

(the last equality is due to Theorem II.1), and we see that condition i) of Definition II.2 is satisfied. Next, if f does not depend on $y_1, \dots, y_{j-1}$ then we have

$$\begin{aligned} r_j f &= f(\overset{j}{l_j}, \dots, \overset{n}{l_n})(y_j) = f(\overset{j}{l_j}, \dots, \overset{n}{l_n})\,(l_j 1) \\ &= (y_j f)(\overset{j}{l_j}, \dots, \overset{n}{l_n})(1) = y_j f(y_j, \dots, y_n), \end{aligned}$$

so that condition ii) of Definition II.2 is also satisfied. □

2 Some Examples

Note that a given Feynman tuple of operators does not necessarily have a left ordered representation. Really, why should it exist at all? We have seen that the principal usefulness of the ordered representation is to move the operators in products (or even more complicated aggregates such as composite functions) to the places rigidly prescribed by the Feynman indices. All in all, such an operation requires (perhaps implicit) permutations of the operators involved, and so we may suggest that the existence of a sufficiently rich set of commutation relations is a necessary condition for the existence of the left ordered representation. This guess is supported by the observation that in the example considered in Section 1 (creation-annihilation operators) it was the commutation relation $[a^+, a^-] = I$ that enabled us to construct the left ordered representation.

Here we consider some extremely simple, but at the same time useful examples of calculating the ordered representation operators. In the next section we distinguish some classes of Feynman tuples (more precisely, of commutation relations) and present general methods for computing the regular representation.

2.1 Functions of the Operators x and $-ih\partial/\partial x$

In Section 1 we have mentioned that differential, pseudodifferential, and Fourier integral operators can be considered as functions of the position and momenta operators $x_1, \dots, x_n$ and $-i\partial/\partial x_1, \dots, -i\partial/\partial x_n$. Since the main problems of the theory of differential equations involve computation of products of such operators, it would be useful to calculate the ordered representation for the tuples consisting of these operators. In order to simplify the notation, we make all the calculations for $n = 1$. Since the operators x_j and $-i\partial/\partial x_j$ commute with x_k and $-i\partial/\partial x_k$ whenever $j \neq k$, the result for $n > 1$ can easily be obtained from the result for $n = 1$ by attaching the indices.

First of all, consider the Feynman tuple

$$\left(\overset{2}{x}, -i\overset{1}{\frac{\partial}{\partial x}}\right).$$

Clearly, we have the commutation relation

$$\left[-i\frac{\partial}{\partial x}, x\right] = -i$$

(we drop the identity operator I in our notation); this relation differs from that for the creation-annihilation operators a^+ and a^- only by the factor i. Hence it is not surprising that the computations and the result are quite similar to what we have for a^+ and a^-. Thus, we will just formulate the result, leaving the computation to the reader. Namely, the left ordered representation of the tuple $\left(\overset{2}{x}, \overset{1}{-i\partial/\partial x}\right)$ has the form

$$l_x = q, \quad l_{-i\partial/\partial x} = p - i\frac{\partial}{\partial q}.$$

A similar computation yields an expression for the right ordered representation operators:

$$r_x = q - \frac{\partial}{\partial p}, \quad r_{-i\partial/\partial x} = p.$$

One can also consider the reverse-ordered tuple

$$\left(\overset{1}{x}, -i\overset{2}{\frac{\partial}{\partial x}}\right).$$

For this tuple the ordered representation operators have the form

$$\begin{aligned} l_x &= q + i\frac{\partial}{\partial p}, \quad l_{-i\partial/\partial x} = p; \\ r_x &= q, \quad r_{-i\partial/\partial x} = p + i\frac{\partial}{\partial q}. \end{aligned}$$

Let $P\left(\overset{2}{x}, \overset{1}{-i\partial/\partial x}\right)$ and $Q\left(\overset{2}{x}, \overset{1}{-i\partial/\partial x}\right)$ be pseudodifferential operators. By Theorem II.6, the product $[\![P\left(\overset{2}{x}, \overset{1}{-i\partial/\partial x}\right)]\!][\![Q\left(\overset{2}{x}, \overset{1}{-i\partial/\partial x}\right)]\!]$ is again a pseudodifferential operator,

$$[\![P\left(\overset{2}{x}, -i\overset{1}{\frac{\partial}{\partial x}}\right)]\!][\![Q\left(\overset{2}{x}, -i\overset{1}{\frac{\partial}{\partial x}}\right)]\!] = H\left(\overset{2}{x}, -i\overset{1}{\frac{\partial}{\partial x}}\right),$$

where

$$H(p,q) = P\left(\overset{2}{q}, \overset{1}{\overline{p - i\frac{\partial}{\partial q}}}\right) Q(q,p).$$

The last formula immediately implies the classical formula [109], [84] for the symbol of the product of pseudodifferential operators. In order to obtain this formula, we expand the operator $P\left(\overset{2}{q}, \overset{1}{\overline{p - i\partial/\partial q}}\right)$ into a Taylor series in powers of $-i\partial/\partial q$. This poses no difficulties since p and $-i\partial/\partial q$ commute, and we obtain

$$H(p,q) \cong \sum_{k=0}^{\infty} \frac{(-i)^k}{k!} \frac{\partial^k P}{\partial p^k}(q,p) \frac{\partial^k Q}{\partial q^k}(q,p).$$

Of course, this expansion needs justification by estimating the decay as $|p| \to \infty$ of the remainder terms, but we omit these routine computations.

2.2 Perturbed Heisenberg Relations

The operators $A_0 = -i\partial/\partial x$ and $B_0 = x$ satisfy the commutation relation

$$B_0 A_0 - A_0 B_0 = i.$$

Consider the "perturbed" relation

$$BA - \alpha AB = i,$$

where α is a constant, $\alpha = 1 + \varepsilon$, $\varepsilon \to 0$ valid for some "perturbed" operators A, B; we do not need to use below the concrete form of these operators. Let us compute the left ordered representation for the tuple $(\overset{2}{A}, \overset{1}{B})$. We have

$$\overset{3}{A}[\![f(\overset{2}{A}, \overset{1}{B})]\!] = \overset{3}{A} f(\overset{2}{A}, \overset{1}{B}) = \overset{2}{A} f(\overset{2}{A}, \overset{1}{B}),$$

so that $l_A = x$ (here x and y are the arguments of f, $f = f(x, y)$). Furthermore, by analogy with Subsection 2.1 we obtain

$$\begin{aligned} B[\![f(\overset{2}{A}, \overset{1}{B})]\!] &= \overset{3}{B} f(\overset{2}{A}, \overset{1}{B}) \\ &= \overset{3}{B} f(\alpha \overset{4}{A}, \overset{1}{B}) + \overset{3}{B}(\overset{2}{A} - \alpha \overset{4}{A}) \frac{\delta f}{\delta x}(\overset{2}{A}, \alpha \overset{4}{A}, \overset{1}{B}). \end{aligned}$$

Moving the indices apart and introducing the autonomous brackets, we obtain

$$B[\![f(\overset{2}{A}, \overset{1}{B})]\!] = \overset{1}{B} f(\alpha \overset{2}{A}, \overset{1}{B}) + [\![\overset{3}{B}(\overset{2}{A} - \alpha \overset{4}{A})]\!] \frac{\delta f}{\delta x}(\overset{2}{A}, \alpha \overset{4}{A}, \overset{1}{B}).$$

But

$$\overset{3}{B}(\overset{2}{A} - \alpha \overset{4}{A}) = BA - \alpha AB = i,$$

so that we obtain

$$B[\![f(\overset{2}{A}, \overset{1}{B})]\!] = \overset{1}{B} f(\alpha \overset{2}{A}, \overset{1}{B}) + i \frac{\delta f}{\delta x}(\overset{2}{A}, \alpha \overset{2}{A}, \overset{1}{B}).$$

In contrast to the preceding example, the arguments of the difference derivative do not coincide, so that it does not reduce to the usual derivative.

Let I_α denote the dilatation operator

$$I_\alpha \varphi(x, y) = \varphi(\alpha x, y).$$

We obtain the left ordered representation operators in the form

$$l_A = x, \quad l_B = y I_\alpha + \frac{i}{(1-\alpha)x}(1 - I_\alpha).$$

Let us demonstrate how this ordered representation can be applied. Assume that we wish to invert an operator of the form $f(\overset{2}{A}, \overset{1}{B})$, i.e., to solve the equation

$$[\![f(\overset{2}{A}, \overset{1}{B})]\!]\,[\![\chi(\overset{2}{A}, \overset{1}{B})]\!] = 1,$$

where $\chi(x, y)$ is an unknown symbol. By Theorem II.1, we obtain the following equation for the symbol χ:

$$f\left(\overset{2}{x}, \overset{1}{[\![}yI_\alpha + \frac{i}{(1-\alpha)x}(1 - I_\alpha)]\!]\right) \chi(x, y) = 1.$$

If $f(x, y)$ is a polynomial with respect to y, then this is a difference equation and can be solved explicitly; it may seem more convenient to reduce this equation to a difference equation on a uniform grid, which can be accomplished by the change of variables $x = e^\theta$. Under this change of variables, I_α is carried into the operator $e^{a\partial/\partial\theta}$ of translation by $a = \ln\alpha$, and the equation takes the form

$$f\left(\overset{2}{e^\theta}, \overset{1}{[\![}ye^{a\partial/\partial\theta} + \frac{ie^{-\theta}}{1-\alpha}(1 - e^{a\partial/\partial\theta})]\!]\right) \chi(e^\theta, y) = 1.$$

This is a difference equation on a uniform grid with mesh width a.

2.3 Examples of Nonlinear Commutation Relations

Let A, B, and C be operators satisfying the relations

$$[A, B] = 1, \quad [A, C] = \alpha(C)A + \beta(C)B + \gamma(C),$$
$$[B, C] = \varepsilon(C)A + \delta(C)B + \sigma(C),$$

where $\alpha(x)$, $\beta(x)$, $\gamma(x)$, $\delta(x)$, $\varepsilon(x)$, and $\sigma(x)$ are some symbols. Let us find the left ordered representation for the Feynman tuple $(\overset{1}{A}, \overset{2}{B}, \overset{3}{C})$. The arguments of symbols will be denoted by (x, y, z), $x \mapsto \overset{1}{A}$, $y \mapsto \overset{2}{B}$, and $z \mapsto \overset{3}{C}$.

The last two relations satisfied by A, B, and C can be written in the matrix form

$$\begin{pmatrix} A \\ B \end{pmatrix} C = S(C) \begin{pmatrix} A \\ B \end{pmatrix} + \begin{pmatrix} \gamma(C) \\ \sigma(C) \end{pmatrix},$$

where $S(z)$ is the matrix symbol

$$S(z) = \begin{pmatrix} z + \alpha(z) & \beta(z) \\ \varepsilon(z) & z + \delta(z) \end{pmatrix}.$$

Let $F(x, y, z)$ be an arbitrary symbol. By Theorem I.6,

$$\begin{pmatrix} A \\ B \end{pmatrix} [\![F(\overset{1}{A}, \overset{2}{B}, \overset{3}{C})]\!] = \overline{\begin{pmatrix} A \\ B \end{pmatrix}}^{\,4} F(\overset{1}{A}, \overset{2}{B}, \overset{3}{C})$$

$$= F(\overset{1}{A}, \overset{2}{B}, S(\overset{4}{C})) \overline{\begin{pmatrix} A \\ B \end{pmatrix}}^{\,3} + \frac{\delta F}{\delta z}(\overset{1}{A}, \overset{2}{B}, \overset{3}{C}, S(\overset{3}{C})) \overline{\begin{pmatrix} \gamma(C) \\ \sigma(C) \end{pmatrix}}^{\,3}.$$

In the second term of the last expression the ordering of operators is just the required one, whereas in the first term $\overset{3}{A}$ still needs to be moved to the first position. This, by analogy with the previous example, can be performed by applying the operator $x + \partial/\partial y$ to F. Finally, we obtain

$$\begin{pmatrix} l_A \\ l_B \end{pmatrix} = I_{S(z)} \begin{pmatrix} x + \frac{\partial}{\partial y} \\ y \end{pmatrix} + \frac{1 - \overset{1}{I}_{S(z)}}{\overset{2}{z} - S(\overset{2}{z})} \begin{pmatrix} \gamma(\overset{2}{z}) \\ \sigma(\overset{2}{z}) \end{pmatrix},$$

where $I_{S(z)}$ is the operator of substituting the matrix $S(z)$ instead of z into a function,

$$I_{S(z)} F(z) = \exp\left\{ (S(\overset{2}{z}) - \overset{2}{z}) \frac{\overset{1}{\partial}}{\partial z} \right\} F(z) \stackrel{\text{def}}{=} F(S(z)).$$

since the operator C acts last, it is clear that $l_C = z$.

Our next example is concerned with the phenomenon that the ordered representation may or may not exist depending on the chosen ordering in a Feynman tuple.

Let A and B be operators satisfying the permutation relation

$$AB = B^n A,$$

where $n > 1$ is an integer. Let us first construct the left ordered representation for the tuple $(\overset{1}{A}, \overset{2}{B})$. We have

$$B [\![f(\overset{1}{A}, \overset{2}{B})]\!] = \overset{2}{B} f(\overset{1}{A}, \overset{2}{B})$$

and

$$\begin{aligned} A [\![f(\overset{1}{A}, \overset{2}{B})]\!] &= \overset{3}{A} f(\overset{1}{A}, \overset{2}{B}) = \overset{3}{A} f(\overset{1}{A}, \overset{4}{B^n}) + [\![\overset{3}{A} (\overset{3}{B} - \overset{2}{B^n})]\!] \frac{\delta f}{\delta y}(\overset{1}{A}, \overset{2}{B}, \overset{4}{B^n}) \\ &= \overset{1}{A} f(\overset{1}{A}, \overset{2}{B^n}), \end{aligned}$$

and so

$$l_A = x e^{\overline{(y^n - y)}^{\,2} \overset{1}{\frac{\partial}{\partial y}}}, \quad l_B = y,$$

if we choose the ordering $(\overset{1}{A}, \overset{2}{B})$. On the other hand, one can easily check that the ordered representation does not exist at all for the ordering $(\overset{2}{A}, \overset{1}{B})$.

Indeed, consider the quotient $\mathcal{A}$ of the free algebra with generators A and B modulo the ideal generated by $AB - B^n A$. Each element of $\mathcal{A}$ can be uniquely represented in the form $\sum a_{kl} B^k A^l$, where a_{kl} are complex numbers and the sum is finite. The existence of the representation follows from the existence of the left ordered representation operators l_A and l_B (see above). The uniqueness is also clear: since

$$l_A l_B = l_B^n l_A,$$

it follows that the algebra generated by l_A and l_B is a homomorphic (in fact, isomorphic) image of $\mathcal{A}$; now the coefficients a_{kl} can be determined by applying the operator $\sum a_{kl} l_B^k l_A^l$ to the function 1:

$$\left[\sum a_{kl} l_B^k l_A^l\right](1) = \sum a_{kl} y^k x^l.$$

Now consider the element $BA \in \mathcal{A}$. It has no representation of the form

$$\sum c_{kl} A^k B^l,$$

for if such a representation could be found, we would have

$$BA = \sum c_{kl} A^k B^l = \sum c_{kl} B^{n^k l} A^k,$$

which is impossible since the system $n^k l = k$ has no integer solutions.

This implies that the left ordered representation simply does not exist for the ordering $(\overset{2}{A}, \overset{1}{B})$.

Let us consider two more general examples concerned with Lie algebras.

2.4 Lie Commutation Relations

Generators of a Lie algebra, special case. Let $A = (\overset{1}{A_1}, \dots, \overset{m}{A_m})$ be a Feynman tuple satisfying the following conditions:

i) for any $j, k \in \{1, \dots, m\}$ there exists an index $s = s(j, k) \in \{1, \dots, m\}$ such that

$$[A_j, A_k] = -ic_s A_s,$$

where c_s is some constant;

ii) there exists a positive integer N such that any commutator of length $\geq N$ composed of $A_1, \dots, A_n$ is equal to zero,

$$[A_{j_1}, [A_{j_2}, [\dots [A_{j_{N-1}}, A_{j_N}] \dots]]] = 0$$

for any $j_1, j_2, \dots, j_N \in \{1, \dots, N\}$.

Condition i) means that the operators $A_1, \dots, A_m$ define a representation of a Lie algebra of a rather special form (with a general Lie algebra, the right-hand sides would contain linear combinations of all A_l, $l \in \{1, \dots, n\}$; see the next example). The factor $-i$ is introduced so as to ensure that c_s are real in case $A_1, \dots, A_s$ are self-adjoint operators.

Condition ii) is simply the requirement that this Lie algebra is nilpotent; N is referred to as the *nilpotency rank*.

This class of commutation relations was considered in [129]. Note that *any* nilpotent Lie algebra can be described as a homomorphic image of a nilpotent algebra of the type described above.

Let us describe the construction of the left ordered representation for this case. The main (and the only) idea is the same as in Subsection 2.1. Assume that we need to compute the operator l_{A_j}, i.e., in fact, the product

$$A_j f(A) = \overset{m+1}{A_j} f(\overset{1}{A_1}, \dots, \overset{m}{A_m})$$

for an arbitrary symbol $f = f(x_1, \dots, x_n)$.

To this end, the product should be transformed in such a way that the Feynman indices of $A_{j+1}, \dots, A_m$ become greater than $m + 1$. Using the permutation formula (see Corollary I.1)

$$\overset{a}{A}\varphi(\overset{b}{B}) = \overset{b}{A}\varphi(\overset{a}{B}) + [\overset{a}{A}, B]\frac{\delta\varphi}{\delta x}(\overset{b}{B}, \overset{c}{B}),$$

where $c > a > b$ and φ may have additional operator arguments, we obtain, by successive commutation,

$$A_j f(A) = (x_j f)(A) + \sum_{l=j+1}^{m} \overset{l+1}{[A_j, A_l]} \frac{\delta f}{\delta x_l}(\overset{1}{A_1}, \dots, \overset{l}{A_l}, \overset{l+2}{A_l}, \dots, \overset{m}{A_m}).$$

We now use the expression for the commutator and obtain

$$A_j f(A) = (x_j f)(A) - i \sum_{l=j+1}^{m} c_{s(j,l)} \overset{l+1}{A}_{s(j,l)} \frac{\delta f}{\delta x_l}(\overset{1}{A_1}, \dots, \overset{l}{A_l}, \overset{l+2}{A_l}, \dots, \overset{m}{A_m}).$$

The first summand already has the desired form; as for the other summands, let us move the operator $A_{s(j,l)}$ in each of them to its place, thus obtaining remainder terms with second-order difference derivatives and second-order commutators, then we proceed to the remainders, etc. By virtue of nilpotency, the process will terminate after the Nth step. Clearly, at this stage all difference derivatives disappear and transform into the usual ones. Thus we have proved the following theorem.

Theorem II.6 ([129]) *Under conditions* i) *and* ii) *the Feynman tuple*

$$(\overset{1}{A_1}, \dots, \overset{n}{A_n})$$

possesses the left ordered representation $l_1, \dots, l_n$, *and each of the operators* l_j *is a differential operator of order* $\leq N - 1$ *with linear homogeneous coefficients.*

In fact, it still remains to prove that

$$f(\overset{1}{l_1}, \dots, \overset{m}{l_m})\, 1 = f(x_1, \dots, x_m)$$

for any symbol $f(x)$. By Lemma II.1, it suffices to prove that

$$l_{A_j} g(x) = x_j g(x)$$

whenever $g(x)$ is independent of $x_{j+1}, \dots, x_m$. However, this follows by construction.

Generators of a Lie algebra, general case. Let us now consider the case in which the operators $A_1, \dots, A_n$ are generators of a representation of an arbitrary Lie algebra. This means that

$$[A_i, A_j] = \sum_{k=1}^{n} c_{ij}^k A_k, \quad i, j = 1, \dots, n,$$

where c_{ij}^k are some (complex) numbers called the *structural constants* of the Lie algebra.

Let us calculate the left ordered representation $(l_1, \dots, l_n)$ for the tuple of operators $(\overset{1}{A_1}, \dots, \overset{n}{A_n})$. In order to avoid nonalgebraic difficulties, we consider only polynomial symbols $f(x)$, $x = (x_1, \dots, x_n)$.

Remark II.3 As shown below, for polynomial symbols we can accomplish the construction by purely algebraic means. For more general symbols this is not the case; as is shown in the Appendix, in the latter situation it is more convenient to pass to Fourier transforms, which takes the left ordered representation operators into right-invariant vector fields on the corresponding Lie group.

Remark II.4 The constants c_{ij}^k are structural constants of a Lie algebra, and therefore they must satisfy the antisymmetry condition and the Jacobi identity. However, our construction does not use these properties. These conditions are discussed in detail in Section 4 of this chapter.

Denote by $\mathcal{A}$ the associative subalgebra generated by $A_1, \dots, A_n$ and by $\mathcal{L} \subset \mathcal{A}$ the linear span of $A_1, \dots, A_n$. In the algebra $\mathcal{A}[[t_1, \dots, t_n]]$ of formal power series in $t = (t_1, \dots, t_n)$ with coefficients in $\mathcal{A}$ consider the element

$$U(t) = e^{t_n A_n} \dots e^{t_1 A_1}.$$

Clearly, for any polynomial $p(x)$, $x = (x_1, \dots, x_n)$, we have[1]

$$P(\overset{1}{A}_1, \dots, \overset{n}{A}_n) = \left\{ p\left(\frac{\partial}{\partial t}\right) U(t) \right\}\Big|_{t=0}$$

(it suffices to check this identity for monomials, which is trivial).

Let us find operators $\hat{L}_j$ such that

$$\hat{L}_j U(t) = A_j U(t), \quad j = 1, \dots, n.$$

To this end, we compute the derivative $(\partial/\partial t_q)U(t)$ using the relationship

$$e^{Bt} D = e^{\operatorname{ad}_B t}(D)\, e^{Bt}.$$

Differentiating $U(t)$ and applying the last equation successively, we obtain

$$\begin{aligned} \frac{\partial}{\partial t_q} U(t) &= e^{t_n A_n} \dots e^{t_{q+1} A_{q+1}} A_q e^{t_q A_q} \dots e^{t_1 A_1} \\ &= [e^{t_n ad_{A_n}} \dots e^{t_{q+1} \operatorname{ad}_{A_{q+1}}}(A_q)] U(t). \end{aligned} \qquad \text{(II.3)}$$

For each $j \in \{1, \dots, n\}$ denote by c_j the matrix with the elements $(c_j)_{kl} = c^k_{jl}$. Evidently, the operator ad_{A_j} acts from $\mathcal{L}$ into $\mathcal{L}$, and if $A = \sum \lambda_s A_s$, then $\operatorname{ad}_{A_j}(A) = \sum \mu_s A_s$, where $\mu_s = c_j \lambda_s$.

Note that the last assertion is valid regardless of whether or not the operators $(\overset{1}{A}_1, \dots, \overset{n}{A}_n)$ are linearly independent. Therefore, the element in square brackets on the right-hand side of (II.3) belongs to $\mathcal{L}$, and moreover, we have

$$\frac{\partial}{\partial t_q} U(t) = \sum_j A_{qj}(t) A_j U(t), \qquad \text{(II.4)}$$

where[2]

$$A_{qj}(t) = \begin{cases} [e^{t_n c_n} \dots e^{t_{q+1} c_{q+1}}]_{jq}, & \text{if } \ q < n, \\ \delta_{kj} & \text{if } \ q = n. \end{cases}$$

In particular, $A_{qj}(0) = E$, and we see that the matrix $A(t) = (A_{qj}(t))$ is invertible in the algebra of formal power series. Multiplying (II.4) by $A^{-1}(t)$ on the left, we obtain

$$A_j \circ U(t) = \sum_q A^{-1}_{jq}(t) \frac{\partial}{\partial t_q} U(t),$$

[1]Recall that the substitution $t = 0$ is a well-defined operation in the algebra of formal power series; this operation assigns to each formal series its constant term.

[2]Here $[\ ]_{jq}$ stands for the (j, q)th entry of the matrix inside the square brackets.

i.e., we can set

$$\hat{L}_j = L_j \left(\overset{2}{t}, \overset{1}{\frac{\partial}{\partial t}} \right),$$

where

$$L_j(t, x) = \sum_q A^{-1}_{jq}(t)\, x_q.$$

Now, for any polynomial $p(x)$ we have

$$\begin{aligned} A_j [\![p(\overset{1}{A_1}, \ldots, \overset{n}{A_n})]\!] &= A_j \left\{ p\left(\frac{\partial}{\partial t}\right) U(t) \right\}\bigg|_{t=0} \\ &= \left\{ p\left(\frac{\partial}{\partial t}\right) A_j U(t) \right\}\bigg|_{t=0} \\ &= \left\{ [\![p\left(\frac{\partial}{\partial t}\right)]\!] [\![L_j \left(\overset{2}{t}, \overset{1}{\frac{\partial}{\partial t}} \right)]\!] U(t) \right\}\bigg|_{t=0}. \end{aligned} \tag{II.5}$$

To compute the composition of the two autonomic brackets in the latter expression we use the right ordered representation

$$r_t = t + \frac{\partial}{\partial x}, \quad r_{\frac{\partial}{\partial t}} = x$$

for the tuple $(\overset{2}{t}, \overset{1}{\partial/\partial t})$ computed above in Subsection 2.1. With the help of this representation the above mentioned composition can be written in the form

$$[\![p\left(\frac{\partial}{\partial t}\right)]\!] [\![L_j(\overset{2}{t}, \overset{1}{\frac{\partial}{\partial t}})]\!] = q \left(\overset{2}{t}, \overset{1}{\frac{\partial}{\partial t}} \right)$$

where the function $q(t, x)$ equals

$$q(t, x) = L_j \left(t + \overset{1}{\frac{\partial}{\partial x}}, \overset{2}{x} \right) p(x).$$

Substituting this equality in the right-hand side of formula (II.5) we represent the left-hand side of this formula in the form $h(\overset{1}{A_1}, \ldots, \overset{n}{A_n})$, where the function $h(x)$ is given by

$$h(x) = q(0, x) = \sum_q \overset{2}{x_q} A^{-1}_{jq} \left(\overset{1}{\frac{\partial}{\partial x}} \right) p(x)$$

(we used here the above expression of the functions $L_j(t, x)$ via the matrix A^{-1}).

Thus, we have obtained the left ordered representation operators in the form

$$l_j = \sum_q \overset{2}{x}_q A^{-1}_{jq} \begin{pmatrix} \overset{1}{\partial} \\ \overline{\partial x} \end{pmatrix}, \quad j = 1, \ldots, n.$$

It is easy to check that these operators are well-defined on polynomials and satisfy condition ii) of Definition II.1.

2.5 Graded Lie Algebras

We again consider the creation-annihilation operators, as in Subsection 1.1.

This time we assume that there are different types of particles in the system, namely, bosons and fermions. Let $b_j^{\pm}$ be the creation-annihilation operators for bosons, and $f_j^{\pm}$ be the corresponding operators for fermions. Then the following relations must be valid:

$$[\overset{+}{b}_j, \overset{+}{b}_k] = [\overset{-}{b}_j, \overset{-}{b}_k] = [\overset{\pm}{b}_j, \overset{\pm}{f}_k] = 0,$$
$$[\overset{+}{b}_j, \overset{-}{b}_k] = \delta_{jk} I,$$
$$[\overset{+}{f}_j, \overset{-}{f}_k]_+ \equiv \overset{+}{f}_j \overset{-}{f}_k + \overset{-}{f}_k \overset{+}{f}_j = \delta_{jk} I.$$

These relations are different from the usual Lie relations in that one uses anticommutators instead of commutators for fermion creation-annihilation operators. Thus, the model symmetric with respect to bosons and fermions leads to the consideration of a new mathematical notion, namely, of *graded Lie algebras* that involve both commutators and anticommutators (see [26]).

Formally, the underlying linear space of a graded Lie algebra L is the direct sum $L = L_0 \oplus L_1$.

The space L is equipped with a bilinear operation, $[\cdot, \cdot]$, that respects the $\mathbb{Z}_2$ gradation, that is,

$$[L_0, L_0] \subset L_0, \quad [L_0, L_1] \subset L_1, \quad [L_1, L_1] \subset L_0,$$

and if a, b, c are graded elements of L (that is, they each belong either to L_0 or to L_1), then the graded Jacobi identity holds:

$$(-1)^{|a|\,|c|}[[a, b], c] + (-1)^{|b|\,|a|}[[b, c], a] + (-1)^{|c|\,|b|}[[c, a], b] = 0.$$

Furthermore,

$$[a, b] = -(-1)^{|a|\,|b|}[b, a]$$

(here $|a|$ is the gradation of a: we write $|a| = j$ if $a \subset L_j$).

A *representation* of a graded Lie algebra L is a linear mapping of L into an associative algebra such that $[a, b]$ is carried into $ab - (-1)^{|a||b|}ba$; in particular, the bracket of two elements of L_1 is realized as the anticommutator and the brackets of all other combinations of graded elements are taken into commutators.

Let us consider a simple example in which we construct the ordered representation for some operators forming a representation of a graded Lie algebra. Let A and B be operators satisfying the relation

$$AB + BA = 1.$$

This can be regarded as a representation of a graded Lie algebra with $A, B \in L_1$, and $1 \in L_0$.

Note that this graded algebra is likely to be infinite-dimensional since the above relation does not form a closed system of graded Lie relations by itself: we have not supplied any data for the anticommutators $[A, A] = 2A^2$ and $[B, B] = 2B^2$. Formally, this can be considered as a special case (for $\alpha = -1$) of the perturbed Heisenberg relations considered in Subsection 2.2.

Let us consider functions of the Feynman tuple $(\overset{1}{A}, \overset{2}{B})$. Then for any symbol $f(x, y)$ we have

$$B[\![f(\overset{2}{B}, \overset{1}{A})]\!] = \overset{3}{B}\, f(\overset{2}{B}, \overset{1}{A}) = \overset{2}{B}\, f(\overset{2}{B}, \overset{1}{A}),$$

which implies that $l_B = x$.

Proceeding as in the cited example, we obtain

$$\begin{aligned} A[\![f(\overset{2}{B}, \overset{1}{A})]\!] &= \overset{3}{A}\, f(\overset{2}{B}, \overset{1}{A}) = \overset{3}{A}\, f(-\overset{4}{B}, \overset{1}{A}) \\ &\quad + \overset{3}{A}\,(\overset{2}{B} + \overset{4}{B}) \frac{f(\overset{2}{B}, \overset{1}{A}) - f(-\overset{4}{B}, \overset{1}{A})}{\overset{2}{B} + \overset{4}{B}} \\ &= f(-\overset{2}{B}, \overset{1}{A})\, \overset{1}{A} + [\![\overset{2}{\overset{3}{A}}\,(\overset{2}{B} + \overset{4}{B})]\!] \frac{\delta f}{\delta x}(\overset{1}{B}, -\overset{5}{B}, \overset{0}{A}). \end{aligned}$$

Since $\overset{3}{A}\,(\overset{2}{B} + \overset{4}{B}) = 1$, it follows that

$$A[\![f(\overset{2}{B}, \overset{1}{A})]\!] = l_A f(\overset{2}{B}, \overset{1}{A}),$$

where

$$l_A f(x, y) = y f(-x, y) + \frac{f(x, y) - f(-x, y)}{2x}.$$

Thus, the left ordered representation of the Feynman tuple $(\overset{1}{A}, \overset{2}{B})$ is

$$l_B = x, \quad l_A = y I_x + \frac{1 - I_x}{2x},$$

where I_x is the spatial inversion operator $I_x f(x, y) = f(-x, y)$.

3 Evaluation of the Ordered Representation Operators

In Section 2 we have considered several examples of computation for the operators of left and right ordered representation. Here we present a general method to compute these operators for some "distinguished" classes of commutation relations.

3.1 Equations for the Ordered Representation Operators

We consider a special class of commutation relations of the following form. Let $k \leq n$ be a positive integer. Consider the relations

$$[A_i, A_j] = 0, \quad j = k+1, \ldots, n, \quad i = 1, \ldots, n, \tag{II.6}$$

$$A_i A_j = \sum_{r=1}^{n} \varphi_{ij}^r (A_j, A_{k+1}, \ldots, A_n) A_r, \quad i, j = 1, \ldots, k, \tag{II.7}$$

where $\varphi_{ij}^r(y_j, y_{k+1}, \ldots, y_n)$ are some symbols; since, according to (II.6), the operators $A_j, A_{k+1}, \ldots, A_n$ are pairwise commuting, their Feynman indices are inessential and therefore omitted. The relations (II.6) and (II.7) can be rewritten in the matrix form

$$AA_j = \varphi_j(A_j, A_{k+1}, \ldots, A_n)A, \quad j = 1, \ldots, n, \tag{II.8}$$

where A is the vector operator $A = {}^t(A_1, \ldots, A_n)$, and $\varphi_j(A_j, A_{k+1}, \ldots, A_n)$ is the $n \times n$ matrix operator with entries

$$(\varphi_j(A_j, A_{k+1}, \ldots, A_n))_{ir} = \begin{cases} \varphi_{ij}^r(A_j, A_{k+1}, \ldots, A_n) & \text{if } \ j \leq k \ \text{ and } \ i \leq k, \\ A_j \delta_{ir} & \text{otherwise.} \end{cases}$$

Example II.1 Let $k = n$ and

$$\varphi_j(y_j) = \begin{pmatrix} \sigma_{1j} & & 0 \\ & \ddots & \\ 0 & & \sigma_{nj} \end{pmatrix} y_j + \Lambda_j, \tag{II.9}$$

where Λ_j are constant matrices. Then, if all $\sigma_{ij} = +1$, we obtain Lie commutation relations; if $\sigma_{ij} = \pm 1$, we obtain graded Lie algebras, if σ_{ij} are close to 1, we obtain "perturbations" of Lie algebras.

One can also consider the more general case

$$\varphi_j(y_j) = M_j y_j + \Lambda_j, \tag{II.10}$$

where the matrix M_j is not diagonal, or even the case in which $\varphi_j(y_j)$ is a nonlinear function (such relations will be referred to as strongly nonlinear).

First of all, note that it follows from (II.6) that $L_{k+1} = y_{k+1}, \dots, L_n = y_n$; without loss of generality it can be assumed that the operators L_j, $j = k+1, \dots, n$, commute with all the other left ordered representation operators. Thus we can assume that $k = n$, which is done throughout the sequel.

Along with (II.7) one can also consider the "dual" relations

$$A_k A_j = \sum_{s=1}^{n} A_s \Omega_{kj}^s (A_k). \tag{II.11}$$

They transform into (II.7) and vice versa if we pass from the algebra

$$\mathcal{A} \ni A_1, \dots, A_n$$

to the opposite algebra $\mathcal{A}^{op}$, whose multiplication is given by the formula

$$A^{op} B = BA.$$

It follows that the left ordered representation of $\mathcal{A}$ corresponds to the right ordered representation of $\mathcal{A}^{op}$, and the same is true for the right ordered representation of $\mathcal{A}$ and left ordered representation of $\mathcal{A}^{op}$. Let us first consider the commutation relations (II.7). Introduce the auxiliary operators L_{ij}, $i = 1, \dots, n$, $j = 0, \dots, n$ satisfying the relations

$$(L_{ij} f)(\overset{1}{A_1}, \dots, \overset{n}{A_n}) = \overset{j+1}{A}_i f(\overset{1}{A_1}, \dots, \overset{j}{A_j}, \overset{j+2}{A}_{j+1}, \dots, \overset{n+1}{A}_n)$$

for any symbol f. In particular,

$$L_{in} = L_{A_i} \quad \text{and} \quad L_{i0} = R_{A_i}$$

are the operators of left and right ordered representation respectively.

From the moving indices apart rule it follows that

$$L_{jj} = y_j, \quad j = 1, \dots, n.$$

We seek the operators L_{ij} in the form

$$L_{ij} = L_{ij}\left(\overset{2}{y}, \overset{1}{\frac{\partial}{\partial y}}\right).$$

Our aim is to derive equations for the symbols $L_{ij}(x, p)$. Introduce the vector operator[1]

$$\mathcal{L}_j = {}^t(L_{1j}, \dots, L_{nj}), \quad j = 0, \dots, n,$$

[1] Here t stands for the transposed matrix.

with symbol

$$\mathcal{L}_j(x,p) = {}^t(L_{1j}(x,p), \ldots, L_{nj}(x,p)).$$

By Theorem I.4 we conclude that

$$\begin{aligned} &\overset{j+1}{A_i}\, f(\overset{1}{A_1}, \ldots, \overset{j}{A_j}, \overset{j+2}{A}_{j+1}, \ldots, \overset{n+1}{A}_n) \\ &= \overset{j}{A_i}\, f(\overset{1}{A_1}, \ldots, \overset{j-1}{A}_{j-1}, \overset{j+1}{\varphi}_j(A_j), \overset{j+2}{A}_{j+1}, \ldots, \overset{n+1}{A}_n) \end{aligned} \tag{II.12}$$

(the operators $A_{j+1}, \ldots, A_n$ on the right-hand side of the latter equation are tensored by the identity matrix).

Using the substitution operator (see Subsection 2.3 above) one can write

$$f(y_1, \ldots, \varphi(y_j), \ldots, y_n) = e^{(\varphi(y_j)-y_j)\partial/\partial y_j} f(y_1, \ldots, y_n).$$

Thus it follows that the operators corresponding to the symbols

$$L_{ij} f(y_1, \ldots, y_n)$$

and

$$\sum_{r=1}^{n} L_{r,j-1} \left\{ e^{(\varphi_j(y_j)-y_j)\partial/\partial y_j} \right\}_{ir} f(y_1, \ldots, y_n)$$

coincide. We require that the symbols of these operators coincide and use the right regular representation of the tuple $(\overset{2}{y}, \overset{1}{\partial/\partial y})$ constructed in the preceding section. We obtain

$$\mathcal{L}_j(x,p) = e^{\overset{2}{p}_j F_j\left(\overset{1}{x_j+\partial/\partial p_j}\right)} \mathcal{L}_{j-1}(x,p), \tag{II.13}$$

where

$$F_j(y) = \varphi_j(y) - y.$$

Using the relation $L_{jj} = y_j$, we can write out the following system of equations to define $\mathcal{L}_0(x,p)$:

$$\sum_k \left\{ e^{\overset{2}{p}_j F_j\left(\overset{1}{x_j+\partial/\partial p_j}\right)} \cdots e^{\overset{2}{p}_1 F_1\left(\overset{1}{x_1+\partial/\partial p_{j1}}\right)} \right\}_{jk} L_{k0}(x,p) = x_j. \tag{II.14}$$

If this system is solvable, its solution gives the right ordered representation operators; then we can also obtain the left ordered representation operators. Of course, it remains to prove that the solution obtained indeed gives the ordered representation operators. This requires some additional assumptions, and we will return to this question later; the corresponding theorem will be stated only for relations (II.11), since the statements are essentially the same. Let us now consider relations (II.11) in more detail. We define matrix functions $\Lambda_k(y)$ by the equations

$$(\Lambda_k(y))_{sj} = \Omega^s_{kj}(y) - y\delta_{sj}, \tag{II.15}$$

where δ_{sj} is the Kronecker delta, and introduce the matrix operator

$$U_k = \exp\left(\Lambda_k(\overset{2}{y}_k)\frac{\overset{1}{\partial}}{\partial x_k}\right).$$

The operator U_k acts on scalar functions of y_k, and the result is $n \times n$ matrix functions of y_k,

$$[U_k f(y_k)]_{sj} = f(y_k + \Lambda_k(y_k))_{sj} \equiv f(\Omega_k(y_k))_{sj}, \tag{II.16}$$

where $\Omega_k(y_k)$ is the matrix with elements

$$(\Omega_k(y_k))_{sj} = \Omega^s_{kj}(y_k),$$

and $f(\Omega_k(y_k))_{sj}$ stands for the (s, j)th entry of $f(\Omega_k(y_k))$.

Define operators $D_{sj}, s, j = 1, \ldots, n$, acting on functions of n variables $y_1, \ldots, y_n$ by the formula

$$\begin{aligned} D_{sj} &= [U_n \times \cdots \times U_{j+1}]_{sj}, \quad j < n, \\ D_{sn} &= \delta_{sn}. \end{aligned}$$

Theorem II.7 *Suppose that the operators $l_1, \ldots, l_n$ of the left ordered representation for system* (II.11) *exist and are uniquely defined. Suppose also that all the operators U_j are invertible. Then the operators $l_1, \ldots, l_n$ satisfy the system of equations*

$$\sum_s l_s D_{sj} = y_j. \tag{II.17}$$

Proof. Let $\varphi(y_1, \ldots, y_n)$ be a given symbol. Relations (II.11) can be rewritten in the form

$$A_k A = A\Omega_k(A_k), \quad k = 1, \ldots, n,$$

where as above $A = {}^t(A_1, \ldots, A_n)$. Thus, using successively relations (II.11) and (II.16) we obtain

$$\begin{aligned} &\varphi(\overset{1}{A}_1, \ldots, \overset{k-1}{A}_{k-1}, \overset{k+1}{A}_k, \ldots, \overset{n+1}{A}_n)\overset{k}{A} \\ &= \varphi(\overset{1}{A}_1, \ldots, \overset{k-1}{A}_{k-1}, \overset{k}{\Omega_k(A_k)}, \overset{k+2}{A}_{k+1}, \ldots, \overset{n+1}{A}_n) \overset{k+1}{A} \\ &= (U_k\varphi)(\overset{1}{A}_k, \ldots, \overset{k-1}{A}_{k-1}, \overset{k}{A}_k, \ldots, \overset{k+2}{A}_{k+1}, \ldots, \overset{n+1}{A}_n) \overset{k+1}{A} \ldots \\ &= (l\, U_n \ldots U_k\varphi)(\overset{1}{A}_k, \ldots, \overset{n}{A}_n) \overset{\text{def}}{=} (l_{(k)}\varphi)(A_1, \ldots, A_n), \end{aligned}$$

where

$$l = (l_1, \dots, l_n), \quad l_{(k)} = (l_{(k)1}, \dots, l_{(k)n}).$$

Since all operators U_j are invertible and l is uniquely determined, it follows that $l_{(k)}$ is also uniquely determined by the property

$$\varphi(\overset{1}{A_1}, \dots, \overset{k-1}{A}_{k-1}, \overset{k+1}{A}_k, \dots, \overset{n+1}{A}_n)\overset{k}{A} = (l_{(k)}\varphi)(\overset{1}{A_k}, \dots, \overset{n}{A_n}),$$

that is, $l_{(k)j} = l_{jk}$. From this it follows that

$$\sum_s l_s[U_n \dots U_{j+1}]_{sj} = l_{ij} = y_j, \quad 1 \le j \le n-1.$$

Taking into account that $l_n = y_n$, we find that the operators l_s satisfy system (II.17). The theorem is proved. □

3.2 How to Obtain the Solution

Let us now solve the equations for the ordered representation operators. Equations (II.14) and (II.17) admit explicit solution in a variety of cases. We shall now analyze these cases and then state a general theorem concerning the solvability of system (II.14).

(a) Let the operators $A_1, \dots, A_n$ form a Lie algebra, i.e., satisfy the relations

$$[A_i, A_j] = \sum \lambda_{ij}^k A_k, \tag{II.18}$$

where λ_{ij}^k are structural constants. These relations can be rewritten in the form (II.8) by setting

$$\begin{aligned} \varphi_j(y_j) &= y_j + \Lambda_j, \\ F_j(y_j) &= \varphi_j(y_j) - y_j = \Lambda_j, \quad (\Lambda_j)_{ik} = \lambda_{ij}^k, \end{aligned}$$

and we obtain from (II.13)

$$\mathcal{L}_j(x, p) = e^{p_j \Lambda_j} \mathcal{L}_{j-1}(x, p).$$

Let us compute the left ordered representation $\mathcal{L}_n$. We have

$$\mathcal{L}_j(x, p) = e^{-p_{j+1}\Lambda_{j+1}} \cdot e^{-p_{j+2}\Lambda_{j+2}} \dots e^{-p_n\Lambda_n} \mathcal{L}_n(x, p).$$

Thus we can write out the system of equations

$$(e^{-p_{j+1}\Lambda_{j+1}} \dots e^{-p_n\Lambda_n} \mathcal{L}_n(x, p))_j = x_j$$

and obtain the following familiar expression for $\mathcal{L}_n$ (cf. Subsection 2.4):

$$\mathcal{L}_n = A^{-1} \begin{pmatrix} \overset{1}{\partial} \\ \overline{\partial y} \end{pmatrix} \overset{2}{y},$$

where

$$(A(p))_{ij} = (e^{-p_{j+1}\Lambda_{j+1}} \ldots e^{-p_n \Lambda_n})_{ij}.$$

Note that

$$A(0) = \begin{pmatrix} 1 & & 0 \\ & \ddots & \\ 0 & & 1 \end{pmatrix}$$

and consequently, the obtained operator is well-defined as a series in powers of $\partial/\partial y$. Simple though cumbersome calculations show that the components of $\mathcal{L}_n$ satisfy (II.18) if and only if the structural constants satisfy the Jacobi identity

$$\sum_k (\lambda_{ij}^k \lambda_{kl}^s + \lambda_{li}^k \lambda_{kj}^s + \lambda_{jl}^k \lambda_{ki}^s) = 0$$

for all tuples (i, j, l, s) (cf. Section 4).

(b) Let us now analyze a much less trivial case in which one manages to solve system (II.17) with nonlinear functions Ω_{kj}^s.

Definition II.3 The system of commutation relations (II.11) is said to be *solvable* if (a) all matrices $\Lambda_k(y)$ in (II.15) are lower-triangular, that is, $\Lambda_{kj}^s \equiv 0$ for $s < j$; (b) the function $\varphi_{ks}^s(y)$ have inverses $(\Omega_{ks}^s)^{-1}(y)$ for all k and s.

Theorem II.8 *If system* (II.11) *is solvable, then system* (II.17) *can be solved for* l_s, $s = 1, \ldots, n$ *(the explicit form of the solution is given below in the proof of the theorem).*

Proof. First of all, let us transform (II.17) to a more convenient form. Let us seek l_s in the form

$$l_s = \sum_{k=1}^n y_k M_{ks},$$

where M_{ks} are now unknown operators. Inserting this into (II.17), we obtain

$$\sum_{k,s} y_k M_{ks} D_{sj} = y_j, \quad j = 1, \ldots, n,$$

so that it suffices to solve the system of equations

$$\sum_s M_{ks} D_{sj} = \delta_{kj} \quad j, k = 1, \ldots, n. \tag{II.19}$$

Let us introduce the operators R_j by setting

$$(R_j f)(y_1, \dots, y_n) = f(y_1, \dots, y_j, (\Omega^j_{j+1,j})^{-1}(y_{j+1}), \dots, (\Omega^j_{nj})^{-1}(y_n),$$

$j = 1, \dots, n-1$, $R_n = 1$. Now set

$$\begin{aligned} M_{ss} &= R_s, \quad s = 1, \dots, n, \\ M_{sj} &= 0, \quad s < j, \\ M_{sj} &= -\sum_{l=1}^{s-1} R_s D_{sl} M_{lj}, \quad s > j. \end{aligned}$$

These equations permit us to determine the operators M_{sj} for all s and j. We claim that they satisfy (II.19). To prove this, note that if A is a lower-triangular matrix, then so is $f(A)$, and $f(A)_{jj} = f(A_{jj})$ for all $j = 1, \dots, n$. We conclude that $D_{sj} = 0$ for $s < j$ and $D_{jj} \circ R_j = 1$, $j = 1, \dots, n$. Inserting M_{sj} into (II.19), we obtain, by the above,

$$\begin{aligned} \sum_{s=1}^{n} M_{ks} D_{sj} &= \sum_{s \le k, s \ge j} M_{ks} D_{sj} \\ &= M_{kk} D_{kj} - \sum_{s=j}^{k-1} \sum_{l=s}^{k-1} R_k D_{kl} M_{ls} D_{sj} \\ &= M_{kk} D_{kj} - \sum_{l=1}^{k-1} R_k D_{kl} \left\{ \sum_{s=j}^{l} M_{ls} D_{sj} \right\}. \end{aligned}$$

If $j = k$, then only the term $M_{kk} D_{kk} = R_j D_{jj} = 1$ occurs on the right-hand side of the last equation. Let us show by induction over $k - j$ that

$$\sum_{s=j}^{k} M_{ls} D_{sj} = \delta_{lj}.$$

There is nothing to prove for $k \le j$. If $k > j$, then, by the induction hypothesis,

$$\begin{aligned} M_{kk} D_{kj} &- \sum_{l=j}^{k-1} R_k D_{kl} \left\{ \sum_{s=j}^{l} M_{ls} D_{sj} \right\} \\ &= M_{kk} D_{kj} - \sum_{l=j}^{k-1} R_k D_{kl} \delta_{lj} \\ &= M_{kk} D_{kj} - R_k D_{kj} = R_k D_{kj} - R_k D_{kj} = 0. \end{aligned}$$

The theorem is proved. □

Let us now prove that system (II.14) is solvable in the class of formal power series provided that the functions $\varphi_j(y_j)$ have the form (II.10).

We seek the solution of (II.14) in the form

$$L_{s0}(x, p) = \sum_{|\alpha|=0}^{\infty} C_{s,\alpha}(x) p^{\alpha}.$$

Then we obtain the following infinite system of equations for the functions $C_{s,\alpha}$:

$$\sum_{s=1}^{n} \sum_{|\alpha|=0}^{\infty} C_{s,\alpha}(x) \frac{\nu_n!}{(\nu_n - \alpha_n)!} \cdots \frac{\nu_{j+1}!}{(\nu_{j+1} - \alpha_{j+1})!} x_{j+1}^{\nu_{j+1}-\alpha_{j+1}}$$
$$\times \cdots \times x_n^{\nu_n - \alpha_n} \left[\frac{\partial^{\alpha_j}}{\partial x_j^{\alpha_j}} (\varphi_j(x_j))^{\nu_j} \cdots \frac{\partial^{\alpha_1}}{\partial x_1^{\alpha_1}} (\varphi_1(x_1))^{\nu_1} \right]_{js} \tag{II.20}$$
$$= x_1^{\nu_1} \cdots x_j^{\nu_j+1} \cdots x_n^{\nu_n},$$

$(\nu_i = 0, 1, \ldots; i = 1, \ldots, n)$.

Indeed, let us make the substitution

$$x \to y, \quad p \to \frac{\partial}{\partial y}$$

in (II.14) and apply both sides of this equation to the monomial $y_1^{\nu_1} \ldots y_n^{\nu_n}$. Then we will have $y_1^{\nu_2} \ldots y_j^{\nu_j+1} \ldots y_n^{\nu_n}$ on the right-hand side of this equation and on the left-hand side we can use the fact that $(p, x + \partial/\partial p)$ is the right regular representation of the tuple $\left(\overset{1}{\partial/\partial y}, \overset{2}{y}\right)$. We obtain

$$\sum_k [\![L_{k0}\left(\overset{2}{y}, \overset{1}{\frac{\partial}{\partial y}}\right)]\!] [\![\left\{ e^{F_j(\overset{2}{y_j})\overset{1}{\partial/\partial y_j}} \cdots e^{F_1(\overset{2}{y_1})\overset{1}{\partial/\partial y_1}} \right\}_{jk}]\!] (y_1^{\nu_1} \cdots y_n^{\nu_n})$$
$$= \sum_k L_{k0}\left(\overset{2}{y}, \overset{1}{\frac{\partial}{\partial y}}\right) \left[y_{j+1}^{\nu_{j+1}} \cdots y_n^{\nu_n} \right] \left[(\varphi_j(y_j))^{\nu_j} \cdots (\varphi_1(y_1))^{\nu_1} \right]_{jk}$$
$$= \sum_{s=1}^{n} \sum_{|\alpha|=0}^{\infty} C_{s\alpha}(y) \frac{\nu_n!}{(\nu_n - \alpha_n)!} \cdots \frac{\nu_{j+1}!}{(\nu_{j+1} - \alpha_{j+1})!} y_{j+1}^{\nu_{j+1}-\alpha_{j+1}} \cdots y_n^{\nu_n - \alpha_n}$$
$$\times \left[\frac{\partial^{\alpha_j}}{\partial y_j^{\alpha_j}} (\varphi_j(y_j))^{\nu_j} \cdots \frac{\partial^{\alpha_1}}{\partial y_1^{\alpha_1}} (\varphi_1(y_1))^{\nu_1} \right]_{j,s},$$

and we arrive directly at (II.20). Let $\varphi_j(y_j)$ have the form (II.10). Then the summation in (II.20) ranges over $\alpha \leq \nu$, and (II.20) can be rewritten in the form

$$\nu_1! \ldots \nu_n! \sum_{s=1}^{n} \left[M_j^{\nu_j} \ldots M_1^{\nu_1} \right]_{js} C_{s,\nu}(x)$$

$$= x_1^{\nu_1} \dots x_j^{\nu_j+1} \dots x_n^{\nu_n} + \sum_{\alpha<\nu} \sum_{l=1}^{n} P_{jl\alpha\nu}(x) C_{l,\alpha}(x),$$

where $P_{jl\alpha\nu}(x)$ is a known polynomial of degree $\leq |\nu| - |\alpha|$. The last equation immediately implies the following theorem.

Theorem II.9 *Let the functions* $\varphi_j(x)$ *have the form* (II.10). *Then system* (II.14) *has a solution provided that for any multi-index* $\nu = (\nu_1, \dots, \nu_n)$ *we have*

$$\det ||(M_j^{\nu_j} \dots M_1^{\nu_1})_{js}|| \neq 0.$$

Under the conditions of Theorem II.9, $C_{s,\alpha}(x)$ is a polynomial of degree $\leq |\alpha| + 1$ and consequently, the operator $L_{s0}\ (y, \partial/\partial y)$ is well-defined on the space of polynomial symbols. In particular, the hypotheses of Theorem II.9 are necessarily satisfied if φ is a function of the form (II.9) and $\sigma_{ij} \neq 0$ for all i and j.

3.3 Semilinear Commutation Relations

Let us now consider a generalization of the commutation relations defining a Lie algebra. It turns out that the method used can be generalized to a wider class of relations, namely, to that of *quasilinear* commutation relations.

Consider a Feynman tuple

$$\overset{1}{A}_1, \dots, \overset{n}{A}_n, \quad \overset{n+1}{B}_1, \dots, \overset{n+1}{B}_m$$

of operators satisfying the following relations:

$$\begin{aligned} &[A_j, A_k] = -i \sum_{l=1}^{n} c_{ij}^l(B) A_l, \quad j, k = 1, \dots, n; \\ &[A_j, B_s] = -\operatorname{id}_{js}(B), \quad j = 1, \dots, n, \quad s = 1, \dots, m; \\ &[B_s, B_r] = 0, \quad s, r = 1, \dots, m. \end{aligned} \tag{II.21}$$

Here $c_{jk}^l(z_1, \dots, z_m)$ and $d_{js}(z_1, \dots, z_m)$ are given m-ary symbols. We see that the operators B_s commute with one another so it is quite appropriate to assign the same Feynman index to all these operators, Relations (II.21) generalize the Lie commutation relations in the sense that the structural constants c_{jk}^l are now allowed to depend on the additional operator arguments B. However, these arguments all commute with one another and satisfy a rather special commutation relations with the operators A_j (see the second line in (II.21).

Let us outline the method that we follow to construct the left ordered representation; the reader will see that in principle it differs only slightly from that used for Lie commutation relations.

(a) Consider the Feynman-ordered exponential

$$U(t,\tau) = e^{i\tau_m B_m} \dots e^{i\tau_1 B_1} e^{it_n B_n} \dots e^{it_1 B_1},$$

where $t \in \mathbb{R}^n$ and $\tau \in \mathbb{R}^m$. We will find operators

$$\hat{Q}_j = Q_j\left(\overset{2}{t}, \overset{2}{\tau}, -i\overset{1}{\frac{\partial}{\partial t}}, -i\overset{1}{\frac{\partial}{\partial \tau}}\right)$$

such that

$$\hat{Q}_j U(t,\tau) = A_j\, U(t,\tau).$$

(b) For any symbol $f(y,z)$, $y \in \mathbb{R}^n$, $z \in \mathbb{R}^m$, we have

$$f(\overset{1}{A_1}, \dots, \overset{n}{A_n}, \overset{n+1}{B}) = \left[f\left(-i\frac{\partial}{\partial t}, -i\frac{\partial}{\partial \tau}\right) U(t,\tau)\right]\Big|_{t=\tau=0}.$$

Combining this with the preceding formula, we obtain

$$\begin{aligned} A_j[\![f(\overset{1}{A_1}, \dots, \overset{n}{A_n}, \overset{n+1}{B})]\!] &= A_j\left[f\left(-i\frac{\partial}{\partial t}, -i\frac{\partial}{\partial \tau}\right) U(t,\tau)\right]_{t=\tau=0} \\ &= \left[f\left(-i\frac{\partial}{\partial t}, -i\frac{\partial}{\partial \tau}\right) A_j U(t,\tau)\right]\Big|_{t=\tau=0} \\ &= \left[\left(f\left(-i\frac{\partial}{\partial t}, -i\frac{\partial}{\partial \tau}\right)\hat{Q}_j\right) U(t,\tau)\right]\Big|_{t=\tau=0}. \end{aligned} \tag{II.22}$$

The composition

$$f\left(-i\frac{\partial}{\partial t}, -i\frac{\partial}{\partial \tau}\right)\hat{Q}_j = [\![f\left(-i\frac{\partial}{\partial t}, -i\frac{\partial}{\partial \tau}\right)]\!] [\![Q_j\left(\overset{2}{t}, \overset{2}{\tau}, -i\overset{1}{\frac{\partial}{\partial t}}, -i\overset{1}{\frac{\partial}{\partial \tau}}\right)]\!]$$

can be computed with the help of the right ordered representation for the tuple $\left(\overset{2}{t}, \overset{2}{\tau}, -i\overset{1}{\partial/\partial t}, -i\overset{1}{\partial/\partial \tau}\right)$.

The computations similar to those carried out in the end of Subsection 2.4 lead us to the formula

$$l_j = Q_j\left(-i\overset{1}{\frac{\partial}{\partial y}}, -i\overset{1}{\frac{\partial}{\partial z}}, \overset{2}{y}, \overset{2}{z}\right), \quad j = 1, \dots, n,$$

since the operators B_l act last and commute with one another, we have

$$B_l[\![f(\overset{1}{A_1}, \dots, \overset{n}{A_n}, \overset{n+1}{B})]\!] = z_j\, \overset{n+1}{B_l}\, f(\overset{1}{A_1}, \dots, \overset{n}{A_n}, \overset{n+1}{B}),$$

so that the corresponding representation operators are very simple,

$$l_{B_l} = z_l, \quad l = 1, \dots, m.$$

Let us now proceed to the implementation of the outlined scheme. Since, in contrast to Example II.1, we do not assume that the symbol space $\mathcal{F}$ consists only of polynomials, we should make some other assumptions so as to ensure the rigorousness of our considerations. Specifically, we assume the following.

Condition II.1 The exponential e^{itx} belongs to $\mathcal{F}$ for any $t \in \mathbb{R}$; moreover, there exists a subalgebra Φ of the algebra $C^\infty(\mathbb{R}; \mathcal{F})$ of all infinitely differentiable mappings from $\mathbb{R}$ to $\mathcal{F}$ such that e^{itx} (more precisely, the mapping $t \mapsto e^{itx}$) belongs to $\mathcal{F}$ and the operator $-i\partial/\partial t$ is an $\mathcal{F}$-generator in Φ.

Lemma II.2 *Under Condition A, for any tuple* $C = (\overset{1}{C}_1, \dots, \overset{s}{C}_s)$ *of* $\mathcal{F}$*-generators in an operator algebra* $\mathcal{A}$ *and for any symbol* $f \in \mathcal{F}_s$ *we have*

$$f(C) = \left[f\left(-i\frac{\partial}{\partial t}\right) e^{itC} \right]\bigg|_{t=0}, \tag{II.23}$$

where e^{itC} *is the Feynman-ordered exponential*

$$e^{itC} = e^{it_sC_s} \dots e^{it_1C_1}.$$

Proof. Since factorable symbols are dense in $\mathcal{F}_s$ and both sides of (II.23) depend on f linearly and continuously, it suffices to prove the lemma for symbols of the form

$$f(y) = f_1(y_1) \dots f_s(y_s).$$

Furthermore, we see that for such symbols the statement of the lemma can easily be reduced to the case $s = 1$. Thus, here is what we have to prove: if C is an $\mathcal{F}$-generator and Condition II.1 holds, then equation (II.23) is valid. Let $\tilde{\Phi} \subset C^\infty(\mathbb{R}; \mathcal{A})$ be the image of Φ under the mapping $\varphi(t, x) \mapsto \varphi(t, C)$ and let $\tilde{\Phi}_0 \subset \tilde{\Phi}$ be the subalgebra of elements independent of t. Evidently, the diagram

$$\begin{array}{ccc}
\tilde{\Phi}_0 & \xrightarrow{L_{e^{itC}}} & \tilde{\Phi} \\
{\scriptstyle L_C}\downarrow & & \downarrow{\scriptstyle -i\frac{\partial}{\partial t}} \\
\tilde{\Phi}_0 & \xrightarrow{L_{e^{itC}}} & \tilde{\Phi}
\end{array}$$

commutes (here L_C stands for the operator of left multiplication by C etc.) Thus $L_{e^{itC}}$ is an intertwining operator for L_C and $-i\partial/\partial t$ in the above-mentioned spaces, and, consequently,

$$L_{e^{itC}} f(L_C) = f\left(-i\frac{\partial}{\partial t}\right) L_{e^{itC}}.$$

Apply both sides of the last equation to the identity operator $1 \in \tilde{\Phi}_0$. We obtain

$$e^{itC} f(C) = f\left(-i\frac{\partial}{\partial t}\right) e^{itC}.$$

The desired relation can now be derived from this one by setting $t = 0$. The lemma is proved. □

Let us now try to find the operators $\hat{Q}_j$ (item (a) of the outline). The idea is quite simple. We have

$$\begin{aligned} -i\frac{\partial}{\partial t_j}U(t,\tau) &= e^{i\tau B} e^{it_n A_n} \dots e^{it_{j+1}A_{j+1}} A_j e^{it_j A_j} \dots e^{it_1 A_1}, \\ j &= 1, \dots, n; \\ -i\frac{\partial}{\partial t_j}U(t,\tau) &= B_l U(t,\tau), \quad l = 1, \dots, m. \end{aligned} \tag{II.24}$$

We would like to represent the first group of equations in the form

$$-i\frac{\partial}{\partial t_j}U(t,\tau) = [\![\chi_j(\overset{1}{A}_1, \dots, \overset{n}{A}_n, B, T, \tau)]\!] U(t,\tau)$$

and then construct the $\hat{Q}_j$ by choosing appropriate "quasilinear combination" of the equations obtained.

To transform (II.24) into the desired form, we should move A_j to the last place in the product on the right-hand side, i.e., make it act after all other operators in this expression.

Lemma II.3 *We have*

$$e^{it_n A_n} \dots e^{it_{j+1}A_{j+1}} A_j = \sum_{r=1}^{n} \varphi_j^r(B,t) A_r e^{it_n A_n} \dots e^{it_{j+1}A_{j+1}},$$

where $\varphi_j^r(z,t)$ are symbols determined in the proof of the lemma (in fact, $\varphi_j^r(z,t)$ is independent of $t_1, \dots, t_j$).

Proof. Let

$$X = \sum_{r=1}^{n} \omega^r(B) A_r;$$

let us compute

$$X(t_k) = e^{it_k A_k} X e^{-it_k A_k} = e^{it_k \operatorname{ad}_{A_k}}(X), \quad t_k \in \mathbb{R}.$$

One has

$$-i\frac{d}{dt}(X(t_k)) = [A_k, X(t_k)], \quad X(0) = X. \tag{II.25}$$

We seek $X(t_k)$ in the form

$$X(t_k) = \sum_{r=1}^{n} \omega^r(B, t_k) A_r,$$

where $\omega^r(z, t_k)$ will be determined later. On substituting this expression into (II.25) we obtain in the commutator term

$$\begin{aligned} [A_k, X(t_k)] &= \sum_{r=1}^{n} \{[A_k \omega^r(B, t_k)] A_r + \omega^r(B, t_k)[A_k, A_r]\} \\ &= -i \sum_{r=1}^{n} \left\{ \sum_{s=1}^{m} d_{ks}(B) \frac{\partial \omega^r(B, t_k)}{\partial z_s} A_r + \omega^r(B, t_k) \sum_{l=1}^{n} c_{kr}^{l}(B) A_l \right\} \end{aligned}$$

(we have used relations (II.21) and the commutation formulas from Chapter I). On the whole, we get

$$\sum_{r=1}^{n} \sum_{s=1}^{m} \frac{\partial \omega^r}{\partial t_s}(B, t_k) A_r = \sum_{r=1}^{n} \left\{ \sum_{s=1}^{m} d_{ks}(B) \frac{\partial \omega^r}{\partial z_s} + \sum_{l=1}^{n} \omega^l(B, t_k) c_{kl}^{l}(B) \right\} A_r$$

(in the second term on the right-hand side we have interchanged the notation for the dummy indices r and l). Clearly, it suffices to require that the symbols $\omega^r(B, t_k)$, $r = 1, \dots, n$, satisfy the system

$$\begin{cases} \dfrac{\partial \omega^r}{\partial t_k} - \sum\limits_{s=1}^{m} d_{ks}(z) \dfrac{\partial \omega^r}{\partial z_s} = \sum\limits_{l=1}^{n} c_{kl}^{r}(z) \omega^l, \\ \omega^r(z, t_k)|_{t_k=0} = \omega^r(z), \end{cases}$$

which is a system of ordinary differential equations along the trajectories of the vector field[1]

$$\frac{d}{dt_k} = \frac{\partial}{\partial t_k} - \sum_{s=r}^{m} d_{ks}(z) \frac{\partial}{\partial z_s}.$$

Let $C_k(z)$ be the matrix with the elements

$$(C_k^{(z)})_{rl} = c_{kl}^{r}(z).$$

[1]We assume this field to generate a global phase flow.

We have

$$\omega(z, t_k) = T\text{-}\exp\left(\int_0^{t_k} C_k(z_{(k)}(z_0, \tau))d\tau\right)\omega(z_0),$$

where the integral is taken along the trajectory $z = z_{(k)}(z_0, t)$ of the field $\frac{d}{dt_k}$ such that $z_{(k)}(z_0, t) = z$, and $\omega = {}^t(\omega^1, \dots, \omega^n)$.

We apply this formula successively for $k = j+1,\ j+2, \dots, n$ and obtain the following expression for the coefficients $\varphi_j^r(z, t)$:

$$\begin{aligned}\varphi_j(z, t) &= T\text{-}\exp\left(\int_0^{t_n} C_k(z_{(n)}(z_{0,n}, \tau))d\tau\right)\\ &\quad\times T\text{-}\exp\left(\int_0^{t_{n-1}} C_{n-1}(z_{(n-1)}(z_{0,n-1}, \tau))d\tau\right)\\ &\quad\times\cdots\times T\text{-}\exp\left(\int_0^{t_{j+1}} C_{j+1}(z_{(j+1)}(z_{0,j+1}, \tau))d\tau\right)1_j, \qquad \text{(II.26)}\end{aligned}$$

where $\varphi_j = (\varphi_j^1, \dots, \varphi_j^n)$, 1_j is the vector with jth component 1 and other components zero, and the integrals are taken along the trajectories of the related vector fields. The lemma is proved. □

Note that (II.26) can also be rewritten in the form

$$\begin{aligned}\varphi_j(z, t) &= \exp t_n\left(C_n(z) + \sum_s d_{ns}(z)\frac{\partial}{\partial z_s}\right)\\ &= \exp t_{j+1}\left(C_{j+1}(z) + \sum_s d_{j+1,s}(z)\frac{\partial}{\partial z_s}\right)\cdot 1_j. \qquad \text{(II.27)}\end{aligned}$$

We have obtained the equation

$$-i\frac{\partial}{\partial t_j}U(t, \tau) = \sum_{r=1}^{n}\varphi_j^r(B, t)A_r e^{i\tau B}e^{it_nA_n}\dots e^{it_1A_1},$$

where the functions $\varphi_j^r(B, t)$ are given by (II.26) or (II.27).

It remains to permute A_r and the exponential $e^{i\tau B}$. We have

$$e^{i\tau_mB_m}\dots e^{i\tau_1B_1}A_r = \left[e^{i\tau_m \operatorname{ad} B_m}\dots e^{i\tau_1 \operatorname{ad} B_1}\right](A_r)\cdot e^{i\tau_mB_m}\dots e^{i\tau_1B_1},$$

so that our aim is to evaluate the first factor on the left-hand side.

However, this is quite simple. We have

$$\begin{aligned}\mathrm{ad}_{B_k}(A_r) &= [B_k, A_r] = \mathrm{id}_{rk}(B),\\ (\mathrm{ad}_B)^\alpha(A_r) &= 0 \quad \text{for} \quad |\alpha| > 1;\end{aligned}$$

here $\alpha(\alpha_1, \dots, \alpha_n)$ is a multi-index, $|\alpha| = \alpha_1 + \cdots + \alpha_m$,

$$(\mathrm{ad}_B)^\alpha = (\mathrm{ad}_{B_m})^{\alpha_m} \dots (\mathrm{ad}_{B_1})^{\alpha_1}.$$

Consequently, we can expand the exponential $e^{i\tau\,\mathrm{ad}_B}$ into the Taylor series, retaining only the first two terms:

$$e^{i\tau\,\mathrm{ad}_B}(A_r) = (1 + i\tau\,\mathrm{ad}_B)(A_r) = A_r - \sum_{l=1}^{m} \tau_l d_{rl}(B).$$

Finally, we get

$$\begin{aligned}-i\frac{\partial}{\partial t_j}U(t,\tau) &= \sum_{t=1}^{n} \varphi_j^r(B,t)\left(A_r - \sum_{l=1}^{m} \tau_l d_{rl}(B)\right)U(t,\tau)\\ &= \left[\sum_{r=1}^{n} \varphi_j^r(B,t)A_r - \sum_{r,l=1}^{m} \tau_l \varphi_j^r(B,t)d_{rl}(B)\right]U(t,\tau) \quad \text{(II.28)}\\ -i\frac{\partial}{\partial \tau_l}U(t,\tau) &= B_l U(t,\tau).\end{aligned}$$

(We tacitly assume that the functions $\varphi_j^r(z,t)$ are well-defined for all t and belong to the symbol space $\mathcal{F}_m$ for each t.)

Set

$$\sum_{r,l=1}^{m} \tau_l \varphi_j^r(z,t)d_{rl}(z) = \omega_j(z,t,\tau).$$

Then we have

$$-i\frac{\partial}{\partial t_j}U(t,\tau) = \left[\sum_{r=1}^{n} \varphi_j^r(B,t)A_r - \omega_j(B,t,\tau)\right]U(t,\tau).$$

Assume that the matrix $||\varphi_j^r(z,t)||$ is invertible for each (z,t), and the entries of the inverse $||\psi_j^r(z,t)||$ belong to the symbol space $\mathcal{F}_m$ for each t. Since

$$\chi\left(-i\frac{d}{d\tau}\right)U(t,\tau) = \chi(B)U(t,\tau)$$

for any symbol χ by virtue of the second equation in (II.28), it follows that

$$\begin{aligned}&-i\sum_{j=1}^{n} \psi_k^j\left(-i\overset{1}{\frac{\partial}{\partial\tau}}, \overset{2}{t}\right)\overset{1}{\frac{\partial}{\partial t_j}}U(t,\tau)\\ &= \sum_{j=1}^{n}\left[\sum_{r=1}^{n} \varphi_j^r(B,t)A_r\psi_k^j(B,t) - \omega_j(B,t,\tau)\psi_k^j(B,t)\right]U(t,\tau).\end{aligned}$$

Furthermore, permuting A_r and $\psi_k^j(B, t)$, we obtain the relation

$$Q_k\left(\overset{1}{t}, \overset{2}{\tau}, -i\overset{1}{\frac{\partial}{\partial t}}, -i\overset{1}{\frac{\partial}{\partial \tau}}\right) U(t, \tau) = A_k U(t, \tau),$$

where

$$Q_k(t, \tau, y, z) = \sum_{j=1}^{n} \psi_k^j(z, t) y_j + \mu_k(z, t, \tau).$$

and the functions $\mu_k(z, t, \tau)$ are given by

$$\mu_k(z, t, \tau) = \sum_{j=1}^{n} \left(i \sum_{r=1}^{n} \sum_{l=1}^{m} \varphi_j^r(z, t) d_{rl}(z) \frac{\partial \psi_k^j}{\partial z_l}(z, t) - \omega_j(z, t, \tau) \psi_k^j(z, t) \right).$$

Thus we obtain the formulas for the left regular representation of the relations (II.21):

$$l_{B_s} = z_s, \quad s = 1, \dots, m,$$

$$l_{A_k} = \sum_{j=1}^{n} \psi_k^j \left(\overset{2}{z}, -i\overset{1}{\frac{\partial}{\partial y}}\right) \overset{2}{y}_j + \mu_k \left(\overset{2}{z}, -i\overset{1}{\frac{\partial}{\partial y}}, -i\overset{1}{\frac{\partial}{\partial z}}\right),$$

where the functions $\psi_k^j(z, t)$ and $\mu_k(z, t, \tau)$ were defined above.

Note that, in analogy with the case of Lie algebras, the left ordered representation operators are linear in y, but no longer homogeneous: there appears a constant term.

As with Lie algebras, the condition that the matrix $||\varphi_j^r(z, t)||$ be "good" (i.e., everywhere defined and invertible) can be guaranteed if we assume that the matrices $C_k(z)$ are lower-triangular and the derivatives of functions $d_{ks}(z)$ are bounded. We encourage the reader to check these facts for himself.

4 The Jacobi Condition and Poincaré–Birkhoff–Witt Theorem

The examples given in Section 2 and somewhat more general computation methods presented in Section 3 show that it is not the specific form of the operators $A_1, \dots, A_n$ that is essential to constructing the ordered representation. In fact, we use only the relations satisfied by these operators, and it is quite evident that in the lack of such relations there would be no regular representation at all. Thus, the ordered representation actually "represents" the relations that exist between the operators, which motivates the definitions given in the following.

4.1 Ordered Representation of Relation Systems and the Jacobi Condition

First, let us fix a class $\mathcal{F}$ of unary symbols and the corresponding classes $\mathcal{F}_k = \mathcal{F}^{\hat{\otimes} k}$ of k-ary symbols to work with. We assume that $A_1, \dots, A_n$ are $\mathcal{F}$-generators in some operator algebra $\mathcal{A}$. A *relation* between $A_1, \dots, A_n$ is an equation of the form

$$\omega(A) \equiv \omega(\overset{i_1}{A}_{j_1}, \dots, \overset{i_r}{A}_{j_r}) = 0, \tag{II.29}$$

where $\omega \in \mathcal{F}_r$, and the sequences $(j_1, \dots, j_r)$ and $(i_1, \dots, i_r)$ satisfy the conditions:

$$j_s \in \{1, \dots, n\}, \quad s = 1, \dots, r$$

(it is not prohibited that some of the j_s coincide);

$$i_l \neq i_k \quad \text{whenever} \quad [A_{i_l}, A_{i_k}] \neq 0,$$

so that the left-hand side of (II.29) makes sense.

It is easy to see that the relations considered in the preceding sections all fall under our definition.

For example, the commutation relation

$$\left[-i\frac{\partial}{\partial x}, x\right] = -iI$$

can be written in the form

$$\omega\left(\overset{2}{x}, -i\overset{1}{\frac{\partial}{\partial x}}, -i\overset{3}{\frac{\partial}{\partial x}}\right) = 0,$$

where

$$\omega(y_1, y_2, y_3) = y_2(y_3 - y_1) + i.$$

Let Ω be a (possibly, infinite) set of relations of the form (II.29). Strictly speaking, such a relation is determined by a triple $(\omega, \{i_s\}, \{j_s\})$, where ω is an r-ary symbol for some r and $\{i_s\}$ and $\{j_s\}$ are two sequences satisfying the cited conditions. However, for brevity we will simply write $\omega \in \Omega$; this cannot lead to misunderstanding, since the sequences $\{i_s\}$ and $\{j_s\}$ will be supplied by the context where necessary.

Definition II.4 A tuple $l = (\overset{1}{l}_1, \dots, \overset{n}{l}_n)$ of continuous linear operators on the space $\mathcal{F}_q$ of n-ary symbols is called a *left ordered representation of the system* Ω if the following conditions hold:

i) for any Feynman tuple $A = (\overset{1}{A}_1, \dots, \overset{n}{A}_n)$ of $\mathcal{F}$-generators satisfying all the relations $\omega \in \Omega$ and any $f \in \mathcal{F}_n$ we have

$$A_j f(A) = (l_j f)(A), \quad j = 1, \dots, n$$

ii) if $f \in \mathcal{F}_n$ is a symbol independent of $x_{j+1}, \ldots, x_n$, then

$$(l_j f)(y) = y_j\, f(y_1, \ldots, y_j).$$

In other words, we require that l be a left ordered representation for any tuple $A = (\overset{1}{A_1}, \ldots, \overset{n}{A_n})$ of $\mathcal{F}$-generators satisfying Ω (see Definition II.1 above). One can easily give the definition of the right ordered representation by himself or herself.

Definition II.5 We say that Ω is a system of permutation relations if it has a left regular representation.

Thus, Ω is a system of permutation relations if one can always move $\overset{n+1}{A_j}$ into the jth place in the expression $\overset{n+1}{A_j} f(\overset{1}{A_1}, \ldots, \overset{n}{A_n})$.

Let Ω be a system of permutation relations and l a left ordered representation of Ω. Suppose that each of the operators l_j is an $\mathcal{F}$-generator in $\mathcal{F}_n$. Then, according to Section 1, we can make $\mathcal{F}_n$ into an algebra by defining the bilinear operation (twisted product)

$$\begin{aligned} * : \mathcal{F}_n \times \mathcal{F}_n &\to \mathcal{F}_n, \\ (f, g) &\mapsto f * g = f(L)(g). \end{aligned}$$

The algebra $(\mathcal{F}_n, *)$ (which will be denoted by $\mathcal{F}_n$ for brevity) is not necessarily associative. However, $\mathcal{F}_n$ contains the identity. It is given by the function identically equal to one.

Indeed, $f * 1 = f$ for any f, and the equation $1 * f = 1(L)(f) = f$ is trivial. We see that 1 is the two-sided identity in $\mathcal{F}_n$.

If $A = (\overset{1}{A_1}, \ldots, \overset{n}{A_n})$ is a Feynman tuple of $\mathcal{F}$-generators in an arbitrary operator algebra $\mathcal{A}$ with left ordered representation $L = (L_1, \ldots, L_n)$, then, as shown in Section 1 (see Theorem II.1 above) the mapping

$$\begin{aligned} \mu_A : \quad & \mathcal{F}_k \to \mathcal{A}, \\ & f \mapsto \mu_A(f) = f(A) \end{aligned}$$

is an algebra homomorphism,

$$(f * g)(A) = f(A)\, g(A).$$

In practice, this implies that one can study the twisted multiplication in $\mathcal{F}_n$ instead of the operator multiplication in $\mathcal{A}$ as long as functions of A are considered. Clearly, the mapping μ_A determines an inclusion of the quotient algebra $\mathcal{F}_n / \operatorname{Ker} \mu_A$, (where $\operatorname{Ker} \mu_A$, the kernel of μ_A, is the set of symbols taken by μ_A into zero) in $\mathcal{A}$. Since $\mathcal{A}$

is an associative algebra, it follows that $\mathcal{F}_n/\operatorname{Ker}\mu_A$ is associative as well. Moreover, consider all possible tuples A of $\mathcal{F}$-generators satisfying Ω and set

$$\mathcal{J}_\Omega \stackrel{\text{def}}{=} \bigcap_A \operatorname{Ker}\mu_A,$$

where the intersection is taken over all such tuples. Since each $\operatorname{Ker}\mu_A$ is a two-sided ideal in $\mathcal{F}_n$, the same is true of $\mathcal{J}_\Omega$. By construction, $f(A) = 0$ for $f \in \mathcal{J}_\Omega$ whenever A is a Feynman tuple of $\mathcal{F}$-generators satisfying Ω. Moreover, the algebra $\mathcal{F}_n/\mathcal{J}_\Omega$ is associative. Indeed,

$$\mu_A(f * (g * h) - (f * g) * h) = 0$$

for any tuple A satisfying Ω. Hence

$$f * (g * h) - (f * g) * h \in \mathcal{J}_\Omega,$$

which implies the associativity of $\mathcal{F}_n/\mathcal{J}_\Omega$.

Actually, $\mathcal{F}_n/\mathcal{J}_\Omega$ is a "natural" symbol space for functions of operators satisfying Ω. The simplest (and most important) is the case in which $\mathcal{J}_\Omega = \{0\}$. It turns out that there is a simple criterion for the triviality of $\mathcal{J}_\Omega$ in terms of the left ordered representation operators.

Theorem II.10 *The following conditions are equivalent:*

(i) $\mathcal{J}_\Omega = \{0\}$.

(ii) *The left ordered representation operators $l_1, \dots, l_n$ of system Ω themselves satisfy the system Ω.*

Under either of these conditions the operators $l_1, \dots, l_n$ are uniquely determined and $(\mathcal{F}_n, *)$ is an associative algebra.

Recall that $l_1, \dots, l_n$ are assumed to be $\mathcal{F}$-generators.

Condition (ii) will be referred to as the *(generalized) Jacobi condition*. It was originally introduced by V. Maslov [130], who also suggested to call the associative algebra $(\mathcal{F}_n, *)$ a *hypergroup*. Theorem II.10 was stated and proved in [130] in somewhat different form.

Proof. (ii) $\Rightarrow$ (i). If the tuple $l = (\overset{1}{l_1}, \dots, \overset{n}{l_n})$ satisfies Ω, then $A = l$ is one of the tuples occurring in the intersection defining J_Ω and so we have $\mathcal{J}_\Omega \subset \operatorname{Ker}\mu_l$. But $\operatorname{Ker}\mu_l = 0$ by Lemma II.1, and we arrive at (i).

(i) $\Rightarrow$ (ii). We prove this implication by *reductio ad absurdum*.

Suppose that

$$\omega(l) \stackrel{\text{def}}{=} \omega(\overset{i_1}{l}_{j_1}, \dots, \overset{i_r}{l}_{j_r}) \neq 0$$

for some $\omega \in \Omega$. Then there exists a symbol $f \in \mathcal{F}_n$ such that

$$\varphi = \omega(l) f \neq 0.$$

Now let $A = (\overset{1}{A_1}, \dots, \overset{n}{A_n})$ be an arbitrary Feynman tuple satisfying Ω. Then,

$$\varphi(A) = (\omega(l)f)(A) = \omega(A)\, f(A) = 0,$$

since the first factor is zero. If follows that $\varphi \in \mathcal{J}_\Omega$ and hence $\mathcal{J}_\Omega \neq \{0\}$, which contradicts (i).

The associativity of $(\mathcal{F}_n, *)$ under condition (i) is clear, since in that case we have

$$\mathcal{F}_n = \mathcal{F}_n / \mathcal{J}_\Omega,$$

and the latter algebra is associative.

Let us prove the uniqueness of l. Suppose that there exist operators $l_1', \dots, l_n'$ with the same properties as $l_1, \dots, l_n$. For any Feynman tuple $A = (\overset{1}{A_1}, \dots, \overset{n}{A_n})$ satisfying Ω we have

$$\varphi_j(A) \overset{\text{def}}{=} (l_j' f)(A) - (l_j f)(A) = A_j f(A) - A_j f(A) = 0$$

for any $f \in \mathcal{F}_n$.

Set $A = l$ in the last equation. Then we obtain $\varphi_j(l) = 0$, and hence

$$(l_j' f)(y) - (l_j f)(y) = \varphi_j(y) = \varphi_j(l)1 = 0.$$

Since $f \in \mathcal{F}_n$ is arbitrary, it follows that $l_j' = l_j$. The theorem is proved. □

In fact, we can prove even more. The statement that follows can best be expressed in the language of category theory, and we do just that. The reader unfamiliar with this language can skip the following theorem without hesitation since it is not used directly in the remaining part of the book.

Consider the category $\mathrm{Alg}(\Omega)$ whose objects are tuples $(\mathcal{A}, A_1, \dots, A_n)$, where $\mathcal{A}$ is an operator algebra and $A_1, \dots, A_n \in \mathcal{A}$ are $\mathcal{F}$-generators satisfying the system Ω. The morphisms in $\mathrm{Alg}(\Omega)$ are continuous algebra homomorphisms taking the distinguished elements into the corresponding distinguished ones, that is, if $(\mathcal{A}, A_1, \dots, A_n)$ and $(\mathcal{B}, B_1, \dots, B_n)$ are elements of $\mathrm{Ob}(\mathrm{Alg}(\Omega))$, then a morphism

$$\psi : (\mathcal{A}, A_1, \dots, A_n) \to (\mathcal{B}, B_1, \dots, B_n)$$

is a continuous algebra morphism

$$\psi : \mathcal{A} \to \mathcal{B}$$

such that

$$\psi(A_i) = B_i, \quad i = 1, 2, \dots, n.$$

We can ask ourselves whether there exists a universal object in $\mathrm{Alg}(\Omega)$, i.e., an algebra $\mathcal{A}^{(0)}$ with elements $A_1^{(0)}, \dots, A_n^{(0)} \in \mathcal{A}$ that are $\mathcal{F}$-generators and satisfy Ω, and moreover, for any $(\mathcal{A}, A_1, \dots, A_n) \in \mathrm{Ob}(\mathrm{Alg}(\Omega))$ there exists a unique continuous morphism $\psi : \mathcal{A}^{(0)} \to \mathcal{A}$ taking each $A_j^{(0)}$ into A_j. The answer is given by the following theorem.

Theorem II.11 *Suppose that the system* Ω *has left ordered representation* $(l_1, \dots, l_n)$, *and moreover, that* $(A_1, \dots, A_n)$ *are* $\mathcal{F}$*-generators and satisfy* Ω. *Then the algebra* $(\mathcal{F}_n, *)$, *where* $*$ *is the twisted product with the elements*

$$A_1^{(0)} = y_1, \dots, A_n^{(0)} = y_n$$

is the universal object in the category $\mathrm{Alg}(\Omega)$.

Proof. By Theorem II.10, the algebra $(\mathcal{F}_n, *)$ is associative. Next, let us assume $(\mathcal{A}, A_1, \dots, A_n) \in \mathrm{Ob}(\mathrm{Alg}(\Omega))$. The mapping

$$\mu_A : \quad \mathcal{F}_n \to \mathcal{A}, \quad f \mapsto f(A)$$

is an algebra homomorphism, and we have

$$\mu_A(y_j) = A_j, \quad j = 1, \dots, n.$$

Since all A_j are $\mathcal{F}$-generators, it follows from Theorem I.1 that the mapping μ_A is uniquely determined by the last property. The theorem is thereby proved. □

We can give another interpretation to this theorem by noting that the algebra $(\mathcal{F}_n, *)$ is isomorphic to the subalgebra of $\mathcal{L}(\mathcal{F}_n)$ consisting of elements of the form $f(\overset{1}{l_1}, \dots, \overset{n}{l_n})$. Denoting this subalgebra by $\tilde{\mathcal{L}}(\mathcal{F}_n)$, we can state Theorem II.11 as follows:

The algebra $\tilde{\mathcal{L}}(\mathcal{F}_n)$ *with distinguished elements* $l_1, \dots, l_n$ *is the universal object in the category* $\mathrm{Alg}(\Omega)$.

In fact the last theorem deals with a far-reaching generalization of the notion of an algebra determined by generators and relations. Indeed, if $\mathcal{F}$ is the set of polynomials, then all relations in Ω are polynomial and the universal object $\mathcal{U} \in \mathrm{Alg}(\Omega)$ is just the algebra determined by the generators $A_1, \dots, A_n$ and the relations $\omega \in \Omega$. In this situation Theorem II.11 says that each element of $\mathcal{U}$ can be represented as an ordered polynomial $p(\overset{1}{A_1}, \dots, \overset{n}{A_n})$ in the generators, and that such polynomials are linearly independent, i.e., $p(\overset{1}{A_1}, \dots, \overset{n}{A_n}) = 0$ implies $p(y_1, \dots, y_n) \equiv 0$. The latter property is known as the Poincaré–Birkhoff–Witt property, or PBW-property for short; this name takes its origin from the Poincaré–Birkhoff–Witt theorem valid for Lie algebras. Let us state this theorem and prove it using the above-stated Theorem II.10. In the course of the proof it will become clear why we use the term “Jacobi condition” for condition (ii) in Theorem II.10.

4.2 The Poincaré–Birkhoff–Witt Theorem

Let us proceed to exact statements. Since the Poincaré–Birkhoff–Witt theorem is an assertion about the enveloping algebra of a Lie algebra, we begin by introducing the latter notion.

Let L be a finite-dimensional Lie algebra determined by its structural constants c_{ij}^k in some linear basis $a_1, \dots, a_n$ of L:

$$[a_i, a_j] = \sum_{k=1}^{n} c_{ij}^k a_k, \quad i, j = 1, \dots, n; \tag{II.30}$$

here $[\cdot\,, \cdot]$ is the Lie bracket on L.

Definition II.6 The *enveloping algebra* of L is the associative algebra $U(L)$ with identity element, determined by the generators $a_1, \dots, a_n$ and the relations (II.30), where $[a_i, a_j]$ now stands for the commutator

$$[a_i, a_j] = a_i a_j - a_j a_i.$$

Thus, the enveloping algebra is the quotient of the free associative algebra generated by $a_1, \dots, a_n$ w.r.t. the minimal ideal containing all elements of the form $[a_i, a_j] - \sum c_{ij}^k a_k$. In other words, the elements of $U(L)$ are finite linear combinations of the products $a_{j_1} a_{j_2} \dots a_{j_k}$, and the factors in such products can be permuted according to (II.30).

Clearly, the linear span of ordered monomials

$$a_n^{\alpha_n} \dots a_1^{\alpha_1} = \overset{1}{a}{}_1^{\alpha_1} \dots \overset{n}{a}{}_n^{\alpha_n}, \quad |\alpha| \equiv \alpha_1 + \dots + \alpha_n = 0, 1, 2, \dots$$

coincides with $U(L)$. Indeed, any unordered monomial can be represented as

$$a_{j_1} a_{j_2} \dots a_{j_N} = P(\overset{1}{a}_1, \dots, \overset{n}{a}_n),$$

where $P(y_1, \dots, y_n)$ is the polynomial

$$P(y_1, \dots, y_n) = l_{j_1} l_{j_2} \dots l_{j_N}(1)$$

and $l_1, \dots, l_n$ are the operators of the left ordered representation constructed for the relations (II.30) in Section 2.

Theorem II.12 (Poincaré–Birkhoff–Witt) *The ordered monomials form a basis in the linear space $U(L)$.*

Remark II.5 Here "basis" means "Hamel basis", i.e., we only consider finite linear combinations.

Proof. Clearly, we only have to show that these monomials are linearly independent. To do this we need the following assertion.

Lemma II.4 *The operators $l_1, \dots, l_n$ satisfy the generalized Jacobi condition, i.e.,*

$$[l_i, l_j] = \sum_k c_{ij}^k l_k, \quad i, j = 1, \dots, n.$$

The proof will be given later in this section; assuming that the lemma is valid, we proceed with the proof of Theorem II.12. Let V be an associative algebra and $A_1, \dots, A_n \in V$ be elements satisfying the relations

$$[A_i, A_j] = \sum_k c_{ij}^k A_k. \tag{II.31}$$

Consider the linear mapping $\tau : U(L) \to V$ given by the formula

$$\tau(a_{j_1} a_{j_2} \dots a_{j_N}) = A_{j_1} A_{j_2} \dots A_{j_N}$$

for any sequence of indices $j_1, \dots, j_N \in \{1, \dots, N\}$. This is well-defined since all relations between $a_1, \dots, a_n$ are corollaries of (II.30) and the operators $A_1, \dots, A_n$ satisfy the same relations[1]. Furthermore, τ is an algebra homomorphism , which can be observed from the fact that

$$\tau(a_{j_1} \dots a_{j_s}) \tau(a_{j_{s+1}} \dots a_{j_N}) = \tau(a_{j_1} \dots a_{j_s} a_{j_{s+1}} \dots a_{j_N}).$$

Suppose that the ordered monomials are linearly dependent, i.e., there exists a nonzero polynomial $P(y_1, \dots, y_n)$ such that $P(\overset{1}{a}_1, \dots, \overset{n}{a}_n) = 0$. But then

$$P(\overset{1}{A}_1, \dots, \overset{n}{A}_n) = P(\tau(\overset{1}{a}_1), \dots, \tau(\overset{n}{a}_n)) = \tau(P(\overset{1}{a}_1, \dots, \overset{n}{a}_n)) = 0.$$

We point out that this conclusion holds for an arbitrary Feynman tuple $(\overset{1}{A}_1, \dots, \overset{n}{A}_n)$ satisfying the relations (II.31). Hence we obtain

$$P \in \bigcap_A \operatorname{Ker} \mu_A = \mathcal{J}_\Omega,$$

where Ω is the set of relations (II.31) and the intersection is taken over all tuples A satisfying Ω. Taking into account the result of Lemma II.4, we can apply Theorem II.10, which asserts that $\mathcal{J}_\Omega = \{0\}$. Thus $P(y) \equiv 0$, which contradicts our assumption. The theorem is thereby proved. □

[1]This conclusion reflects the fact that $U(l)$ is universal with respect to homomorphisms of l into associative algebras.

Proof of Lemma II.4. We proceed by straightforward computation. The operators l_j were computed in Section 2. Recall that they have the form

$$l_j = \sum_q \overset{2}{y}_q B_{jq} \left(\overset{1}{\frac{\partial}{\partial y}} \right), \quad j = 1, \dots, n,$$

where $B(t) = A^{-1}(t)$ and the matrix $A(t)$ is given by

$$A_{qj}(t) = \begin{cases} (e^{t_n C_n} \dots e^{t_{q+1} C_{q+1}})_{jq}, & q < n, \\ \delta_{kj} & q = n. \end{cases}$$

Here δ_{kj} is the Kronecker delta and C_i are the matrices of structural constants,

$$(C_i)_{kj} = c_{ij}^k.$$

The structural constants c_{ij}^k cannot be arbitrary; they satisfy a number of relations [184]. Specifically,

$$c_{ij}^k + c_{ji}^k = 0, \quad i, j, k = 1, \dots, n$$

(antisymmetry), and

$$\sum_k \{c_{ij}^k c_{kl}^s + c_{jl}^k c_{ki}^s + c_{li}^k c_{kj}^s\} = 0, \quad i, j, l, s = 1, \dots, n$$

(the Jacobi identities for structural constants). The latter equations can be rewritten in terms of the matrices C_q, namely,

$$[C_i, C_j] = \sum_k c_{ij}^k C_k. \tag{II.32}$$

Thus we see that the matrices C_q themselves satisfy the relations defining L (in fact, these matrices define the so-called *adjoint representation* [184] of L in the basis $(a_1, \dots, a_n)$).

To compute the commutator $[l_i, l_j]$ we use the right ordered representation of the tuple $\left(\overset{2}{y}, \overset{1}{\partial/\partial y}\right)$. It was computed in Section 2 and has the form

$$r_{\partial/\partial y} = p, \quad r_y = q + \frac{\partial}{\partial p}.$$

We compute the products $l_i l_j$ and $l_j l_i$ according to Theorem II.1 and obtain

$$[l_i, l_j] = H\left(\overset{2}{y}, \overset{1}{\frac{\partial}{\partial y}}\right),$$

where

$$
\begin{aligned}
H(q,p) &= \sum_{k,s}\left[B_{js}(p)\left(q_s+\frac{\partial}{\partial p_s}\right)B_{ik}(p)q_kB_{is}(p)\left(q_s+\frac{\partial}{\partial p_s}\right)B_{jk}(p)q_k\right]\\
&= \sum_{k,s} q_k\left(\frac{\partial B_{ik}}{\partial p_s}B_{js}-\frac{\partial B_{jk}}{\partial p_s}B_{is}\right).
\end{aligned}
$$

In order to prove the lemma it suffices to check that

$$
\sum_s\left(\frac{\partial B_{ik}}{\partial p_s}B_{js}-\frac{\partial B_{jk}}{\partial p_s}B_{is}\right)=\sum_l c^l_{ij}B_{lk}. \tag{II.33}
$$

Multiplying the last identity by A_{kr} and summing over q we obtain[2]

$$
c^r_{ij}=\sum_{s,k}(B_{is}B_{jk}-B_{js})\frac{\partial A_{kr}}{\partial p_s}.
$$

Let us now multiply this by $A_{mj}A_{li}$ and sum over i and j. This yields

$$
\sum_{i,j} c^r_{ij}A_{mj}A_{li}=\frac{\partial A_{mr}}{\partial p_l}-\frac{\partial A_{lr}}{\partial p_m}. \tag{II.34}
$$

To compute the derivatives on the right-hand side of this equation, we make use of (II.32). We obtain

$$
\frac{\partial}{\partial p_k}(e^{p_nC_n}\dots e^{p_1C_1})=\sum_j A_{kj}C_je^{p_nC_n}\dots e^{p_1C_1}.
$$

Hence it follows that

$$
\begin{aligned}
\frac{\partial A_{mr}}{\partial p_l} &= \frac{\partial}{\partial p_l}(e^{p_nC_n}\dots e^{p_{m+1}C_{m+1}})_{rm}\\
&= \begin{cases} 0, & \text{if } l\le m,\\ \left[\sum_j A_{lj}C_je^{p_nC_n}\dots e^{p_{m+1}C_{m+1}}\right]_{rm} \sum_{j,s} A_{lj}(C_j)_{rs}A_{ms}, & \text{if } l>m,\end{cases}
\end{aligned}
$$

or

$$
\frac{\partial A_{mr}}{\partial p_l}=\begin{cases} 0, & \text{if } l\le m,\\ \sum_{i,j} c^r_{ij}A_{li}A_{mj}, & \text{if } l>m.\end{cases}
$$

---

[2] In doing so we have used the relation

$$
\sum_k\left(\frac{\partial B_{ik}}{\partial p_s}A_{kr}+B_{jk}\frac{\partial A_{kr}}{\partial p_s}B_{is}\right)=\frac{\partial}{\partial p_s}(\delta_{ir})=0.
$$

Thus we have proved (II.34) for $l > m$ (and, by antisymmetry, for $m < l$). If $m = l$, then both sides of (II.34) are zero, so that everything is all right in this case, too. It remains to note that the passage from (II.33) to (II.34) was an equivalence. The lemma is proved.

Remark II.6 The proof of Lemma II.4 could be substantially shorter if we used some standard facts from the theory of Lie algebras. However, we have preferred to give a direct proof since it shows with full clarity that there is a direct algebraic relationship between the Jacobi condition for the operators $l_1, \dots, l_n$ of the left regular representation and the Jacobi condition for the structural constants of the Lie algebra. It is because of this relationship that the term "generalized Jacobi condition" has been introduced.

4.3 Verification of the Jacobi Condition: Two Examples

In the preceding subsection we have checked the generalized Jacobi condition for commutation relations. Here we accomplish such a verification for the examples given in Subsections 2.2, 2.3, and 2.5.

The perturbed Heisenberg relations. The left ordered representation constructed in Subsection 2.2 for the Feynman tuple (A, B) satisfying the relations $BA - \alpha AB = i$ has the form

$$l_A = x, \quad l_B = yI_\alpha + \frac{i}{(1-\alpha)x}(1 - I_\alpha),$$

where I_α is the dilatation operator, $I_\alpha f(x, y) = f(\alpha x, y)$. Let us check that

$$l_B l_A - \alpha l_A l_B = i.$$

The variable y can be considered as a parameter. Therefore, we can assume without loss of generality that our symbols depend only on x. Let $f(x)$ be an arbitrary symbol. We have

$$\begin{aligned} l_B l_A f = l_B(xf) &= \left(yI_\alpha + \frac{i}{(1-\alpha)x}(1 - I_\alpha) \right)(xf(x)) \\ &= \alpha yxf(\alpha x) + \frac{ix(f(x) - \alpha f(\alpha x))}{(1-\alpha)x}, \end{aligned}$$

or, on cancelling out the factor x,

$$l_B l_A f(x) = \alpha yxf(\alpha x) + i\frac{f(x) - \alpha f(\alpha x)}{1-\alpha}.$$

Furthermore,

$$l_A l_B f(x) = xl_B(f(x)) = xyf(\alpha x) + i\frac{f(x) - f(\alpha x)}{1-\alpha}.$$

We obtain, by combining the last two inequalities,

$$\begin{aligned}(l_B l_A - \alpha l_A l_B) f(x) &= \alpha x y f(\alpha x) + i\frac{f(x) - \alpha f(\alpha x)}{1-\alpha} \\ &\quad -\alpha\left(xyf(\alpha x) + i\frac{f(x) - f(\alpha x)}{1-\alpha}\right) = if(x),\end{aligned}$$

that is, the operators l_A and l_B satisfy the same perturbed Heisenberg relation as the operators A and B themselves.

Thus we may conclude that the twisted product

$$(f * g)(x, y) \stackrel{\text{def}}{=} f(\overset{2}{l}_A, \overset{1}{l}_B)(g(x, y))$$

defines the structure of an associative algebra on the symbol space $\mathcal{F}_2$, or, in other words, functions of A and B form a hypergroup.

Of course, this argument applies to the case $\alpha = -1$ as well, so that for the simple example of a graded Lie algebra considered in Subsection 2.5 we can also claim that the functions of A and B form a hypergroup. However, the situation is quite different as soon as a closed system of graded Lie relations defining a finite-dimensional graded Lie algebra is considered. For example, consider the relations

$$\begin{aligned}[A, B]_+ &\equiv AB + BA = 1, \\ [A, A]_+ &\equiv 2A^2 = 0, \\ [B, B]_+ &\equiv 2B^2 = 0.\end{aligned}$$

The last two relations are not needed in calculating the ordered representation operators; nor do l_A and l_B satisfy $l_A^2 = l_B^2 = 0$, and we see that the generalized Jacobi condition, together with the Poincaré–Birkhoff–Witt property is lost. The remedy may be to consider symbols as functions of commuting and anticommuting arguments (see, e.g., [12] and [110]).

The nonlinear commutation relations of Subsection 2.3. Let us check that the operators l_A, l_B, and l_C, constructed there satisfy the relations

$$\begin{aligned}&[l_A, l_B] = 1, \quad [l_A, l_C] = \alpha(l_C)L_A + \beta(l_C)l_B + \gamma(l_C), \\ &[l_B, l_C] = \beta^*(l_C)l_A + \delta(l_C)l_B + \sigma(l_C).\end{aligned}$$

We have

$$\begin{aligned}\begin{pmatrix} l_A \\ l_B \end{pmatrix} l_C \varphi(x, y, z) &= \left[I_{S(z)} \begin{pmatrix} x + \frac{\partial}{\partial y} \\ y \end{pmatrix} + \tfrac{1 - I_{S(z)}}{z - S(z)} \begin{pmatrix} \gamma(z) \\ \sigma(z) \end{pmatrix} \right] - l_C \varphi(x, y, z) \\ &= S(l_C) I_{S(z)} \begin{pmatrix} x + \frac{\partial}{\partial y} \\ y \end{pmatrix} \varphi(x, y, z) + \tfrac{z - S(z) I_{S(z)}}{z - S(z)} \begin{pmatrix} \gamma(z) \\ \sigma(z) \end{pmatrix} \varphi(x, y, z).\end{aligned}$$

By adding and subtracting $S(z)$ in the numerator, we obtain

$$\begin{pmatrix} l_A \\ l_B \end{pmatrix} l_C = S(l_C) \begin{pmatrix} l_A \\ l_B \end{pmatrix} + \begin{pmatrix} \gamma(l_C) \\ \sigma(l_C). \end{pmatrix}.$$

It remains to check that $[l_A, l_B] = 1$.

Exercise II.1 Check that $[l_A, l_B] = 1$ if and only if

$$\operatorname{tr}(S(z) - z) \equiv 0.$$

Thus, the last condition guarantees that functions of $(\overset{1}{A}, \overset{2}{B}, \overset{3}{C})$ form a hypergroup.

5 The Ordered Representations, Jacobi Condition, and the Yang–Baxter Equation

In this section we clarify the relationship between the generalized Jacobi condition for the ordered representation operators introduced in the preceding section and the Yang–Baxter equation playing an important role, e.g. in the theory of quantum groups. In Section 4 we studied the classical Poincaré–Birkhoff–Witt theorem. Let us recall its statement. The theorem claims that if L is a Lie algebra with basis $A_1, \ldots, A_n$ and U the enveloping algebra of L, than the ordered monomials

$$A_n^{\alpha_n}, \ldots, A_1^{\alpha_1}, (\alpha_1, \ldots, \alpha_n) \in (\mathbb{Z}_+ \cup \{0\})^n$$

form a basis in U. This result can be viewed as follows. The algebra U is generated by generators $A_1, \ldots, A_n$ and Lie relations between these generators. The Poincaré–Birkhoff–Witt theorem provides a linear basis in U constructed from $A_1, \ldots, A_n$.

Here we discuss the same problem for general algebras defined by generators and relations.

As we have learned in the above exposition, the main problem which is solved by introducing the ordered representation operators is as follows. Suppose that we are given an algebra $\mathcal{A}$, generated by the operators $A_1, \ldots, A_n$ and the relations

$$\{\omega(\overset{i_1}{A}_{j_1}, \ldots, \overset{i_s}{A}_{j_s}) = 0\}_{\omega \in \Omega} \tag{II.35}$$

(for simplicity, we assume that all symbols $\omega(y_1, \ldots, y_s)$ are polynomials; the number s of arguments of this polynomial is allowed to depend on ω). We are interested in whether any element of $\mathcal{A}$ can be represented in the form $P(\overset{1}{A}_1, \ldots, \overset{n}{A}_n)$ for an appropriately chosen symbol $P(y_1, \ldots, y_n)$. It was shown that this is true if and only if there exist operators $l_1, \ldots, l_n$ in the space of n-ary symbols such that

$$A_j [\![f(\overset{1}{A}_1, \ldots, \overset{n}{A}_n)]\!] = (l_j f)(\overset{1}{A}_1, \ldots, \overset{n}{A}_n) \tag{II.36}$$

for any $j = 1, \ldots, n$ and any symbol $f(y_1, \ldots, y_n)$. With polynomial (or power series) symbols it is not hard to show that provided that the operators $l_1, \ldots, l_n$ exist at all, they can be chosen so that

$$f(\overset{1}{l_1}, \ldots, \overset{n}{l_n})(1) = f(y_1, \ldots, y_n) \tag{II.37}$$

for any symbol $f(y_1, \ldots, y_n)$.

Using the operators $l_1, \ldots, l_n$ we represent any element

$$\hat{\psi} = \psi(\overset{i_1}{A_{j_1}}, \ldots, \overset{i_k}{A_{j_k}}) \in \mathcal{A}$$

as an ordered polynomial,

$$\psi(\overset{i_1}{A_{j_1}}, \ldots, \overset{i_k}{A_{j_k}}) = f(\overset{1}{A_1}, \ldots, \overset{n}{A_n}), \tag{II.38}$$

where

$$f(y_1, \ldots, y_n) = \psi(\overset{i_1}{l_{j_1}}, \ldots, \overset{i_k}{l_{j_k}})(1), \tag{II.39}$$

i.e., we can bring the operator $\hat{\psi}$ into normal form. From this viewpoint, formula (II.37) means that the normal form of an operator already in normal form is just the operator itself.

We can arrange things in a slightly different way. Let $\mathcal{F}_n$ be the space of n-ary symbols. We have the linear mapping

$$\mu = \mu_{\overset{1}{A_1}, \ldots, \overset{n}{A_n}} : \mathcal{F}_n \to \mathcal{A}$$

$$f \mapsto f(\overset{1}{A_1}, \ldots, \overset{n}{A_n}), \tag{II.40}$$

whose image consists of operators in normal form. Since any operator $\hat{\psi} \in \mathcal{A}$ can be brought into normal form (see (II.38) — (II.39)), we see that μ is an epimorphism. Moreover, by (II.38) — (II.39) we have

$$[\![f(\overset{1}{A_1}, \ldots, \overset{n}{A_n})]\!] \, [\![g(\overset{1}{A_1}, \ldots, \overset{n}{A_n})]\!] = h(\overset{1}{A_1}, \ldots, \overset{n}{A_n}), \tag{II.41}$$

where the symbol $h(y_1, \ldots, y_n)$ of the product can be expressed by the formula

$$h(y_1, \ldots, y_n) = [\![f(\overset{1}{l_1}, \ldots, \overset{n}{l_n})]\!] \, [\![g(\overset{1}{l_1}, \ldots, \overset{n}{l_n})]\!](1). \tag{II.42}$$

Since, according to (II.37), we have

$$g(\overset{1}{l_1}, \ldots, \overset{n}{l_n})(1) = g(y_1, \ldots, y_n),$$

(II.42) reduces to

$$h(y_1, \dots, y_n) = f(\overset{1}{l_1}, \dots, \overset{n}{l_n})(g(y)). \tag{II.43}$$

Denote

$$f * g = f(\overset{1}{l_1}, \dots, \overset{n}{l_n})(g(y)). \tag{II.44}$$

This is a bilinear operation on $\mathcal{F}_n$ and the element $1 \in \mathcal{F}$ is the two-sided identity with respect to the operation $*$ (indeed, 1 is the right identity by (II.37) and the left identity since

$$1 * g = 1(\overset{1}{l_1}, \dots, \overset{n}{l_n})(g) = g. \tag{II.45}$$

The mapping (II.40) is an algebra homomorphism, and it is epimorphic. There are two important closely related questions.

(1) Is $(\mathcal{F}_n, *)$ an associative algebra (or, in Maslov's terminology [130], is $(\mathcal{F}_n, *)$ a hypergroup)?

(2) Is the mapping (II.40) a monomorphism?

The implication (II.37) $\Leftrightarrow$ (II.36) is easy, since if (II.40) is a monomorphism, then it is an isomorphism taking $(\mathcal{F}_n, *)$ into an associative algebra. So we can concentrate on condition (II.37). It turns out (see Theorem II.10) that the mapping (II.40) is an isomorphism if and only if the left ordered representation operators $l_1, \dots, l_n$ themselves satisfy relations (II.35), i.e.,

$$\omega(\overset{i_1}{l}_{j_1}, \dots, \overset{i_n}{l}_{j_n}) = 0 \tag{II.46}$$

for all $\omega \in \Omega$. This condition is called the *generalized Jacobi condition.* Hence, we have the equivalence

$$\text{Generalized Jacobi condition} \Rightarrow \mu \text{ is an isomorphism.} \tag{II.47}$$

Let us study the right-hand side of (II.47) in more detail. With polynomial symbols, it can be restated as follows:

The ordered monomials

$$A_n^{k_n} \dots A_1^{k_1} \tag{II.48}$$

form a (linear) basis in the algebra $\mathcal{A}$.

This property is known as the Poincaré–Birkhoff–Witt (PBW) property and the basis (II.48) as the PBW basis. One can even define a finer property as follows. Let $\mathcal{A}^{(j)} \subset \mathcal{A}$ be the subspace spanned by the unordered monomials of order $\leq j$. Clearly,

$$\mathcal{A}^{(0)} \subset \mathcal{A}^{(1)} \subset \dots \subset \mathcal{A}^{(j)} \subset \mathcal{A}^{(j+1)} \subset \dots \tag{II.49}$$

is an increasing filtration on $\mathcal{A}$.

Definition II.7 One says that the PBW property is satisfied in $\mathcal{A}$ up to order s if for any $k \leq s$ the unordered monomials (II.48) with $k_1 + \dots + k_n \leq k$ form a basis in the space $\mathcal{A}^{(k)}$.

The major advantage of the new definition is that the PBW property (= the PBW property up to order ∞ in terms of this definition) can be verified step-by-step, via, say, induction on k. However, this is not so easy, and in practice it is much simpler to invert the emphasis and to find some (necessary) conditions on the constants involved in the commutation relations (II.35) than to prove these conditions to be sufficient for the PBW property for any given order. Let us consider some examples.

Example II.2 Enveloping algebra of a Lie algebra. Let $\mathcal{A}$ be the algebra generated by $A_1, \ldots, A_n$ and the relations

$$[A_j, A_k] \stackrel{\text{def}}{=} A_j A_k - A_k A_j = \sum_l c^l_{jk} A_l. \tag{II.50}$$

Let us assume that $\mathcal{A}$ satisfies the PBW property and infer some consequences on the constants c^l_{jk}. This is quite easy. First of all we have

$$[A_j, A_k] + [A_k, A_j] = 0, \tag{II.51}$$

that is, according to (II.50),

$$\sum_l (c^l_{jk} + c^l_{kj})\, A_l = 0. \tag{II.52}$$

Assuming that the PBW property is valid for $\mathcal{A}$ up to order 1, that is, the A_l, $l = 1, \ldots, n$, are linearly independent, we readily obtain that

$$c^l_{jk} = -c^l_{kj}. \tag{II.53}$$

Furthermore, consider the Jacobi identity

$$[[A_j, A_k], A_l] + \text{c.p.} = 0, \tag{II.54}$$

where c.p. stands for the two summands obtained from $[[A_j, A_k], A_l]$ by cyclic permutations. Using (II.50) twice, we obtain

$$\sum_s c^s_{jk} c^r_{sl} A_r + \text{c.p.} = 0, \tag{II.55}$$

where c.p. denotes the terms obtained by cyclic permutations of the indices j, k, and l. Again using the linear independence of the A^s_j, we obtain

$$\sum_s (c^s_{jk} c^r_{sl} + c^s_{lj} c^r_{sk} + c^s_{kl} c^r_{sj}) = 0. \tag{II.56}$$

The identities (II.53) and (II.56) are known, respectively, as the *antisymmetry condition* and the *Jacobi condition* for the structural constants of a Lie algebra. Let us emphasize that they both follow from the assumed PBW property up to order 1 (!) in $\mathcal{A}$ (the algebra $\mathcal{A}$ is called the enveloping algebra of the Lie algebra determined by the structural constants c^s_{jk}). Let us also point out that it is not obvious at this stage how to infer the PBW property of $\mathcal{A}$ (at least for order 1) from these identities. Nevertheless, the PBW property holds in this situation to arbitrary order (the Poincaré–Birkhoff–Witt theorem). The standard "combinatorial" proof of this theorem can be found in any advanced textbook on Lie algebras. In this book we have given another proof. First, the left ordered representation of the relations (II.50) was constructed (see Section 4), and then it was proved that the identities (II.53) and (II.56) imply that the generalized Jacobi condition holds for the operators of the left ordered representation.

Example II.3 Consider the algebra $\mathcal{A}$ defined by generators A, B, and C and the relations

$$\begin{aligned} BA &= \alpha AB + \sigma C, \\ BC &= \gamma CB + \nu B, \\ CA &= \beta AC + \mu A, \end{aligned} \tag{II.57}$$

where α, γ, β, σ, ν, and μ are some prescribed constants. Let us assume the PBW property and infer some equations involving these constants.

First of all, it is easy to see that $\alpha\beta\gamma \neq 0$. Indeed, assume the opposite, say, $\alpha = 0$. Then

$$BA - \sigma C = 0, \tag{II.58}$$

that is, BA and C are linearly dependent and the PBW property up to order 2 fails.

Next, consider the monomial ABC and transform it into CBA using the relations (II.57) in two different ways.

Method 1. Interchange first B and C. We have

$$ABC = A(\gamma CB + \nu B) = \nu AB + \gamma ACB. \tag{II.59}$$

In the last term we interchange A with C and with B, thus obtaining

$$\begin{aligned} \nu AB + \gamma ACB &= \nu AB + \gamma\left(\frac{1}{\beta}CAB - \frac{\mu}{\beta}AB\right) \\ &= \left(\nu - \frac{\mu\gamma}{\beta}\right)AB + \frac{\gamma}{\beta\alpha}CBA - \frac{\gamma\sigma}{\beta\alpha}C^2. \end{aligned} \tag{II.60}$$

Method 2. Let us first interchange B and A in the product ABC. Then we have, by analogy with the above,

$$ABC = \frac{\gamma}{\alpha\beta}CBA + \frac{\gamma\mu}{\alpha\beta}BA - \frac{\sigma}{\alpha}C^2. \tag{II.61}$$

Comparing (II.60) and (II.61) we see that

$$\frac{\nu-\mu}{\beta}AB+\frac{\sigma(\nu-\mu)}{\alpha\beta}C-\frac{\sigma}{\alpha}C^2=\left(\nu-\frac{\mu\gamma}{\beta}\right)AB-\frac{\gamma\sigma}{\beta\alpha}C^2. \tag{II.62}$$

Assuming the PBW property up to order 2 we see that one should have

$$\frac{\sigma(\nu-\mu)}{\alpha\beta}=\frac{\sigma}{\alpha}\left(1-\frac{\gamma}{\beta}\right)=0$$

and

$$\nu-\mu=\nu\beta-\mu\gamma. \tag{II.63}$$

In other words, either

$$\sigma=0 \quad\text{and}\quad \nu(1-\beta)=\mu(1-\gamma) \tag{II.64}$$

or

$$\beta=\gamma \quad\text{and}\quad \nu=\mu. \tag{II.65}$$

Note that we have only proved that conditions (II.60) or (II.65) are *necessary* for the PBW property to be satisfied and not vice versa. To prove the converse statement, we again calculate the left ordered representation operators. This can easily be done by analogy with numerous other examples that can be found in the preceding sections. Choose the Feynman ordering $\overset{1}{C}, \overset{2}{B}, \overset{3}{A}$.

We omit the calculations and only present the final result. It reads

$$\begin{aligned} l_A &= y_s, \\ l_B &= \sigma y_1 I^{\alpha,\beta}_{y_3} I^{1,\gamma}_{y_2} + y_2\left(I^{\alpha}_{y_3} - \frac{\sigma\nu}{\gamma} I^{\alpha,\beta}_{y_3} I^{1/\gamma,1}_{y_2}\right) + \sigma\mu y_3 I^{\alpha,\beta,1}_{y_3}, \\ l_C &= y_1 I^{\beta}_{y_3} I^{1/\gamma}_{y_2} - \frac{\nu}{\gamma} I^{\beta}_{y_3} I^{1/\gamma,1}_{y_2} + \mu y_3 I^{\beta,1}_{y_3}, \end{aligned} \tag{II.66}$$

where the notation I^{α}_z, $I^{\alpha,\beta}_z$, and $I^{\alpha,\beta,\gamma}_z$ stands for the following combinations of dilatations and difference derivatives:

$$\begin{aligned} (I^{\alpha}_z f)(z) &= f(\alpha z), \\ (I^{\alpha,\beta}_z f)(z) &= \frac{\delta f}{\delta x}(\alpha z, \beta z), \\ (I^{\alpha,\beta,\gamma}_z f)(z) &= \frac{\delta^2 f}{\delta x^2}(\alpha z, \beta z, \gamma z). \end{aligned} \tag{II.67}$$

It is a routine but useful exercise to check that the operators (II.66) satisfy the original commutation relations (II.57) if and only if one of conditions (II.64) and (II.65) is satisfied.

Example II.4 (Graded algebras) Let G be a finite abelian group, and let

$$\mathcal{A} = \bigoplus_{g \in G} \mathcal{A}_g \tag{II.68}$$

be a G-graded algebra with 1. Recall that this means that the underlying linear space $\mathcal{A}$ is the direct sum of linear spaces $\mathcal{A}_g$, $g \in G$, and that

$$\mathcal{A}_g \mathcal{A}_h \subset \mathcal{A}_{g+h} \tag{II.69}$$

for any $g, h \in G$. In particular, it is assumed that $1 \in \mathcal{A}_0$, where $0 \in G$ is the neutral element of G.

Assume that $\mathcal{A}$ is generated (as an algebra) by a finite number of elements $A_1, \ldots,$ A_n. These elements are *homogeneous*, that is,

$$A_i \in \mathcal{A}_{g_i}, \quad i = 1, \ldots, n,$$

and that the following permutation relations are valid:

$$A_i A_j - \omega_{ij} A_j A_i = \sum_k \lambda_{ij}^k A_k \tag{II.70}$$

where ω_{ij} and λ_{ij}^k are some prescribed constants (for homogeneity reasons, the sum on the right-hand side of (II.70) extends only over k such that

$$g_k = g_i + g_j; \tag{II.71}$$

all the other λ_{ij}^k are zero). We shall assume that the ω_{ij} depend not on i and j themselves but rather on g_i and g_j:

$$\omega_{ij} = \omega(g_i, g_j). \tag{II.72}$$

The numbers ω_{ij} and λ_{ij}^k will be referred to as *the scale and structural constants* of the algebra $\mathcal{A}$, respectively. Assume that the PBW property is valid in $\mathcal{A}$ with respect to the tuple $A_1, \ldots, A_n$, and deduce some equations involving the scale and structural constants.

Let

$$\mathcal{A}^{(0)} \subset \mathcal{A}^{(1)} \subset \cdots \subset \mathcal{A}^{(s)} \subset \mathcal{A}^{(s+1)} \subset \cdots \tag{II.73}$$

be the filtration in $\mathcal{A}$ induced by the degree of unordered monomials in $A_1, \ldots, A_n$. Clearly, one has

$$\mathcal{A}^{(i)} = \bigoplus_{g \in G} \mathcal{A}_g^{(i)}, \tag{II.74}$$

where

$$\mathcal{A}_g^{(i)} = \mathcal{A}^{(i)} \cap \mathcal{A}_g. \tag{II.75}$$

Consider the associated graded algebra

$$\mathcal{B} = \bigoplus_{i=0}^{\infty} \mathcal{A}^{(i)} / \mathcal{A}^{(i-1)} \equiv \bigoplus_{i=0}^{\infty} \mathcal{B}^{(i)} \tag{II.76}$$

(here we set $\mathcal{A}^{(-1)} = \{0\}$). Then, clearly, each $\mathcal{B}^{(i)}$ is G-graded, that is,

$$\mathcal{B}^{(i)} = \bigoplus_{g \in G} \mathcal{B}_g^{(i)}, \tag{II.77}$$

and, moreover, $\mathcal{B}$ has the PBW property with respect to the tuple $B_1, \ldots, B_n$, where each B_i is the image of the corresponding A_i in $\mathcal{A}^{(1)}/\mathcal{A}^{(0)}$.

The operators $\mathcal{B}_i$ satisfy the "homogeneous" permutation relations that can be obtained from (II.70) by omitting the linear term, lost in the course of factorization:

$$B_i B_j = \omega(g_i, g_j) B_j B_i. \tag{II.78}$$

Consider the product $B_i B_j B_k$ and transform it into $B_k B_j B_i$ using the relations (II.78). We have

$$\begin{aligned} B_i B_j B_k &= \omega(g_i, g_j) B_j B_i B_k \\ &= \omega(g_i, g_j)\omega(g_i, g_k) B_j B_k B_i \\ &= \omega(g_i, g_j)\omega(g_j, g_k)\omega(g_i, g_k) B_k B_j B_i \end{aligned}$$

and the same factor results if we first interchange B_j and B_k. Hence, the PBW property imposes no restrictions on the constants $\omega(g_i, g_j)$ alone, except for

$$\omega(g_i, g_j) = \omega(g_j, g_i)^{-1} \tag{II.79}$$

In particular, if all the constants λ_{ij}^k are zero, then $\mathcal{B}$ is isomorphic to $\mathcal{A}$ and it is easy to see that the PBW property holds in $\mathcal{A}$ up to any order.

Let us now proceed to the general case of nonzero constants λ_{ij}^k. We assume that the scale constants $\omega(g, h)$ satisfy the following condition:

$$\omega(g, h + k) = \omega(g, h)\omega(g, k). \tag{II.80}$$

If we define a bracket in $\mathcal{A}$ by setting

$$[A_g, A_h] = A_g A_h - \omega(g, h) A_h A_g \tag{II.81}$$

for $A_g \in \mathcal{A}_g$ and $A_h \in \mathcal{A}_h$, then the Jacobi identity

$$[A_g, [A_h, A_k]] = [[A_g, A_h], A_k] + \omega(g, h)[A_h, [A_g, A_k]] \tag{II.82}$$

holds for any homogeneous elements $A_g \in \mathcal{A}_g$, $A_h \in \mathcal{A}_h$, and $A_k \in \mathcal{A}_k$. Applying the identity (II.82) to A_i, A_j, and A_k and taking into account that the third-order terms in this identity necessarily cancel out, we obtain

$$\begin{aligned} \sum_l \lambda_{ij}^l (A_i A_l - \omega(g_i, g_l) A_l A_i) &= \sum_l \lambda_{ij}^l (A_l A_k - \omega(g_l, g_k) A_k A_l) \\ &+ \omega(g_i, g_j) \sum_l \lambda_{ik}^l (A_j A_l - \omega(g_j, g_l) A_l A_j), \end{aligned} \tag{II.83}$$

or

$$\sum_{l,m}(\lambda^l_{jk}\lambda^m_{il} - \lambda^l_{ij}\lambda^m_{lk} - \omega(g_i, g_j)\lambda^l_{ik}\lambda^m_{jl})A_m = 0. \tag{II.84}$$

Assuming the PBW property of order 1, we obtain the following equation tying up the scale and structure constants of the graded algebra:

$$\sum_{l}(\lambda^l_{jk}\lambda^m_{il} - \lambda^l_{ij}\lambda^m_{lk} - \omega(g_i, g_j)\lambda^l_{ik}\lambda^m_{jl}) = 0 \tag{II.85}$$

for all $i, j, k, m = 1, \dots, n$.

It can be shown by constructing the ordered representation and computing the commutators that conditions (II.79), (II.80) and (II.85) guarantee that the PBW property holds in the algebra $\mathcal{A}$.

Example II.5 (The Faddeev–Zamolodchikov algebra) In the quantum inverse scattering problem method the key role is played by the Faddeev–Zamolodchikov algebra, defined as follows.

Let $R = (R^{\beta\beta'}_{\alpha\alpha'})$ be a tensor whose indices range from 1 to n. Consider the algebra $\mathcal{A}$ generated by the operators A^{β}_{α}, where $\alpha, \beta = 1, \dots, n$ and the relations

$$R^{\beta\beta'}_{\alpha\alpha'}A^{\gamma}_{\beta}A^{\gamma'}_{\beta'} = R^{\gamma\gamma'}_{\beta\beta'}A^{\beta'}_{\alpha'}A^{\beta}_{\alpha}, \tag{II.86}$$

where $\alpha, \alpha', \gamma, \gamma' = 1, \dots, n$. Such an algebra is called a *Faddeev–Zamolodchikov algebra*.

It can be proved that the PBW property up to degree 3 holds in a Faddeev–Zamolodchikov algebra $\mathcal{A}$ provided that the tensor R satisfies a certain equation known as the Yang–Baxter equation. This equation can be written out as follows. The tensor R can be viewed as an element of the space

$$R \in \mathrm{Mat}(n, \mathbb{C}) \otimes \mathrm{Mat}(n, \mathbb{C}), \tag{II.87}$$

$$R = \sum \sigma_s \otimes \nu_s, \tag{II.88}$$

where σ_s and ν_s are $(n \times n)$ matrices. Define the elements $R^{12}, R^{13}, R^{23} \in \mathrm{Mat}(n, \mathbb{C}) \otimes \mathrm{Mat}(n, \mathbb{C}) \otimes \mathrm{Mat}(n, \mathbb{C})$ by setting

$$\begin{aligned} R^{12} &= \sum \sigma_s \otimes \nu_s \otimes I, \\ R^{13} &= \sum \sigma_s \otimes I \otimes \nu_s, \\ R^{23} &= \sum I \otimes \sigma_s \otimes \nu_s, \end{aligned} \tag{II.89}$$

where I is the identity matrix.

The *Yang–Baxter equation* reads

$$R^{12}R^{13}R^{23} = R^{23}R^{13}R^{12}. \tag{II.90}$$

Lengthy but routine computations show that (II.90) guarantees the PBW property in the Faddeev–Zamolodchikov algebra up to degree 3. Unfortunately, the ordered representation operators cannot be written out in such a general setting, so that it remains unknown whether the PBW property holds up to arbitrary degree. However, in some particular cases the ordered representation can be computed explicitly.

From now on we use the term "Yang–Baxter equation" for any equations following from the PBW property for the constants in the commutation relations.

Example II.6 (The Sklyanin algebra) Consider the algebra $\mathcal{A}$ with generators A_0, A_1, A_2, and A_3 and the relations

$$\begin{aligned} [A_0, A_j] &= i(A_0A_j + A_jA_0), \\ [A_k, A_l] &= i\mathcal{J}_{kl}(A_kA_l + A_lA_k), \end{aligned} \tag{II.91}$$

valid for any cyclic permutation (i, j, k) of the indices $1, 2, 3$.

The constants $\mathcal{J}_{kl}$ occurring in equation (II.91) satisfy the Yang–Baxter equations provided that

$$\mathcal{J}_{kl} = \frac{\mathcal{J}_l - \mathcal{J}_k}{\mathcal{J}_i},$$

where all $\mathcal{J}_k$ are constants. Sklyanin [168] showed that the PBW property holds in $\mathcal{A}$. However, for general $\mathcal{J}_s$ the ordered representation is not known. Let us consider the particular case in which

$$\mathcal{J}_1 = \mathcal{J}_2 = 1 \quad \text{and} \quad \mathcal{J}_3 = 1 + \omega,$$

where for simplicity we assume $\omega \geq 0$ and $\omega \neq 1$.

The relations (II.91) take the form

$$\begin{aligned} [A_0, A_1] &= i\omega(A_2A_3 + A_3A_2), \\ [A_0, A_2] &= -i\omega(A_1A_3 + A_3A_1), \\ [A_0, A_3] &= 0, \\ [A_j, A_k] &= i(A_0A_l + A_lA_0), \end{aligned} \tag{II.92}$$

where (j, k, l) is an arbitrary cyclic permutation of the indices $(1, 2, 3)$. We choose in $\mathcal{A}$ another set of generators given by the formulas

$$\begin{aligned} a_\pm &= A_1 \pm A_2, \\ b_\pm &= \pm A_0 + \sqrt{\omega}A_3. \end{aligned} \tag{II.93}$$

These generators are linear combinations of the original generators. We choose the Feynman ordering $(\overset{1}{a}_-, \overset{2}{a}_+, \overset{3}{b}_-, \overset{4}{b}_+)$ and construct the regular representation for this ordering. For the new generators the relations (II.92) take the form

$$\begin{aligned} [b_+, a_\pm] &= \pm\sqrt{\omega}(b_+a_\pm + a_\pm b_+), \\ [b_-, a_\pm] &= \mp\sqrt{\omega}(b_-a_\pm + a_\pm b_-), \\ [b_+, b_-] &= 0, \\ [a_+, a_-] &= \sqrt{\omega}(b_+^2 - b_-^2). \end{aligned} \tag{II.94}$$

In a slightly different form, these relations can be rewritten as

$$\begin{aligned} b_+a_+ &= \alpha a_+b_+, & b_-a_+ &= \frac{1}{\alpha}a_+b_-, \\ b_+a_- &= \frac{1}{\alpha}a_-b_+, & b_-a_- &= \alpha a_-b_-, \\ [b_+, b_-] &= 0, & [a_+a_-] &= \frac{\alpha-1}{\alpha+1}(b_+^2 - b_-^2), \end{aligned} \tag{II.95}$$

where we denoted

$$\alpha = \frac{1+\sqrt{\omega}}{1-\sqrt{\omega}}. \tag{II.96}$$

Let us proceed to computing the left ordered representation. First, note that $[b_-, b_+] = 0$, and these operators act last. Hence, no commutations are necessary in computing the composition

$$b_\pm[\![f(\overset{1}{a}_-, \overset{2}{a}_+, \overset{3}{b}_-, \overset{4}{b}_+)]\!], \tag{II.97}$$

and consequently, we obtain

$$l_{b_+} = y_4, \quad l_{b_-} = y_3. \tag{II.98}$$

Furthermore, we have

$$\begin{aligned} a_+[\![f(\overset{1}{a}_-, \overset{2}{a}_+, \overset{3}{b}_-, \overset{4}{b}_+)]\!] &= \overset{5}{a}_+ f(\overset{1}{a}_-, \overset{2}{a}_+, \overset{3}{b}_-, \overset{4}{b}_+) \\ &= \overset{2}{a}_+ f\left(\overset{1}{a}_-, \overset{2}{a}_+, \alpha\overset{3}{b}_-, \frac{1}{\alpha}\overset{4}{b}_+\right) \end{aligned} \tag{II.99}$$

by virtue of (II.95) and the permutation formulas from Chapter I. We arrive at the conclusion that

$$l_{a_+} = y_2 I_{y_3}^{\alpha} I_{y_4}^{1/\alpha}, \tag{II.100}$$

where

$$(I_z^\alpha f)(z) = f(\alpha z), \tag{II.101}$$

that is, I_z^α is the dilatation with respect to the variable z.

Similar but somewhat more complicated computations yield the following formula for the representation operator l_{a_-}. We have

$$l_{a_-} = \left[y_1 + \frac{1}{(1+\alpha)^2} \frac{y_3^2}{y_2} (I_{y_2}^{\alpha^2} - 1) + \alpha^2 \frac{y_4^2}{y_2} (I_{y_2}^{1/\alpha^2} - 1) \right] I_{y_4}^{\alpha} I_{y_3}^{1/\alpha}. \qquad \text{(II.102)}$$

The constructed operators are readily verified to satisfy relations (II.95). For example, let us check the commutator $[l_{a_+}, l_{a_-}]$. We have

$$\begin{aligned} l_{a_+} l_{a_-} - l_{a_-} l_{a_+} &= I_{y_3}^{\alpha} I_{y_4} [y_3^2 (I_{y_2}^{\alpha_2} - 1) + \alpha^2 y_4^2 (I_{y_2}^{1/\alpha^2} - 1)] I_{y_4}^{\alpha} I_{y_3}^{1/\alpha} (1 + \alpha^2) \\ &\quad - \left(\frac{y_3^2}{y_2} (I_{y_2}^{\alpha^2} - 1) + \alpha^2 \frac{y_4^2}{y_2} (I_{y_2}^{1/\alpha^2} - 1) \right) y_2 (1+\alpha)^{-2}. \end{aligned}$$

Performing elementary transformations we find that

$$\begin{aligned} [l_{a_+}, l_{a_-}] &= (1+\alpha)^{-2} \{ \alpha^2 y_3^2 (I_{y_2}^{\alpha^2} - 1) + y_4^2 (I_{y_2}^{1/\alpha^2} - 1) \\ &\quad - y_3^2 \alpha^2 I_{y_2}^{\alpha^2} + y_3^2 - y_4^2 I_{y_2}^{1/\alpha^2} + \alpha^2 y_4^2 \} \\ &= \frac{\alpha - 1}{\alpha + 1} (y_4^2 - y_3^2) = \frac{\alpha - 1}{\alpha + 1} (l_{b_+}^2 - l_{b_-}^2). \end{aligned}$$

The other relations in (II.95) can be checked in a similar way.

Now let us proceed to the general scheme. Suppose that we are given an algebra $\mathcal{A}$ determined by generators $A_1, \ldots, A_n$ and relations (II.35) involving certain parameters $\lambda_1, \ldots, \lambda_n$. We suppose that there are sufficiently many such relations so that any element $\psi \in \mathcal{A}$ can be put into normal form

$$\psi = f(\overset{1}{A_1}, \ldots, \overset{n}{A_n}).$$

How can we study the Poincaré–Birkhoff–Witt property in the algebra $\mathcal{A}$? The usual technique is as follows. First, one considers triple products (cf. Examples II.2 and II.3) and uses permutations performed in various orders to infer the conditions on the parameters for which the PBW property holds up to degree $\leq 1, 2$ or 3. This procedure is considerably simpler than that of computing the ordered representation, and it permits one to eliminate from consideration all the parameter values for which the PBW property fails in low orders. The equations obtained at this stage are the Yang–Baxter equations. Then one assumes that the parameters take only the allowed values and computes the ordered representation (if possible).

Finally, the operators of the ordered representation are substituted into the original commutation relations and this makes it clear whether the PBW property holds in the algebra $\mathcal{A}$.

Hence, the situation is as follows. The Yang–Baxter equation is in general only a necessary condition for the PBW property to hold up to a certain low order. The ordered representation, though often hard to obtain, always gives a precise answer (that is, a necessary and sufficient condition for the PBW property to be valid).

6 Representations of Lie Groups and Functions of Their Generators

As is clear from the preceding considerations, noncommutative analysis provides the following method of dealing with commutation relations:

— Given a set Ω of commutation relations, find some reasonable class of representations and an appropriate symbol class such that the operators

$$f(A) = f(\overset{1}{A_1}, \dots, \overset{n}{A_n})$$

with admissible symbols f are well defined and the composition formula

$$[\![f(\overset{1}{A_1}, \dots, \overset{n}{A_n})]\!] [\![g(\overset{1}{A_1}, \dots, \overset{n}{A_n})]\!] = [\![f(\overset{1}{L_1}, \dots, \overset{n}{L_n})\, g]\!] (\overset{1}{A_1}, \dots, \overset{n}{A_n})$$

is valid;

— Furthermore, try to produce a reasonably explicit expression for the left regular representation operators $L_1, \dots, L_n$;

— Then it becomes possible to use the above calculus for investigating such problems as the Cauchy problem and the inversion problem for the operator $f(A)$.

Here we present a partial realization of this programme for Lie commutation relations, that is, the commutation relations which define finite-dimensional Lie algebras.

Let us now proceed to precise formulations.

6.1 Conditions on the Representation

We intend to study functions of the operator tuple $A = (\overset{1}{A_1}, \dots, \overset{n}{A_n})$ whose components satisfy Lie commutation relations. What does this mean exactly? This means that the operators $A_1, \dots, A_n$ act in some linear space E, and that

$$[A_j, A_k]u = -i \sum_{l=1}^{n} c_{jk}^{l} A_l u$$

for any $u \in E$, where c_{jk}^{l} are the structural constants of some Lie algebra $\mathcal{G}$ in some basis $a_1, \dots, a_n \in \mathcal{G}$. Using notions and notations of representation theory, one may write

$$A_j = -i\,\tau(a_j), \quad j = 1, \dots, n,$$

where

$$\tau : \mathcal{G} \longrightarrow \mathrm{End}(E)$$

is a representation of $\mathcal{G}$ in the algebra of endomorphisms of E (see Section 1). The reasons for introducing the factor $(-i)$ will soon become clear.

Thus, we are able to develop the calculus for polynomial symbols, or in terms of formal series. This might be interesting in itself, but it does not cover even the simplest applications to the theory of differential equations (see Chapter III below).

Thus we have to impose some further conditions which make it possible to consider wider symbol classes.

First, we require that $\mathcal{G}$ be a real Lie algebra (that is, $\mathcal{G}$ is an n-dimensional vector space over $\mathbb{R}$; in particular, $c_{jk}^{l} \in \mathbb{R}$), and that $\mathcal{G}$ be *a nilpotent Lie algebra*, that is, any commutator

$$[A_{j1}, \dots, [A_{js-1}, A_{js}] \dots] = 0$$

once $s > N$, where $N < \infty$ is the *nilpotency rank* of $\mathcal{G}$. Violation of these requirements would lead to considering analytical symbols with certain restrictions on the supports of their Fourier transforms; these topics are beyond the framework of this book.

Since the definition of $f(A)$ for nonpolynomial symbols uses some notions of convergence (see Chapter IV below), our model should necessarily involve conditions of a functional-analytic nature. We introduce the simplest version of conditions of this sort:

— E is a dense linear subset in some Hilbert space H;
— The following *Nelson condition* is satisfied: the operators A_j, $j = 1, \dots, n$ and

$$D = 1 + \sum_{j=1}^{n} A_j^2$$

are essentially self-adjoint on E.

These conditions are very important. Indeed, Nelson's theorem (see Nelson [149]) guarantees that under these conditions there exists a unitary representation

$$T : G \longrightarrow \mathrm{Aut}(H)$$

such that τ is a *derived representation* of T; more precisely,

$$\tau = T_{*}|_{E}.$$

The representation T plays a crucial role in what follows.

6.2 Hilbert Scales

We have stated the conditions we impose on the operators A_j. Our next task is to describe the function space, in which the operators $f(\overset{1}{A_1}, \dots, \overset{n}{A_n})$ act. It is not convenient to construct the calculus in the space H itself, since unbounded symbols (which are of practical interest) give unbounded operators. To avoid this difficulty, we use special Hilbert scales associated with H, whose construction uses the operators $A_1, \dots, A_n$ (see the propositions given below).

We postpone using the nilpotency requirement as far as possible; for now we only use the much weaker requirement that G is *unimodular*, that is, its *Haar measure* dg is *biinvariant* (see Appendix).

Proposition II.1 *Let Δ be a strongly positive essentially self-adjoint operator, defined on a dense linear subset E of a Hilbert space H, invariant with respect to Δ (that is, $\Delta E \subseteq E$). Consider the following family of norms on E:*

$$\|u\|_k = (u, \bar{\Delta}^k u)^{1/2}, \ k \in \mathbb{Z}, \ u \in E$$

(here $\bar{\Delta}$ is the closure of Δ; in the last formula one may take Δ instead of $\bar{\Delta}$ for $k \geq 0$). We denote by H^k the completion of E with respect to the norm $\| \cdot \|_k$. The following statements are valid:

(1) *$\{H^k\}_{k \in Z}$ is a Hilbert scale, $H^0 = H$.*
(2) *For $k \geq 0$, H^k may be naturally identified with the domain of the operator $\bar{\Delta}^{k/2}$.*
(3) *$\bar{\Delta}^{1/2} : H^k \longrightarrow H^{k-1}$ is an isomorphism for $k > 0$ and extends to such isomorphism for $k \leq 0$.*
(4) *The bilinear form $(u, \bar{v})$ may be uniquely extended by continuity to the pairing $H^k \times H^{-k} \longrightarrow \mathbb{C}$, which induces an isomorphism $(H^{-k}) \cong (H^k)^*$.*

Proof. Since $\bar{\Delta}$ is strongly positive and self-adjoint, all powers $(\bar{\Delta})^\alpha$ are well defined (see Dunford, Schwartz [42]). Set $W^k = D(\bar{\Delta}^{k/2})$ for $k \geq 0$, $||u||_{W^k} = ||u||_k$, $W^k = (W^{-k})^*$ for $k \leq 0$, with the identification $W^0 = (W^0)^*$ obtained via the scalar product $(\cdot, \cdot)_H$ in the space H. Clearly, properties (1) — (4) are satisfied for the spaces W^k, and we must only prove that $H^k = W^k$ for $k > 0$ (the case $k < 0$ then follows automatically). It suffices to show that E is dense in W^k in the topology of W^k. Assume the opposite is true. Then there exists a nonzero vector $\xi \in W^k$ such that

$$(\xi, u)_k = 0 \quad \text{for all} \quad u \in E.$$

We have

$$(\xi, u)_k = (\xi, \bar{\Delta}^k u) = (\bar{\Delta}^k \xi, u) = 0$$

for any $u \in E$. Since E is dense in H, $\bar{\Delta}^k \xi = 0$, and thus $\xi = 0$, because Ker $\bar{\Delta}^k$ is trivial. The proposition is proved. □

Proposition II.2 *Let G be a Lie group with Lie algebra $\mathcal{G}$, and let*

$$\tau : \mathcal{G} \longrightarrow \operatorname{End}(E)$$

be a representation of $\mathcal{G}$ in a dense linear subset E of a Hilbert space H, satisfying Nelson's conditions. Let $\mathcal{G}_1 \subset \mathcal{G}$ be an ideal, and let $a_1, \ldots, a_k$ be a (linear) basis of $\mathcal{G}_1$, $A_j = -i\tau(a_j)$, $j = 1, \ldots, k$. Denote by

$$T : G \longrightarrow \operatorname{Aut}(H)$$

the unitary representation of G in H, associated to τ by Nelson's theorem, and set

$$\Delta u = \sum_{j=1}^{k} A_j^2 u + u, \; u \in E.$$

The following statements are valid:

(1) *Δ is a strongly positive essentially self-adjoint operator on E; the corresponding Hilbert scale does not depend on the choice of the basis $(a_1, \ldots, a_k)$ up to a norm equivalence.*

(2) *The representation T extends to a bounded strongly continuous representation in the space H^s for any s; there is an estimate*

$$\|T(g)\, u\|_s \le C_s \, (\|\operatorname{Ad} g\|)_{\mathcal{G}_1}^{|s|} \, \|u\|_s,$$

where C_s is a constant independent of u, and $(\|\operatorname{Ad} g\|)_{\mathcal{G}_1}$ is the norm of the restriction of the operator Ad g *on the invariant subspace $\mathcal{G}_1$.*

Proof. 1) Since $\mathcal{G}_1$ is a subalgebra, it generates the corresponding subgroup $G_1 \subset G$. The operators $A_1, \ldots, A_k$ are generators of the unitary representation $T|_{G_1}$. By Nelson's theorem, the operator Δ is essentially self-adjoint. Positivity of Δ is clear. The fact that the corresponding Hilbert scale does not depend on the choice of the basis up to norm equivalence is a consequence of the following lemma.

Lemma II.5 *Let $\{H^k\}$ be the Hilbert scale described above. For any integer s the norm equivalence*

$$\|u\|_{s+1} \sim \|u\|_s + \sum_{j=1}^{k} \|A_j u\|_s$$

holds for $u \in E$.

We postpone the proof of this lemma until the end of the proof of Proposition II.2.

2) It suffices to prove the estimate for positive s and then use conjugation. We proceed by induction on s. The estimate is valid for $s = 0$, since T is a unitary representation, and therefore, $\|T(g)\| = 1$ for all $g \in G$. To perform the induction step $s \Rightarrow s + 1$, we use Lemma II.5.

By this lemma, it suffices to estimate $\|A_j\, T(g)\, u\|_s$. We have

$$A_j T(g) = T(g)\,(T(g)^{-1} A_j T(g)) = -i\, T(g)\tau(\operatorname{Ad} g(a_j)),$$

thus

$$\begin{aligned} \|A_j T(g) u\|_s &= \|T(g)\tau(\operatorname{Ad} g(a_j)) u\|_s \\ &\le \|T(g)\|_{s\to s}\, \|\operatorname{Ad} g\|_{\mathcal{G}_1} \cdot \|u\|_{s+1}. \end{aligned}$$

The latter inequality together with Lemma II.5 immediately yields the desired estimate for the new value of s. The proposition is proved. □

Proof of Lemma II.5. Case 1. $s \ge 0$.

For $s = 0$ we have

$$\|u\|_1^2 = ((1 + A_1^2 + \cdots + A_n^2)\, u,\, u) = \|u\|^2 + \sum_{j=1}^{n} \|A_j\, u\|^2,$$

and the statement of the lemma holds.

For $s > 0$ we use the identity

$$[A_j, \Delta] = \sum_{k=1}^{n} [A_j, A_k^2] = -i \sum_{k,l=1}^{n} c_{jk}^{l} (A_l A_k + A_k A_l).$$

Using the latter relation, we obtain

$$\begin{aligned} \|u\|_{s+1}^2 &= ((1 + \sum_{j=1}^{n} A_j^2)\, \Delta^s u,\, u) = \|u\|_s^2 + \sum_{j=1}^{n} (A_j \Delta^s u,\, A_j\, u) \\ &= \|u\|_s^2 + \sum_{j=1}^{n} \|A_j\, u\|_s^2 + (P_{2s}(A_1, \ldots, A_n)\, u,\, u) \\ &\quad + \sum_{j,k,l=1}^{n} C_{jk}^{l} ((A_j A_k + A_k A_j)\, A_l \Delta^{s-1} u,\, u), \end{aligned}$$

where $P_{2s}(A_1, \ldots A^n)$ is an unordered polynomial of degree $\le 2s$. The last term equals zero due to antisymmetry, so that

$$\|u\|_{s+1}^2 = \|u\|_s^2 + \sum_{j=1}^{n} \|A_j\, u\|_s^2 + (P_{2s}(A_1, \ldots, A_n)\, u,\, u).$$

By induction on s we prove that

$$\|u\|_{k+1} \sim \|u\|_k + \sum_{j=1}^{k} \|A_j u\|_k \quad \text{for all} \quad k \in \{0, \ldots, s-1\}$$

$$|(P_{2s}(A_1, \dots, A_n)u, u)| \le C \|u\|_s^2.$$

First of all, the first relation implies the second, since $(P_{2s}(A_1, \dots, A_n)u, u)$ is a sum of terms of the form $(A_{j_1} \dots A_{j_{s_1}} u, A_{l_1} \dots A_{l_{s_2}} u)$, where $s_1, s_2 \le s$. Thus, it suffices to verify the first relation. The statement is valid for $s = 1$ (see above). The induction step goes as follows. We have

$$\|u\|_{s+1}^2 \le c(\|u\|_s^2 + \sum_{j=1}^n \|A_j u\|_s^2),$$

$$\|u\|_{s+1}^2 \ge \sum_{j=1}^n \|A_j u\|_s^2 - c\,\|u\|_s^2.$$

Moreover, one has

$$\|u\|_{s+1}^2 \ge \|u\|_s^2.$$

Multiplying this by $(c+1)$ and adding the result to the penultimate inequality, we obtain

$$(c+2)\,\|u\|_{s+1}^2 \ge \|u\|_s^2 + \sum_{j=1}^n \|A_j u\|_s^2,$$

as desired.

Case 2. $s \le 0$.

It is easy to see that

$$\bar{\Delta}^{1/2} : H^s \longrightarrow H^{s-1}$$

is an isomorphism for any s. The proof for $s \le 0$ now goes quite similarly to that for $s > 0$ if we use the relations

$$[A_j, \Delta^{-1}] = -\Delta^{-1}[A_j, \Delta]\Delta^{-1};$$

instead of the polynomial $h_{2s}(A_1, \dots, A_n)$ we get a polynomial in $A_1, \dots, A_n$ and Δ^{-1} of weight $\le 2s$, where the weight of A_j is 1, and the weight of Δ^{-1} is -2. □

Thus, we can associate a scale $\{H^s\}$ to any ideal $\mathcal{G}_1 \subset \mathcal{G}$ (see Proposition II.2 above). Of course, $\mathcal{G}_1 = \mathcal{G}$ would be the simplest choice.

6.3 Symbol Spaces

Let us now define admissible symbols. For this purpose, we introduce a special function space on the Lie group G.

Definition II.8 Let G be a Lie group with Lie algebra $\mathcal{G}$, $\mathcal{G}_1 \subset \mathcal{G}$ an ideal. We denote by $\mathcal{P}^+(G, \mathcal{G}_1)$ the space of distributions φ on G representable both in the form

$$\varphi = \sum_{|\alpha| \le m, \alpha_{k+1}, \dots, \alpha_n = 0} l^\alpha \mu_\alpha$$

and in the form

$$\varphi = \sum_{|\alpha|\le m,\alpha_{k+1},\dots,\alpha_n=0} r^\alpha(\tilde{\mu}_\alpha),$$

where $l = (l_1, \dots, l_n)$ is a tuple of right-invariant vector fields on G, forming a basis in $\mathcal{G}$, with $l_1, \dots, l_k$ being a basis in $\mathcal{G}_1$, $m = m(\varphi)$, μ_α are measures on $\mathcal{G}$ summable with the weight $\| \operatorname{Ad} g\|^s_{\mathcal{G}_1}$ for any $s \ge 0$, $r = r_1, \dots, r_n$ is a basis in the space of left-invariant vector fields on $\mathcal{G}$, with $r_1, \dots, r_k$ being a basis in $\mathcal{G}_1$; $\tilde{\mu}_\alpha$ satisfy the same conditions as μ_α.

If $\varphi \in \mathcal{P}^+(G, \mathcal{G}_1)$, then the operator

$$T(\varphi) = \int_G \varphi(g)\, T(g) dg$$

may be defined as follows: for

$$\varphi(g) = \sum_{|\alpha|\le m,\alpha_{k+1},\dots,\alpha_n=0} l^\alpha \mu_\alpha(g)$$

we set

$$T(\varphi) \stackrel{\text{def}}{=} \sum_{|\alpha|\le m,\alpha_{k+1},\dots,\alpha_n=0} (-1)^{|\alpha|} \int_G \mu_\alpha(g)\, [l^\alpha\, T(g)]\, dg. \tag{II.103}$$

Proposition II.3 *The integral* (II.103) *strongly converges in the scale* $\{H^s\}$ *and defines a continuous operator*

$$T(\varphi) : H^s \longrightarrow H^{s-m(\varphi)}$$

for any $s \in \mathbb{Z}$.

Proof. We have

$$l^\alpha\, T(g)\, u = (i\, A)^\alpha T(g)\, u = T(g)\, (i\, \operatorname{Ad} g(A))^\alpha\, u.$$

Thus,

$$\begin{aligned} \|l^\alpha\, T(g)\, u\|_{s-m(\varphi)} &\le c\, \|T(g)\|_{s-m(\varphi)\longrightarrow s-m(\varphi)} \| \operatorname{Ad} g\|^\alpha\, \|u\|_s \\ &\le c_1\, \| \operatorname{Ad} g\|^s\, \|u\|_s. \end{aligned}$$

Since the measure μ_α is integrable with the weight $\| \operatorname{Ad} g\|^s_{\mathcal{G}_1}$, the integral converges in $H^{s-m(\varphi)}$ when applied to u and defines a bounded operator.

The proposition is proved. □

The following two assertions describe the product for operators of the form $T(\varphi)$.

Proposition II.4 *The space $\mathcal{P}^+(G, \mathcal{G}_1)$ is an algebra with respect to the group mollification*

$$(\varphi_1 * \varphi_2) = \int_G \varphi_1(h)\varphi_2(h^{-1}g)\,dh.$$

Proposition II.5 *Let $\varphi_1, \varphi_2 \in \mathcal{P}^+(G, \mathcal{G}_1)$. Then*

$$T(\varphi_1)\,T(\varphi_2) = T(\varphi_1 * \varphi_2).$$

Proof of Proposition II.4. Denote

$$\xi(g) = \|\operatorname{Ad} g\|_{\mathcal{G}_1};$$

since $\operatorname{Ad}(g_1 g_2) = \operatorname{Ad} g_1 \circ \operatorname{Ad} g_2$, we have

$$\xi(g\,h) \le \xi(g)\,\xi(h).$$

First of all, note that it suffices to prove the statement for the case where both φ_1 and φ_2 are measures on G.

Indeed, if

$$\varphi_1 = l^\alpha \mu, \quad \varphi_2 = r^\beta \tilde{\mu},$$

then

$$\varphi_1 * \varphi_2 = l^\alpha r^\beta\,(\mu * \widetilde{\mu}).$$

Next, let us verify that $\varphi_1 * \varphi_2$ is a well defined measure on G. Let $f \in C_0(G)$ be a test function. We have

$$\begin{aligned} < \varphi_1 * \varphi_2,\ f > &= \int_{G\times G} \varphi_1(h)\,\varphi_2(h^{-1}g)\,f(g)\,dh\,dg \\ &= \int_{G\times G} \varphi_1(h)\,\varphi_2(k)\,f(hk)\,dh\,dk, \end{aligned}$$

since the Haar measure is invariant. Since $f(hk)$ is bounded, and φ_1, φ_2 are summable on G, the latter integral converges. Let us now prove that $\varphi_1 * \varphi_2$ is absolutely integrable with the weight $\xi(g)^s$ for any s.

We have

$$\begin{aligned} < |\varphi_1 * \varphi_2|,\ \xi^s > &= \int_{G\times G} |\varphi_1(h)\,\varphi_2(k)|\,\xi^s(hk)\,dh\,dk \\ &\le \int_{G\times G} |\varphi_1(h)|\,|\varphi_2(k)|\,\xi^s(h)\,\xi^s(k)\,dh\,dk \\ &= \left(\int_G |\varphi_1(h)|\,\xi^s(h)\,dh\right)\left(\int_G |\varphi_2(k)|\,\xi^s(k)\,dk\right). \end{aligned}$$

Both integrals on the last line converge. The proposition is therefore proved. □

Proof of Proposition II.5. This reduces to standard computations:

$$\begin{aligned} T(\varphi_1)\,T(\varphi_2) &= \int_G \varphi_1(g)\,T(g)\,dg \int_G \varphi_2(h)\,T(h)\,dh \\ &= \int_{G\times G} \varphi_1(g)\,\varphi_2(g^{-1}k)T(k)\,dg\,dk \\ &= \int_G (\varphi_1 * \varphi_2)(k)T(k)\,dk. \end{aligned}$$

The above propositions mean that $\mathcal{P}^+(G,\ \mathcal{G}_1)$ is a *group algebra*, and the mapping $\varphi \mapsto T[\varphi]$ is a *group algebra representation.*

We use this representation so as to define functions $f(A) = f(\overset{1}{A}_1, \dots, \overset{n}{A}_n)$ for a certain class of symbols. Recall that the mapping

$$\begin{aligned} &\exp_2 : \mathcal{G} \longrightarrow G \\ &(x_1, \dots, x_n) \mapsto \exp(x_n\, a_n) \cdot \dots \cdot \exp(x_1\, a_1) \end{aligned}$$

(coordinates of second genus) is a local diffeomorphism at least in the neighborhood of the point $0 \in \mathcal{G}$. Suppose that a function $f(x)$, $x \in \mathcal{G}^*$, is given such that the restriction of $\exp_2$ to some neighborhood of the support of its Fourier transform

$$\tilde{f}(p) = \frac{1}{(2\pi)^{n/2}} \int_{\mathcal{G}^*} e^{ipx}\, f(x)\,dx, \quad p \in \mathcal{G}$$

is a diffeomorphism. Then we may define the *group Fourier transform*

$$\check{f}(g) = \tilde{f}(\exp_2^{-1}(g))\,|\det(\exp_{2*})|^{-1}.$$

Denote by $\tilde{\mathcal{P}}^+(G,\ \mathcal{G}_1)$ the space of functions $f(x)$, $x \in \mathcal{G}^*$, such that the group Fourier transform $\check{f}(g)$ is defined and belongs to $\mathcal{P}^+(G,\ \mathcal{G}_1)$. □

Definition II.9 Let $f \in \tilde{\mathcal{P}}^+(G,\ \mathcal{G}_1)$. Then

$$f(A) \equiv f(\overset{1}{A}_1, \dots, \overset{n}{A}_n) \overset{\text{def}}{=} \frac{1}{(2\pi)^{n/2}}\, T(\check{f}). \tag{II.104}$$

The above considerations imply that the following theorem is true:

Theorem II.13 *The operator $f(A)$ is well defined and bounded in the scale $\{H^s\}$, that is, there exists an $m \in \mathbb{Z}$ such that*

$$f(A) : H^s \longrightarrow H^{s-m}$$

is a continuous operator for any $s \in \mathbb{Z}$.

Remark II.7 With the help of the above definitions one can easily rewrite (II.104) in the form

$$f(A) = \frac{1}{(2\pi)^{n/2}} \int_G \tilde{f}(p)\, e^{i\, p_n A_n} \dots e^{i\, p_1 A_1}\, dp.$$

(In the computations one has to use the fact that

$$T(\exp(p_j a_j)) = \exp(i p_j A_j),$$

which is valid due to the commutative diagram whose lower arrow is defined via the solution of the Cauchy problem

$$-i\,\frac{du}{dt} = A\,u, \quad u|_{t=0} = u_0 \in E$$

with a self-adjoint operator A.)

Note that we did not use the nilpotency condition thus far.

We now make use of this condition in order to achieve further results. First of all, in the nilpotent case the mapping $\exp_2$ is a global diffeomorphism with $\det\left|\exp_{2*}\right| \equiv 1$, provided that the matrices of the adjoint representation are triangular in the basis $a_1, \dots, a_n$. It follows that the group Fourier transform is defined for any tempered function f, and the group mollification induces a bilinear operation $\tilde{*}$ in $\tilde{\mathcal{P}}^+(G, \mathcal{G}_1)$:

$$(f \tilde{*} g)^\vee \overset{\text{def}}{=} \check{f} * \check{g}, \quad f, g \in \tilde{\mathcal{P}}^+(G, \mathcal{G}_1).$$

Clearly, we have:

$$(f \tilde{*} g)\,(A) = f(A) g(A).$$

We intend to show that the "twisted product" $(f \tilde{*} g)\,(x)$ may be interpreted as the result of action of the operator $f(l) = f(\overset{1}{l_1}, \dots, \overset{n}{l_n})$ on g, where $l_1, \dots, l_n$ are the operators of the *left ordered representation*, whose explicit expression will be given below.

We have

$$\begin{aligned}
[\check{f} * \check{g}]\,(k) &= \frac{1}{(2\pi)^{n/2}} \int_G \check{f}(k)\, \check{g}(k^{-1} h)\, dk \\
&= \frac{1}{(2\pi)^{n/2}} \int_G \check{f}(k)\, \left[\mathcal{L}_k\, \check{g}\right]\,(h) \\
&= \left(\frac{1}{(2\pi)^{n/2}} \int_G \check{f}(k)\, \mathcal{L}_k\, dk \right) \check{g}(h),
\end{aligned}$$

where the operators $\mathcal{L}_k$ act in $\mathcal{P}^+(G, \mathcal{G}_1)$ according to the formula

$$[\mathcal{L}_k\, \varphi]\,(h) = \varphi(k^{-1} h), \quad k, h \in G.$$

The operators $\mathcal{L}_k$ form a representation of the group G,

$$\mathcal{L}_k \, \mathcal{L}_t = \mathcal{L}_{kt}.$$

The corresponding representation of $\mathcal{G}$ is given by the right-invariant vector fields on G, which have been denoted by $\tilde{l}_j, \; j = 1, \ldots, n$. Thus, we have

$$\tilde{l}_j = \mathcal{L}_*(a_j).$$

If $k = \exp_2(p)$, then obviously we obtain

$$\mathcal{L}_k = \mathcal{L}(\exp_2(p)) = \exp(p_n \tilde{l}_n) \ldots \exp(p_1 \tilde{l}_1).$$

Therefore,

$$\begin{aligned} [\check{f} * \check{g}]\,(k) &= \left(\frac{1}{(2\pi)^{n/2}} \int\limits_G \check{f}(k)\, \mathcal{L}_k \, dk \right) \check{g} \\ &= f(\overset{1}{\tilde{l}_1}, \ldots, \overset{n}{\tilde{l}_n})\, \check{g}. \end{aligned}$$

Denote by $\check{F}$ the group Fourier transformation, $\check{g} = \check{F} g$, and by $\check{F}^{-1}$ the inverse transformation,

$$\check{F} = \left(\exp_2^{-1}\right)^* F,$$

where F is the usual Fourier transformation.

We have

$$\begin{aligned} f \tilde{*} g &= \check{F}^{-1} \left[\check{f} * \check{g} \right] = \check{F}^{-1} \left(f(\overset{1}{\tilde{l}_1}, \ldots, \overset{n}{\tilde{l}_n})\, g \right) \\ &= \left[\check{F}^{-1} f(\overset{1}{\tilde{l}_1}, \ldots, \overset{n}{\tilde{l}_n}) \check{F} \right] g = f \left(\overset{1}{\overline{\check{F}^{-1} \tilde{l}_1 \check{F}}} \ldots \overset{n}{\overline{\check{F}^{-1} \tilde{l}_n \check{F}}} \right) g \\ &= f(\overset{1}{l_1}, \ldots, \overset{n}{l_n})\, g \equiv f(l)\, g, \end{aligned}$$

where

$$l_j = \check{F}^{-1} \tilde{l}_j \check{F} = F^{-1} \, \overset{\circ}{l}_j \, F,$$

and $\overset{\circ}{l}_j$ is the expression for $\tilde{l}_j$ in the coordinates of second genus. It is well known that in the nilpotent case $\overset{\circ}{l}_j$ is a vector field with polynomial coefficients, so the operators l_j of the left ordered representation are differential operators with linear coefficients.

6.4 Symbol Classes: More Suitable for Asymptotic Problems

The symbol class $\tilde{\mathcal{P}}^+(G, \mathcal{G}_1)$, described in the preceding subsection, seems rather exotic from the practical viewpoint. It is well known that in applications to (pseudo) differential equations one mostly deals with symbols homogeneous with respect to a group of variables.

Here we consider classical symbols.

Definition II.10 By $S^m = S^m(G, \mathcal{G}_1)$ denote the space of symbols $f(x) \in \tilde{\mathcal{P}}^+(G, \mathcal{G}_1)$ asymptotically homogeneous of degree $\leq m$ with respect to $(x_1, \ldots, x_k)$. In other words, each symbol $f(x) \in S^m$ possesses an asymptotic expansion

$$f(x) = f_m(x) + f_{m-1}(x) + \cdots$$

where $f_j(x) \in \tilde{\mathcal{P}}^+(G, \mathcal{G}_1)$ is a homogeneous function of degree j with respect to $(x_1, \ldots, x_k)$, that is,

$$f_j(\lambda x_1, \ldots, \lambda x_k, x_{k+1}, \ldots, x_k) = \lambda^j f_j(x_1, \ldots, x_k, x_{k+1}, \ldots, x_n)$$

for $\lambda \geq 1$, $|x_1| + \cdots + |x_k|$ large enough, and the remainder decays rapidly as $|x_1| + \cdots + |x_k| \to \infty$.

The function $f_m(x)$ will be called the *principal symbol* of $f(A)$.

From now on we assume that $A_{k+1}, \ldots, A_n$ commute with one another.

Theorem II.14 *Let $f \in S^m$. Then*

$$f(A) : H^s \to H^{s-m}$$

is a continuous operator for any $s \in \mathbb{Z}$.

For the general version of this theorem, see [134], Theorem II.4.F.2.

To prove the theorem, we first consider the following Lemma.

Lemma II.6 *Let $f \in S^m$, $g \in S^{m'}$. Then the principal symbol of the product $f(A)g(A)$ is equal to $f_m(x)g_{m'}(x)$.*

Proof. Clearly, $\varphi \in \mathcal{P}^+(G, \mathcal{G}_1)$ belongs to S^r iff $\varphi(\lambda x_1, \ldots, \lambda x_k, x_{k+1}, \ldots, x_n)$ possesses an asymptotic expansion of the form

$$\varphi(\lambda x_1, \ldots, \lambda x_k, x_{k+1}, \ldots, x_n) = \lambda^r f_r(x) + \lambda^{r-1} f_{r-1}(x) + \cdots$$

as $\lambda \to \infty$.

The symbol of $f(A)\, g(A)$ is equal to $[f(l)\, g](x)$. Let us perform the change of variables

$$x_1 \to \lambda x_1, \quad x_k \to \lambda x_k, \quad x_{k+1} \to x_{k+1}, \quad x_n \to x_n$$

in the latter expression. Then each operator l_j takes the form

$$l_j = \lambda(x_j + \lambda^{-1}\delta_j(\lambda^{-1})),$$

where $\delta_j(\lambda^{-1})$ is a differential operator whose coefficients are polynomials in λ^{-1}. Moreover, $\delta_j(\lambda^{-1})$ commutes with x_j, since the matrices of the adjoint representation are triangular in the basis $(a_1, \dots, a_k)$ (recall that this basis is assumed to be chosen in a special way). Thus, we obtain the expansion

$$f(l)\, g = \lambda^{m+m'} f_m g_{m'} + \lambda^{m+m'-1} F_1 + \lambda^{m+m'-2} F_2 + \cdots$$

by using the ordinary Taylor expansion of $f(\lambda(x + \lambda^{-1}\delta))$ in powers of λ^{-1}.

The lemma is proved. □

Let us now prove the theorem. Clearly, it suffices to consider the case $m = 0$, $s = 0$; results for the other cases may be obtained via left and right multiplications by appropriate powers of Δ. Consider the operator $(f(A))^*$ (the adjoint operator is taken in $H = H^0$). It is clear that[1]

$$(f(A))^* = \bar{f}(\overset{n}{A_1}, \dots, \overset{1}{A_n}).$$

Therefore, we have

$$f(A)^* f(A) = h(A),$$

where

$$h(x) = \bar{f}(\overset{n}{l_1}, \dots, \overset{1}{l_n})\, f(x).$$

As was done in the proof of the lemma, we can show that the principal symbol of $h(A)$ is equal to $|f(x)|^2$.

Now let M be a constant such that

$$M > \max |f(x)|^2,$$

and set

$$g(x) = \sqrt{M - |f(x)|^2}.$$

Then $g(x) \in S^0$, and

$$|f|^2 + |g|^2 = M.$$

Thus, we have

$$f(A)^* f(A) + g(A)^* g(A) = M + R(A),$$

where $R(x) \in S^{-1}$, and its principal symbol $R_{-1}(x)$ is real-valued since $R(A)$ is self-adjoint.

[1] Here the bar over f stands for complex conjugation.

Choose $g_{-1}(x) = -R_{-1}(x)/2g(x)$ and set $G(A) = g(A) + g_{-1}(A)$. Then we have

$$\begin{aligned} f(A)^* f(A) + G(A)^* G(A) &= M + R(A) + g(A)^* g_{-1}(A) + g_{-1}(A)^* g(A) \\ &\quad + g_{-1}(A)^* g_{-1}(A) = MR'(A), \end{aligned}$$

where $R'(x) \in S^{-2}$. Continuing this procedure, we finally obtain

$$f^*(A)\, f(A) + H^*(A)\, H(A) = M + R''(A),$$

where $R''(x) \in S^{-N}$ for N large enough.

The Fourier transform $R''(x)$ is now a measure on G and, so $R''(A)$ is bounded in H^0. We obtain

$$\|f(A)u\|^2 \le \|f(A)u\|^2 + \|H(A)u\|^2 = M\|u\|^2 + (R''(A)u, u) \le C\|u\|^2,$$

for any $u \in H^0$.

Therefore, $f(A) : H^0 \to H^0$ is bounded.

Application of noncommutative analysis to the Cauchy problem and to the inversion problem with the operator $f(A)$ goes as follows (for detail see Chapter III below). Consider the problem

$$\begin{cases} -i\dfrac{\partial B}{\partial t} + f(A)\, B = 0, \\ B|_{t=0} = B_0 \end{cases} \tag{II.105}$$

or

$$f(A)B = 1.$$

Let us try to solve it by a substitution of the form $B = g(A, t)$ and $B = g(A)$. This leads to the equation for the symbol $g(x, t)$,

$$\begin{cases} -i\dfrac{\partial g}{\partial t} + f(L)\, g = 0, \\ g|_{t=0} = g_0(x) \end{cases} \tag{II.106}$$

or

$$f(L)g = 1,$$

respectively.

The solutions of these equations usually do not belong to the space $\mathcal{P}^+(G, \mathcal{G}_1)$, so the problem arises to define operators whose symbols are of this sort and to prove their boundedness in the scale $\{H^s\}$. □

Definition II.11 Let $f(x) \in C^\infty(\mathbb{R}^n)$. We define the operator $f(A)$ as a strong limit

$$f(A) \stackrel{\text{def}}{=} s\text{-}\lim_{\varepsilon \to 0} (\rho_\varepsilon f)(A),$$

where $\rho_\varepsilon(x) = \rho(\varepsilon x)$, $\rho(x) \in C_0^\infty(\mathbb{R}^n)$, $\rho(0) = 1$, provided that the limit exists. The operator $f(A)$ is a bounded operator in the scale $\{H^k\}$, and does not depend on the choice of ρ.

Symbols arising as solutions of (II.105) and (II.106) possess a special structure, intimately related to the geometry of the cotangent bundle T^*G. Thorough study of this structure eventually implies boundedness theorems (see [129], [101], and [134]). Since we have no opportunity to discuss these topics, which require from the reader a great amount of knowledge of symplectic geometry and the theory of Fourier integral operators (see, e.g., [142] and references cited therein), we present only the following result.

Theorem II.15 *Let $f(x) \in S^1$ be real-valued. Suppose also that $g_0(x) \in S^0$. Then the solution $B(t)$ of the Cauchy problem*

$$\begin{cases} -i\dfrac{\partial B}{\partial t} + f(A)B = 0, \\ B|_{t=0} = g_0(A) \end{cases} \tag{II.107}$$

is bounded in the scale $\{H^s\}$ for any t; more precisely, we have

$$||B(t)||_{H^s \to H^s} \leq C_s(t),$$

where $C_s(t)$ is a locally bounded function.

Proof. Since f is real, we have,

$$(f(A))^* = [(f(\overset{n}{l_1}, \dots, \overset{1}{l_n}))\, 1](A) = f(A) + R(A),$$

where $h(x) \in S^0$ and therefore $R(A)$ is bounded. The problem (II.107) can be rewritten in the form

$$\begin{cases} -i\dfrac{\partial B}{\partial t} = KB + LB, \\ B|_{t=0} = B_0, \end{cases}$$

where $K = K^*$, whereas L and B_0 are bounded operators.

It is well known that the solution of the last problem is a bounded operator. Indeed,

$$\frac{d}{dt}||Bu||^2 = (Bu, LBu) - (LBu, Bu) \leq 2||L||\,||Bu||^2,$$

which implies the desired estimate. □

Chapter III
Noncommutative Analysis and Differential Equations

1 Preliminaries

Let us show how the technique developed in the preceding chapters can be applied to differential equations.

We chiefly consider two sorts of problems:

(a) Find the inverse of a differential operator P.

(b) Find the resolvent operator for a Cauchy problem.

Recall that the resolvent operator is defined as follows. Consider the Cauchy problem

$$\begin{cases} \dfrac{\partial^l u}{\partial t^l} + Pu = 0, \\ \left.\dfrac{\partial^j u}{\partial t^j}\right|_{t=0} = v_j, \quad j = 0, \dots, l-1, \end{cases}$$

where $u = u(x,t)$, $x \in \mathbb{R}^n$, $t \in \mathbb{R}^1$; $P = P(x, t, \partial/\partial x, \partial/\partial t)$ is a partial differential operator containing t-derivatives of order at most $l-1$; $v_j = v_j(x)$ are given functions (the *Cauchy data*). Then the resolvent operator of this problem is the (unique) vector operator

$$Q(t) = (Q_0(t), \dots, Q_{l-1}(t))$$

such that the solution is given by the formula

$$u(t) = \sum_{j=0}^{l-1} Q_j(t) v_j$$

for arbitrary initial data $v_0, \dots, v_{l-1}$.

We shall illustrate the approach to these problems suggested by noncommutative analysis on an example of problem (a). One expresses the operator P as a function of a Feynman tuple $\overset{1}{A}_1, \dots, \overset{m}{A}_m$. It is assumed that the left ordered representation $\overset{1}{l}_1, \dots, \overset{m}{l}_m$ of this tuple exists. If so, one can seek for the inverse of P (problem (a)) in the form

$$P^{-1} = Q(\overset{1}{A}_1, \dots, \overset{m}{A}_m).$$

By definition, we should have

$$PP^{-1} = [\![f(\overset{1}{A_1}, \dots, \overset{m}{A_m})]\!] [\![Q(\overset{1}{A_1}, \dots, \overset{m}{A_m})]\!] = 1$$

(to be completely rigorous, we now speak of the *right* inverse, which is often not the same as the left inverse in the context of differential operators).

We obtain an equation for the unknown symbol $Q(y_1, \dots, y_m)$ by passing to the left ordered representation, as is described in Chapter II:

$$f(\overset{1}{l_1}, \dots, \overset{m}{l_m})(Q(y_1, \dots, y_m)) = 1.$$

It remains to solve the latter equation, and one obtains the answer (Problem (b) can be treated in a similar way).

The most evident (though useless) representation is given by the choice $m = 1$, $A = P$ (we omit the subscript 1). We have

$$P = f(A), \quad f(y) = y; \quad L = y;$$ [1]

and obtain the equation

$$y\, Q(y) = 1$$

for the symbol $Q(y)$ of the inverse. The solution

$$Q(y) = 1/y$$

is obvious, and we obtain $P^{-1} = P^{-1}$, which gives no information at all.

Another representation (used very often) follows merely from the fact that P is a differential operator,

$$P = f\left(\overset{2}{x}, -i\overset{1}{\frac{\partial}{\partial x}}\right),$$

so it is automatically a function of the tuple

$$-i\overset{1}{\frac{\partial}{\partial x_1}}, \dots, -i\overset{1}{\frac{\partial}{\partial x_n}}, \quad \overset{2}{x_1}, \dots, \overset{2}{x_n}.$$

Let $y_1, \dots, y_n, y_{n+1}, \dots, y_{2n}$ be the corresponding arguments of symbols. Then the left ordered representation has the form

$$L_1 = y_1 - i\frac{\partial}{\partial y_{n+1}}, \dots, \quad L_n = y_n - i\frac{\partial}{\partial y_{2n}},$$
$$L_{n+1} = y_{n+1}, \dots, \quad L_{2n} = y_{2n},$$

[1] Note that the left ordered representation for a tuple consisting of a single operator A always has the form $L = y$.

and so the equation for the symbol Q of the inverse of P takes the form

$$f\left(\overset{1}{y_1} - i\frac{\overset{1}{\partial}}{\partial y_{n+1}}, \dots, y_n - i\frac{\overset{1}{\partial}}{\partial y_{2n}}, \overset{2}{y}_{n+1}, \dots, \overset{2}{y}_{2n}\right) Q(y) = 1.$$

In particular problems, according to their specific structure, it is often useful to choose the operators $A_1, \dots, A_m$ in some other way. In any case, the main idea is that *an equation in the algebra of operators is reduced to an equation in the symbol space.*

Clearly, the choice of $A_1, \dots, A_m$ is a matter of the investigator's skill and experience rather than of any standard procedure; indeed, this choice should meet two requirements: first the operator P should be a function of $\overset{1}{A}_1, \dots, \overset{m}{A}_m$, and, second, the tuple $A_1, \dots, A_m$ should satisfy a sufficiently rich system of commutation relations, so that functions of $\overset{1}{A}_1, \dots, \overset{m}{A}_m$ form an algebra. Also, one should take care that the passage to an equation for $Q(y)$ be a simplification rather that a complication.

Moreover, the following phenomenon is frequently encountered: *the symbol $Q(y)$ of the operator that gives the solution of the problem considered lies in a function space larger than the original one. The operators with symbols in this space itself do not form an algebra, but only a module over the original algebra.*

We point out that this phenomenon is not a disadvantage of the method but is essentially related to the nature of the problems considered. For example, if P is an lth order differential operator of principal type with real principal symbol, then its right almost-inverse P^{-1} acts as follows:

$$P^{-1} : H_{\circ}^{s+1} \to H_{\text{loc}}^{s+l}, \quad s \in \mathbb{R},$$

where $H_{\circ}^{s+1}$ is the subspace of the Sobolev space H^{s+1} consisting of compactly supported distributions and H_{loc}^{s+l} is the space of distributions belonging to H^{s+l} locally[2]. Products of such operators cannot be considered in general: an application of such an operator to a function with compact support yields a function whose support need not be compact, and the second application thus becomes impossible. Such operators, however, form a module over the algebra of (pseudo)differential operators with proper support.

The method cited applies to problems of various levels of difficulty, and in Section 1.1 we consider the simplest case of ordinary differential operators with constant coefficients. It turns out that in this case the above-described scheme leads to the conventional Heaviside method. Clearly, this case is trivial from the standpoint of noncommutative analysis: no noncommutativity is involved, and the details of the method, in particular, the extension of symbol classes, employ a technique completely different from the one used in the subsequent sections. However, this example displays some characteristic features of the method, and that is why it is included.

[2]For specialists, let us note that P^{-1} is a Fourier integral operator on a manifold *with boundary* that is *not* the graph of a canonical transformation even locally (see [173]).

The table given below shows some differences between the cases of ordinary differential equations with constant coefficients and of partial differential equations.

The main features of the method	The type of equations considered	
	ordinary differential equations with constant coefficients	partial differential equations with variable coefficients
Equation for the symbol of the inverse operator	algebraic	(pseudo) differential
The choice of the tuple $A_1, \dots, A_m$	$m = 1$, $A_1 = \frac{d}{dx}$	depends on a particular problem
The solution obtained	exact	asymptotic (in most cases)
The method of extending the class of symbols and justifying the correctness of definition for operators with symbols from this class	Algebraic (passage to the ring of quotients)	Function-theoretic

1.1 Heaviside's Operator Method for Differential Equations with Constant Coefficients

Consider the equation

$$P(D)\,u(x) = f(x),$$

where $x \in (a, b)$, $f(x) \in C^\infty(a, b)$ is a given function and

$$P(D) = \sum_{k=0}^{m} c_k D^k$$

is a differential operator of order m with constant coefficients c_k (here $D = \frac{d}{dx}$). The function

$$P(y) = \sum_{k=0}^{m} c_k y^k$$

will be called the *symbol* of $P(D)$. It is a polynomial with complex coefficients.

Thus, $P(D)$ is already represented as a function of a tuple of operators.This "tuple", however, consists of a single operator D. According to the general scheme, we seek a right inverse P^{-1} in the form $P^{-1} = Q(D)$, where $Q(x)$ is some unknown function.

(Note that one cannot find the two-sided inverse of f for the obvious reason that $P(D)$ has a nontrivial kernel provided that $m > 0$). Whichever function $f(y)$ we take, it is obviously true that

$$D\,f(D) = f_1(D),$$

where $f_1(y) = y\,f(y)$. Thus, left multiplication by D in the algebra of operators corresponds to multiplication by y in the algebra of their symbols. In accordance with the general theorems provided in Chapter II, we have

$$P(D)\,Q(D) = (P\,Q)\,(D),$$

(which is obvious in itself). We have to find a function $Q(y)$ such that

$$P(D)\,Q(D) = 1,$$

and so we should require that

$$P(y)\,Q(y) = 1.$$

The solution is quite obvious:

$$Q(y) = 1/P(y).$$

In order to calculate the operator $Q(D)$ explicitly, let us expand $Q(y)$ into partial fractions,

$$Q(y) = \frac{1}{P(y)} = \sum_{k=1}^{l}\sum_{j=1}^{r_k} \frac{a_{kj}}{(y-\lambda_k)^j},$$

where $\lambda_1, \dots, \lambda_l$ are the roots of $P(y)$, $r_1, \dots, r_l$ are their multiplicities, and a_{kj} are constants, which can be found from a system of linear equations. We see that if suffices to compute operators of the form $1/(D-\lambda)^j$. To this end, we use the following trick. Let U_λ denote the operator of multiplication by $e^{-\lambda x}$. Clearly, the conjugation mapping

$$A \mapsto U_\lambda^{-1} A U_\lambda$$

is an automorphism of the algebra of operators acting on functions on the real axis. Moreover, we have

$$U_\lambda^{-1}(D+\lambda)U_\lambda = D.$$

By Theorem I.4 we have

$$U_\lambda^{-1} f(D+\lambda) U_\lambda = f(U_\lambda^{-1}(D+\lambda)U_\lambda) = f(D)$$

for any symbol $f(y)$. In particular,

$$\frac{1}{(D-\lambda)^j} = U_\lambda^{-1}\frac{1}{D^j}U_\lambda,$$

and we have reduced our problem to the computation of $1/D^j$. The operator $1/D^j$ should be the right inverse of D^j, that is,

$$D^j \cdot 1/D^j = 1.$$

However, this condition does not uniquely determine $1/D^j$; to get rid of the ambiguity, we assume, completely at random, that $1/D^j$ is the resolvent operator of the problem

$$\begin{cases} \dfrac{d^j u}{dx^j} = f, \\ u|_{x=0} = u'\big|_{x=0} = \cdots = u^{(j-1)}\big|_{x=0} = 0. \end{cases}$$

Then a simple calculation yields

$$\frac{1}{D^j} f = \frac{1}{(j-1)!} \int_0^x (x-\xi)^{j-1} f(\xi)\, d\xi,$$

and consequently,

$$\frac{1}{(D-\lambda)^j} f = e^{\lambda x} \frac{1}{D^j} e^{-\lambda x} f = \frac{1}{(j-1)!} \int_0^x e^{\lambda(x-\xi)} (x-\xi)^{j-1} f(\xi)\, d\xi.$$

The formula for $Q(D)$ follows immediately.

The above argument is extremely simple; however, still there is a point to be clarified. As was mentioned above, our definition of $1/D^j$ is not the only possible one (we have posed the zero Cauchy data at $x = 0$, which is by no means necessary). Therefore, the extension of the mapping *symbol* $\mapsto$ *operator* given here needs further investigation and verification.

Let us show that our method gives a well-defined result for any rational functions taken as symbols. Let $\mathcal{P} = \mathbb{C}[z]$ be the ring of polynomials in z with complex coefficients, and let $\mathrm{Op} = \mathrm{End}(C^\infty(a,b))$ be the ring of continuous linear operators in the space $C^\infty(a,b)$ equipped with the natural topology.

Consider the ring homomorphism

$$\mu : \mathcal{P} \to \mathrm{Op},$$
$$P(z) = \sum_{k=0}^{n} c_k z^k \mapsto P(D) = \sum_{k=0}^{m} c_k \left(\frac{d}{dx}\right)^k.$$

We wish to solve is the inversion problem for the operator $P(D) = \mu(P)$ and, at first glance, its solution from the algebraic point of view is an extension of the above homomorphism to a homomorphism

$$\mu' : \mathcal{R} \to \mathrm{Op},$$

where $\mathcal{R}$ is the *ring of quotients*[3] of $\mathcal{P}$.

However, we know that such an extension is not possible, since the two-sided inverse of $P(D)$ does not exist. We shall seek a right inverse, that is, an operator $\hat{Q}$ that assigns a particular solution $u = \hat{Q}f$ of the equation $P(D)u = f$ to each right-hand side f. To this end, we consider $\mathcal{R}$ as a left $\mathcal{P}$-module and Op as a left module over itself; our aim is to extend μ to a homomorphism

$$\tilde{\mu} : \mathcal{R} \to \mathrm{Op}$$

of left modules over μ.[4]

Let us realize the described program. To simplify the notation we assume that the origin is contained in the interval (a, b) under consideration; clearly, this assumption does not lead to loss of generality.

First of all we present the definition of $\tilde{\mu}$.

Definition III.1 Let

$$R(z) = P(z)/Q(z)$$

be a rational function. Then the *operator*

$$\tilde{\mu}(R) = \frac{P(D)}{Q(D)} : C^\infty(a, b) \to C^\infty(a, b)$$

is defined by the formula

$$\frac{P(D)}{Q(D)} f = P(D)v$$

for any function $f \in C^\infty(a, b)$, where v is the unique solution of the Cauchy problem

$$\begin{cases} Q(D)v = f, \\ \left.\dfrac{d^j v}{dx^j}\right|_{x=0} = 0, \quad j = 0, 1, \dots, \deg Q - 1 \end{cases} \tag{III.1}$$

with zero Cauchy data.

---

[3]The ring $\mathcal{R}$ is the set of pairs $(P, Q) \in \mathcal{P} \times \mathcal{P}$ factorized by the following congruence: $(P, Q) \sim (P_1, Q_1)$ is and only if $PQ_1 = QP_1$. The operations on $\mathcal{R}$ are defined as follows:

$$(P, Q) + (R, W) = (PW + QR, QW); \quad (P, Q)(R, W) = (PR, QW).$$

Since $\mathcal{P}$ is an integral domain, this definition is correct. Each element $(P, Q) \in \mathcal{R}$ is naturally interpreted as a rational function $\frac{P(z)}{Q(z)}$ and $\mathcal{P}$ is embedded in $\mathcal{R}$ by the mapping $P \mapsto P/1$.

[4]A homomorphism of modules over a homomorphism μ of rings is a morphism $\tilde{\mu}$ of abelian groups such that

$$\tilde{\mu}(pr) = \mu(p)\tilde{\mu}(r)$$

for each $p \in \mathcal{P}$ and $r \in \mathcal{R}$.

Since the Cauchy problem is uniquely solvable for an ordinary differential operator, we see that the operator $P(D)/Q(D)$ is uniquely determined by the operators $P(D)$ and $Q(D)$.

However, to verify the correctness of Definition III.1 one must check that the result of acting by the operator $P(D)/Q(D)$ depends not on the polynomials P and Q themselves but only on the equivalence class of the pair (P, Q) in the quotient ring $\mathcal{R}$. Therefore, it is necessary to check the relation

$$\frac{P(D)S(D)}{Q(D)S(D)} = \frac{P(D)}{Q(D)}$$

for any polynomials P, Q, and S. By definition, we obtain

$$\frac{P(D)S(D)}{Q(D)S(D)} f = P(D)S(D)v \tag{III.2}$$

where v is the solution of the Cauchy problem

$$\begin{cases} Q(D)S(D)v = f, \\ \left.\dfrac{\partial^j v}{\partial x^j}\right|_{x=0} = 0, \quad j = 0, 1, \dots, \deg Q + \deg S - 1. \end{cases}$$

Denote $\tilde{v} = S(D)v$. Then we have

$$\frac{P(D)S(D)}{Q(D)S(D)} f = P(D)\tilde{v}$$

and it is quite simple to check that the function $\tilde{v}$ satisfies the Cauchy problem (III.1). Thus, the relation (III.2) is valid and the correctness of Definition III.1 is proved.

Now we must verify that the defined mapping

$$\tilde{\mu} : \mathcal{R} \to \mathrm{Op}$$

is a homomorphism of left modules over the ring homomorphism μ. To do this we must verify the following three assertions.

1°. $P(D)/1(D) = P(D)$ for $P \in \mathcal{P}$, that is, $\tilde{\mu}$ is an extension of μ.

2°. $(R_1 + R_2)(D) = R_1(D)R_2(D)$ for any $R_1, R_2 \in \mathcal{R}$.

3°. $(PR)(D) = P(D)R(D)$ for $R \in \mathcal{R}$ and $P \in \mathcal{P}$.

The verification of these three assertions is quite simple and therefore is left to the reader.

Now we shall justify the naturality property

$$U_\lambda^{-1} f(A) U_\lambda = f(U_\lambda^{-1} A U_\lambda)$$

for $U_\lambda = e^{\lambda x}$ and $A = D$ used above for description of the explicit computational procedure of the operator $P(D)/Q(D)$.

Proposition III.1 *For any rational function $R \in \mathcal{R}$ we have*

$$e^{-\lambda x} R(D) e^{\lambda x} = R(D + \lambda).$$

Proof. If the rational function R is the quotient of the polynomials P and Q, namely, $R = P/Q$, then the function

$$u(x) = R(D)\{e^{\lambda x} f(x)\}$$

is defined by the equality

$$u(x) = P(D)v(x)$$

where $v(x)$ is the solution of the following Cauchy problem:

$$\begin{cases} Q(D)v(x) = e^{\lambda x} f(x), \\ \left.\dfrac{d^j v(x)}{dx^j}\right|_{x=0} = 0, \quad j = 0, 1, \dots, \deg Q - 1. \end{cases} \tag{III.3}$$

Let us seek for the solution $v(x)$ of the latter problem in the form

$$v(x) = e^{\lambda x} \tilde{v}(x).$$

Substituting this equation in the Cauchy problem (III.3), we obtain that the function $\tilde{v}(x)$ is the solution of the problem

$$\begin{cases} Q(D + \lambda)\tilde{v}(x) = f(x), \\ \left.\dfrac{d^j \tilde{v}}{dx^j}\right|_{x=0} = 0, \quad j = 0, 1, \dots, \deg Q - 1. \end{cases} \tag{III.4}$$

Here we used the obvious fact that

$$Q(D)e^{\lambda x}\tilde{v} = e^{\lambda x} Q(D + \lambda)\tilde{v}$$

for any differential operator $Q(D)$. Thus, we obtain

$$R(D)\{e^{\lambda x} f(x)\} = P(D)\{e^{\lambda x}\tilde{v}(x)\} = e^{\lambda x} P(D + \lambda)\tilde{v}(x).$$

The latter relation together with the Cauchy problem (III.4) proves the required property of the operator $R(D)$. □

1.2 Nonstandard Characteristics and Asymptotic Expansions

We now proceed to considering partial differential equations with variable coefficients. As was mentioned above, the theory is asymptotic in this case.

In this subsection we consider the question of which Hamilton–Jacobi equations correspond to a given differential equation if various types of asymptotic expansions are considered. The first subsection is chiefly devoted to nonstandard characteristics (see [131], [134]). Our main topic, noncommuting operators, remains in the background. In the second subsection we compare asymptotic expansions with respect to a large parameter and smoothness and draw the conclusion that noncommuting operators play a leading role in the theory of arbitrary asymptotic expansions. In the following subsections we use noncommutative analysis to develop the main stages in the construction of asymptotic solutions to differential equations of various types, and here noncommuting operators exhibit their full strength.

The theory of characteristics of differential equations has been developing since the very origin of the study of partial differential equations. It gained a strong impetus with the appearance of physical problems requiring asymptotic expansions with respect to a small parameter. Among such problems, we note the WKB method for constructing asymptotic solutions of the Schrödinger equation in quantum mechanics and the method of geometric optics applied to the propagation of high-frequency electromagnetic waves. As soon as the investigation of such problems started, the theory of characteristics of differential equations split into two parallel branches developing almost independently. The first of these branches deals with singularities of solutions to differential equations (and with asymptotic expansions with respect to smoothness), whereas the second one treats asymptotic expansions with respect to a small parameter. It should be noted that the second branch was at first developed chiefly in treatises on mathematical physics.

Having both branches in mind, let us study the following question: what is the "correct" characteristic equation for, say, the Klein–Gordon equation

$$h^2\frac{\partial^2 u}{\partial t^2} - h^2\frac{\partial^2 u}{\partial x^2} + m^2c^4u = 0$$

(here m is the mass of the particle described by the equation, c the velocity of light, and h the Planck constant).

A physicist would probably claim that the characteristic equation is the Hamilton–Jacobi equation for free motion of a relativistic particle:

$$\left(\frac{\partial S}{\partial t}\right)^2 - \left(\frac{\partial S}{\partial x}\right)^2 - m^2c^4 = 0.$$

However, a mathematician specializing in hyperbolic equations could argue that the correct characteristic equation is another Hamilton–Jacobi equation, namely,

$$\left(\frac{\partial S}{\partial t}\right)^2 - \left(\frac{\partial S}{\partial x}\right)^2 = 0. \tag{III.5}$$

We postpone answering the question of which version is correct and note that the problem is even less trivial for the Helmholtz equation

$$\frac{\partial^2 u}{\partial x^2} + \frac{\partial^2 u}{\partial y^2} + k^2 n^2(x, y) u = 0.$$

Indeed, in the previous example we only discussed the *form* of the corresponding Hamilton–Jacobi equation, whereas for the Helmholtz equation there would be different opinions as to whether there exist (real) characteristics at all. A mathematician could reasonably assert that the equation is *elliptic*, so that it makes little sense to speak about its real characteristics. A physicist, in turn, could argue that the equation describes propagation of electromagnetic waves in some (inhomogeneous) medium and hence is a wave equation. Moreover, the physicist would readily write out the Hamilton–Jacobi equation

$$\left(\frac{\partial S}{\partial x}\right)^2 + \left(\frac{\partial S}{\partial y}\right)^2 - n^2(x, y) = 0, \tag{III.6}$$

which is none other than the classical *eikonal* equation.

Which of the described standpoints is correct? So far, the answer seems evident: neither is correct (or both are correct, if you like that better). The point is that the reason of the argument is lies in the terminology. Each side has its own interpretation of the notion of characteristics, and so each point of view is valid. An attentive reader has necessarily noticed that the key question is *what is called a wave.* Let us give this question a somewhat different setting: *what propagates along the characteristic rays corresponding to the Hamilton–Jacobi equation?*

The mathematician specializing in hyperbolic equations gives the following answer: it is the discontinuities of solutions of the differential equation considered that propagate along the trajectories of the Hamiltonian system (more precisely, along the projections of these trajectories to the configuration space). Hence, an appropriate asymptotic solution to the Klein–Gordon equation may have the following form:

$$u(x, t) = \theta(\Phi(x, t))\, \varphi_0(x, t) + \Phi(x, t)\, \theta(\Phi(x, t))\, \varphi_1(x, t) + \cdots, \tag{III.7}$$

where $\Phi, \varphi_0, \varphi_1, \ldots$ are smooth functions of the variables (x, t) and $\theta(z)$ is the Heaviside function

$$\theta(z) = \begin{cases} 1, & z > 0, \\ 0, & z < 0. \end{cases}$$

The solution is discontinuous on the surface $\Phi(x, t) = 0$, and we monitor the evolution of this surface.

The physicists' answer is also evident. He would say that it is the wave, i.e., the oscillations of solutions that propagate along the rays of geometric optics. In particular, the asymptotic expansion of the solution may have the form of an electromagnetic wave,

$$u(x, y) = e^{ikS(x,y)}(\varphi_0(x, y) + k^{-1}\varphi_1(x, y) + \cdots) \tag{III.8}$$

(here we consider the Helmholtz equation). The level surfaces $S(x, y) = \text{const}$ of the phase $S(x, y)$ play the role of wave fronts, and we should analyze the evolution of these surfaces.

It is now completely evident that the same differential equation can have different characteristic equations according to which asymptotic expansions we are interested in. Indeed, if we substitute the asymptotic expansion (III.7) into the Klein–Gordon equation and equate the coefficient of the leading singularity $\delta'(\Phi)$ to zero, we naturally obtain the characteristic equation (III.5), whereas the substitution of the asymptotic expansion (III.8) into the Helmholtz equation yields the eikonal equation (III.6).

We arrive at the following conclusion: there is no natural characteristic equation associated with a given differential equation; the characteristic equation (or, what is the same, the Hamiltonian function) is determined not only by the differential equation, but also by the form of the asymptotic expansion we intend to obtain.

Following the existing tradition, we say that the characteristics associated with expansions with respect to smoothness (e.g., (III.7)) are *standard*; all other types of characteristics (including oscillatory expansions of the form (III.8)) are said to be *nonstandard* (as regards this terminology, see the paper [131], which was extensively used in this section).

One should note that there are numerous types of various asymptotic expansions that could be constructed for a given differential equation: with respect to smoothness, large (or small) parameter, growth at infinity, etc. We also mention the so-called *synchronous* asymptotic expansions involving two or more parameters (say, smoothness and a large parameter). The preceding discussion shows that each type of asymptotic expansion leads to a different characteristic equation (or Hamiltonian function) and to different equations for the amplitudes $\varphi_0, \varphi_1, \dots$ occurring in asymptotic expansions of the form (III.7), (III.8), etc.

It would be illogical if one had to develop the theory from the very beginning for each new asymptotic problem. Therefore, in the next subsection we try to find common features of asymptotic expansions with respect to smoothness and small (large) parameters and thus to guess the outlines of the general theory.

1.3 Asymptotic Expansions: Smoothness vs Parameter

The problem of obtaining asymptotic expansions with respect to a small parameter can easily be reduced to the problem of constructing asymptotics with respect to smoothness. Here we use the Helmholtz equation to illustrate this reduction.

Physically, the idea is quite obvious. The point is that the Helmholtz equation is the stationary equation corresponding to the wave equation

$$\frac{\partial^2 U}{\partial t^2} = \frac{c^2}{n^2(x, y)} \left(\frac{\partial^2 U}{\partial x^2} + \frac{\partial^2 U}{\partial y^2} \right).$$

In other words, the Helmholtz equation is an equation for the Fourier transform of $U(x, t)$ with respect to t taken at frequency ω. The wave number k is equal to ω/c and the ratio $c/n(x, y)$ is the phase velocity of the wave in the medium ($n(x, y)$ is known as the *reflection coefficient* in geometric optics). Since the decay of the Fourier transform $u(x, y, k)$ in k is known to be equivalent to the smoothness of $U(x, y, t)$ with respect to t, we see that the problem of finding asymptotic expansions as $k \to \infty$ of solutions to the Helmholtz equation is reduced to the problem of finding asymptotics expansions with respect to smoothness for the wave equation[1]. The reverse transition (from smoothness asymptotics to parameter asymptotics) is not as simple as the direct one. First, asymptotic expansions with respect to smoothness are usually considered in all variables rather then a single variable. Second, the transition to the Fourier transform in t in the wave equation is simplified by the fact that the coefficients of this equation do not depend on this variable. If we tried to pass to the Fourier transform in the variable x, we would not obtain a differential equation for it unless the coefficients of the equation are polynomials.

To deal with each of these difficulties separately, we first consider the equation

$$\hat{L}u(x, t) = \sum_{j+|\alpha|\leq m} a_{j\alpha}(x, t) \left(\frac{\partial}{\partial t}\right)^j \left(\frac{\partial}{\partial x}\right)^\alpha u(x, t) = 0 \qquad \text{(III.9)}$$

and construct asymptotic solutions with respect to smoothness in t alone. According to what was said above, we perform the Fourier transform $t \to k$ in equation (III.9) and obtain the equation

$$\sum_{j+|\alpha|\leq m} a_{j\alpha}\left(x, -\frac{\partial}{\partial k}\right) (ik)^j \left(\frac{\partial}{\partial x}\right)^\alpha \tilde{u}(x, k) = 0.$$

We seek its solutions in the form

$$\tilde{u}(x, k) = e^{ikS(x)}(\varphi_0(x) + k^{-1}\varphi_1(x) + \cdots) = e^{ikS(x)}\varphi(x, k).$$

The asymptotic solution of the original equation has the form

$$u(x, t) = F^{-1}_{k\to t}\left\{e^{ikS(x)}\varphi(x, k)\right\},$$

where $F^{-1}_{k\to t}$ is the inverse Fourier transform.

Note that the latter formula can be rewritten as

$$u(x, t) = e^{iS(x)(-i\partial/\partial t)}\varphi\left(x, -i\frac{\partial}{\partial t}\right)\delta(t). \qquad \text{(III.10)}$$

It is somewhat difficult to substitute this expression into the equation directly, since the coefficients depend on t and thus do not commute with the operator $(-i\partial/\partial t)$

[1]Note that in spite of the evident relationship described here both theories were developed independently for a long time. The reason probably resides in the difficulties which we discuss below.

involved in the asymptotic expansion (in fact, $-i\partial/\partial t$ is the "large parameter" of our expansion). This indicates that noncommuting operators may be useful here. Clearly, the fact that t does not commute with $-i\partial/\partial t$ directly affects the form of the corresponding Hamiltonian function.

In conclusion, let us try to find the general form of asymptotic expansions with respect to smoothness of solutions to differential equations. We use formula (III.10) as a starting point. Clearly, one should use derivatives with respect to all variables. To simplify the notation, we consider independent variables $(x_1, \dots, x_n)$ (i.e., we do not distinguish t explicitly). Analyzing formula (III.10), we arrive at the conclusion that the general form of the asymptotic expansion with respect to smoothness is[2]

$$u(x) = e^{iS\left(\overset{2}{x}, -i\overset{1}{\frac{\partial}{\partial x}}\right)} \varphi\left(\overset{2}{x}, -i\overset{1}{\frac{\partial}{\partial x}}\right) f(x),$$

where $f(x)$ is an arbitrary function of x, $S(x, p)$ is first-order homogeneous in $p = (p_1, \dots, p_n)$, and $\varphi(x, p)$ is given by a formal series

$$\varphi(x, p) = \varphi_0(x, p) + \varphi_{-1}(x, p) + \cdots.$$

of homogeneous functions (the subscript denotes the degree of homogeneity).

We see that

i) the operators

$$\left(-i\frac{\partial}{\partial x^1}, \dots, -i\frac{\partial}{\partial x^n}\right)$$

play the role of "large parameters" in the expansions with respect to smoothness;

ii) the terminology of noncommutative analysis is convenient when dealing with expansions of the form (III.10).

In more general cases, the operators defining the asymptotic expansion can be chosen in some other way. For example, if we deal with simultaneous asymptotic expansions with respect to smoothness and growth at infinity, we will usually introduce the set of operators

$$\left(-i\partial/\partial x_1, \dots, -i\partial/\partial x_n, x_1, \dots, x_n\right),$$

which itself is not commutative.

In the next subsection we give a sketch of the general theory of asymptotic expansions with respect to a tuple of (noncommuting) operators.

---

[2]We ignore focal (caustic) points for the sake of simplicity.

1.4 Asymptotic Expansions with Respect to an Ordered Tuple of Operators

As was explained in the preceding section, in order to construct the general theory of asymptotic expansions, it is necessary, first of all, to choose and fix a tuple of operators $A = (A_1, \dots, A_n)$, which will serve as large parameters for asymptotic expansions in question. Clearly, this tuple cannot be chosen completely arbitrarily (but rather should satisfy some conditions we do not discuss here; see [133]). We assume that our operators are self-adjoint operators in some Hilbert space[1] H. The choice of the tuple A determines the type of asymptotic expansion and it is now appropriate to give a precise definition of this basic notion.

For this purpose, we introduce a scale of spaces $H^s(A)$ associated with the tuple A in the following way.

Let $D \subset H$ be a linear subset of the common domain of all possible products of powers of operators $A_1, \dots, A_N$. The set D is assumed to be dense in H. We consider the following family of norms on D:

$$\| u \|_s^2 = \left\| (1 + A_1^2 + \dots + A_N^2)^{s/2} u \right\|_H .$$

The space $H^s(A)$ is defined as the completion of D with respect to the norm $\| \cdot \|_s$.

Definition III.2 A function $\tilde{u}$ is called an *asymptotic approximation of order s* of a function u if $u - \tilde{u} \in H^s(A)$.

Thus, if $H = L_2(\mathbb{R}^1)$, $N = n$, and

$$A = (A_1, \dots, A_n) = \left(-i \frac{\partial}{\partial x_1}, \dots, -i \frac{\partial}{\partial x_n} \right),$$

then we can take $D = C_0^\infty(\mathbb{R}^n)$, and the scale $H^s\,(-i\partial/\partial x)$ is none other than the usual Sobolev scale. Asymptotic expansions obtained in this way are those with respect to smoothness. If we choose another tuple of operators, say,

$$A = \left(-i \frac{\partial}{\partial x_1}, \dots, -i \frac{\partial}{\partial x_n}, x^1, \dots, x^n \right),$$

then the norms in the spaces $H^s\,(-i\partial/\partial x, x)$ are defined by the formulas

$$\| u(x) \|_s = \int | (1 + |x|^2 - \Delta)^{s/2} u(x) \, | \, dx.$$

An advantage of these norms is that they are Fourier-invariant. Along with smoothness, these norms allow for the decay at infinity and prove to be useful in studies dealing with pseudodifferential equations (see Shubin [163]).

[1] For example, one can often take $H = L_2$.

The second stage in the construction of asymptotic expansions with respect to the tuple $(A_1, \dots, A_N)$ is to represent the differential operator of the original problem as a function of the (generally noncommuting) tuple $(A_1, \dots, A_N)$, that is, to represent the original equation in the form

$$H(\overset{1}{A_1}, \dots, \overset{N}{A_N})\, u = f. \tag{III.11}$$

Of course, such a representation is generally *ambiguous* and sometimes *impossible*. In constructing asymptotic expansions with respect to smoothness, along with the operators $(-i\partial/\partial x)$ defining the scale H^s, one needs to introduce the operators $x_1, \dots, x_n$ not involved in the definition of the asymptotic expansion. Even at this stage one can obtain different representations of the original equation by choosing different ordering of operators. For simplicity, we do not dwell upon the fact that not all of the operators $A_1, \dots, A_N$ can be involved in the definition of the scale $H^s(A)$.

The first two stages described are not algorithmized as yet. The choice of the tuple A and the representation of the original equation must be done in a special way for each equation and is a matter of art rather than craft. However, we try to show in the following that as soon as the tuple A is chosen and the representation obtained, the remaining stages are automatically determined by the calculus of noncommuting operators.

1.5 Reduction to Pseudodifferential Equations

With the asymptotic expansion constructed in Section 2 as an example, let us try to find asymptotic solutions, in the sense of Definition III.2, of equation (III.11) in the form

$$u = F(\overset{1}{A_1}, \dots, \overset{N}{A_N})\, f.$$

A sufficient condition that u be a solution of equation (III.11) is that the operator $F(\overset{1}{A_1}, \dots, \overset{N}{A_N})$ be a solution of the operator equation

$$[\![H(\overset{1}{A_1}, \dots, \overset{N}{A_N})]\!]\,[\![F(\overset{1}{A_1}, \dots, \overset{N}{A_N})]\!] = 1.$$

This is an equation for the symbol $F(z_1, \dots, z_n)$ of the operator $F(\overset{1}{A_1}, \dots, \overset{N}{A_N})$. Here we consider the case, studied in Chapter II, in which the following condition is satisfied:

The product of any two functions of the ordered tuple $A = (\overset{1}{A_1}, \dots, \overset{N}{A_N})$ *can be represented as a function of the same tuple, i.e., the functions of* $(\overset{1}{A_1}, \dots, \overset{N}{A_N})$ *form an operator algebra.*

As was shown in Chapter II, this is the case if the tuple $A = (A_1, \dots, A_N)$ has a left ordered representation on the space of symbols. Let $l = (l_1, \dots, l_N)$ be this representation. Then

$$[\![H(\overset{1}{A}_1, \dots, \overset{N}{A}_N)]\!]\,[\![F(\overset{1}{A}_1, \dots, \overset{N}{A}_N)]\!] = \varphi(\overset{1}{A}_1, \dots, \overset{N}{A}_N),$$

where

$$\varphi(y_1, \dots, y_N) = H(\overset{1}{l}_1, \dots, \overset{N}{l}_N)\,(F).$$

Thus we obtain the equation

$$H(\overset{1}{l}_1, \dots, \overset{N}{l}_N)\,(F) = 1$$

for the symbol $F(y_1, \dots, y_N)$. Of course, this equation has little practical value unless we can calculate the left ordered representation operator $H(\overset{1}{l}_1, \dots, \overset{N}{l}_N)$ explicitly; we have discussed these problems in Chapter II, and we shall also find numerous examples in the following sections. Here we only recall that "triangular" commutation relations lead to differential operators $l_1, \dots, l_N$ of the left ordered representation; if the symbol H is a polynomial in the variables whose associated operators are nontrivial, then $H(\overset{1}{l}_1, \dots, \overset{N}{l}_N)$ is a differential operator as well.

So, what conclusion can we derive from our considerations? The problem of solving equation (III.11) has been reduced successively to the operator equation and then to the equation for the symbol. At first glance it seems that this "symbolic" problem is no simpler than the original one. This impression would be true if we considered *exact* solutions of equation (III.11) rather that asymptotic ones. But in order to obtain *asymptotic* solutions we should solve the operator equation modulo operators of sufficiently low negative order in the scale $H^s(A)$. Accordingly, the "symbolic" equation should be solved up to functions of $z_1, \dots, z_n$ that decay at infinity sufficiently rapidly (this clearly requires "good" estimates for the operators $F(\overset{1}{A}_1, \dots, \overset{N}{A}_N)$ in the scale $H^s(A)$ and concerns the choice of appropriate symbol classes).

Thus, *the operational calculus permits us to reduce the problem of constructing asymptotic solutions to equation* (III.11) *with respect to an arbitrary tuple of operators to asymptotic expansion of a solution to the symbolic equation with respect to growth at infinity.* On passing in the latter equation to the Fourier transform, *we reduce an "arbitrary" asymptotic problem to a problem of constructing asymptotic expansions with respect to smoothness.*

We can now return (at a new level of understanding) to the problem posed in Subsection 1.2: *What is the "correct" characteristic equation (or, what is the same, the Hamiltonian) for a given differential equation if we consider asymptotic solutions with respect to a tuple* $A = (A_1, \dots, A_N)$? With the above information in mind, we should perform the following steps in order to construct the Hamiltonian.

1) Compute the operators of the left ordered representation.

2) Compute the operator $H(\overset{1}{l}_1, \ldots, \overset{N}{l}_N)$ and pass to the Fourier transform by $(z_1, \ldots, z_N)$ (or only the part of these variables that corresponds to the type of asymptotic expansion considered).

The "standard" Hamiltonian function of the operator obtained is the Hamiltonian function corresponding to the original equation for the given type of asymptotic expansions.

Let us consider the construction of the characteristic equation for the Helmholtz equation. To represent this equation in the form (III.11), we introduce the following set of operators:

$$A_1 = -i\frac{\partial}{\partial x}, \quad A_2 = -i\frac{\partial}{\partial y}, \quad A_3 = k, \quad B_1 = x, \quad B_2 = y.$$

The Helmholtz equation now becomes

$$[\overset{1}{A}{}_1^{2} + \overset{1}{A}{}_2^{2} - \overset{1}{A}{}_3^{2} n^2(\overset{2}{B}_1, \overset{2}{B}_2)]u = 0$$

(the assignment of Feynman indices uses the fact that some of these operators commute with each other).

Let us write out the left ordered representation operators, using the following correspondence between the operators and the arguments of symbols:

$$z_1 \leftrightarrow A_1, \; z_2 \leftrightarrow A_2, \; z_3 \leftrightarrow A_3, \; w_1 \leftrightarrow B_1, \; w_2 \leftrightarrow B_2.$$

We have

$$l_{B_1} = w_1, \quad l_{B_2} = w_2,$$
$$l_{A_3} = z_3, \quad l_{A_1} = z_1 - i\frac{\partial}{\partial w_1}, \quad l_{A_2} = z_2 - i\frac{\partial}{\partial w_2}.$$

Hence the operator $H(\overset{1}{l}_1, \ldots, \overset{N}{l}_N)$ has the following form

$$H(\overset{1}{l}_1, \ldots, \overset{N}{l}_N) = \left(z_1 - i\frac{\partial}{\partial w_1}\right)^2 + \left(z_2 - i\frac{\partial}{\partial w_2}\right)^2 - z_3^2 n^2(w_1, w_2).$$

The subsequent argument depends on which operators we include in the definition of the asymptotic expansions considered.

If we consider asymptotic solutions with respect to smoothness, then the operators A_1 and A_2 serve as large parameters.

Accordingly, the Hamiltonian function is equal to $(z_1 + p_1)^2 + (z_2 + p_2)^2$ (where p_1 and p_2 are the variables dual to w_1 and w_2, respectively), and there are no real characteristics.

On the other hand, if we consider asymptotic expansions with respect to a parameter, then only the operator A_3 is to be included in the definition of the scale. Then the Hamiltonian function equals

$$p_1^2 + p_2^2 - z_3^2 n^2(w_1, w_2),$$

which is in full agreement with the Hamilton–Jacobi equation (III.6).

1.6 Commutation of an h^{-1}-Pseudodifferential Operator with an Exponential

The WKB method of constructing asymptotic expansions of solutions to differential equations, as well as several related approaches, uses Ansätze of the form

$$u(x,h) = e^{iS(x)/h}\varphi(x,h)$$

for the solutions of the differential equation

$$P\left(\overset{2}{x}, -ih\overset{1}{\frac{\partial}{\partial x}}\right) u(x,h) = 0.$$

Here $\varphi(x,h)$ depends regularly on the small parameter h, and the method is to substitute $u = \exp(iS/h)\varphi$ into the equation, permute P and $\exp(iS/h)$, and then solve the obtained equation for φ by means of the regular perturbation theory with respect to h.

Thus the point is to expand the operator

$$e^{-iS(x)/h}[\![P\left(\overset{2}{x}, -ih\overset{1}{\frac{\partial}{\partial x}}\right)]\!]e^{iS(x)/h} \tag{III.12}$$

in a series in powers of h.

The conventional technique used in this situation is the stationary phase method. We present here a different technique that uses noncommutative functional calculus.

Denote $U = e^{iS(x)/h}$. The mapping $A \mapsto U^{-1}AU$ is an algebra homomorphism; it follows from Theorem I.4, 1° that

$$U^{-1}[\![P\left(\overset{2}{x}, -ih\overset{1}{\frac{\partial}{\partial x}}\right)]\!]U = P(\overset{2}{[\![U^{-1}xU]\!]}, \overset{1}{[\![U^{-1}}\left(-i\frac{\partial}{\partial x}\right)U]\!]).$$

Substituting $U = e^{iS(x)/h}$ into the latter formula and taking into account that

$$e^{-iS(x)/h}xe^{iS(x)/h} = x,$$

and

$$e^{-iS(x)/h}\left(-ih\frac{\partial}{\partial x}\right)e^{iS(x)/h} = -ih\frac{\partial}{\partial x} + \frac{\partial S}{\partial x},$$

we obtain the expression for the operator (III.12) in the form

$$e^{-iS(x)/h}[\![P\left(\overset{2}{x}, -ih\overset{1}{\frac{\partial}{\partial x}}\right)]\!]e^{iS(x)/h} = [\![P\left(\overset{2}{x}, \overset{1}{[\![} -ih\frac{\partial}{\partial x} + \frac{\partial S}{\partial x}]\!]\right)]\!].$$

Thus, it suffices to calculate the operator

$$P\left(\overset{2}{x}, \overset{1}{[\![} -ih\frac{\partial}{\partial x} + \frac{\partial S}{\partial x}]\!]\right).$$

In doing so, we may pay no attention to the argument $\overset{2}{x}$ and consider simply

$$f\left([\![-ih\frac{\partial}{\partial x} + \frac{\partial S}{\partial x}]\!]\right),$$

with a given function $f(z)$ (the general case is no more complicated).

In order to obtain the expansion up to a certain power of the small parameter h, we use the Taylor formula (see the discussion following Theorem I.12). Using this formula for $f(B+\varepsilon C)$ with $B = \partial S/\partial x$, $C = -\partial/\partial x$, $\varepsilon = h$, we obtain[2]

$$\begin{aligned} f\left([\![-ih\frac{\partial}{\partial x} + \frac{\partial S}{\partial x}]\!]\right) &= f\left(\frac{\partial S}{\partial x}\right) \\ &+h\left[f'\left(\frac{\partial S}{\partial x}\right)\left(-i\frac{\partial}{\partial x}\right) - \frac{1}{2}f''\left(\frac{\partial S}{\partial x}\right)\mathrm{ad}_{\partial S/\partial x}\left(-i\frac{\partial}{\partial x}\right)\right] + O(h^2), \end{aligned}$$

since $\mathrm{ad}_B^2(C) = 0$.

Furthermore, we have

$$\mathrm{ad}_{\partial S/\partial x}\left(-i\frac{\partial}{\partial x}\right) = \left[\frac{\partial S}{\partial x}, -i\frac{\partial}{\partial x}\right] = i\frac{\partial^2 S}{\partial x^2},$$

which implies that

$$f\left([\![-i\frac{\partial}{\partial x} + \frac{\partial S}{\partial x}]\!]\right) = f\left(\frac{\partial S}{\partial x}\right) - ih\left\{f'\left(\frac{\partial S}{\partial x}\right)\left(\frac{\partial}{\partial x}\right) + \frac{1}{2}\frac{\partial^2 S}{\partial x^2}f''\left(\frac{\partial S}{\partial x}\right)\right\} + O(h^2).$$

Let us also mention an elegant *exact* expression for $f([\![-ih\partial/\partial x + \partial S/\partial x]\!])$ in terms of functions of noncommuting operators (see [98]):

$$f\left([\![-i\frac{\partial}{\partial x} + \frac{\partial S}{\partial x}]\!]\right) = f\left(-ih\overset{2}{\frac{\partial}{\partial x}} + \delta S(\overset{1}{x}, \overset{3}{x})\right). \tag{III.13}$$

---

[2]For simplicity, we write down only two main terms of the expansion; the subsequent terms can be written down easily with the help of Taylor's formula and we leave this computation to the reader.

We are going to prove this formula with the use of the definition of functions of operators via the Fourier transform (this is quite natural if one deals with h^{-1}-pseudodifferential operators). Therefore, we first consider the particular case in which $f(z) = e^{itz}$; the general case follows by integration.

Taking into account the relation

$$\hat{p} + \frac{\partial S}{\partial x} = e^{-iS(x)/h}\hat{p}e^{iS(x)/h}$$

where $\hat{p} = -i\partial/\partial x$, we obtain the formula

$$\begin{aligned} e^{it[\![\hat{p}+\partial S/\partial x]\!]} &= e^{-iS(x)/h}e^{it\hat{p}}e^{iS(x)/h} \\ &= e^{i[t\overset{2}{\hat{p}}+(S(\overset{1}{x})-S(\overset{3}{x}))/h]} = e^{i[t\overset{2}{\hat{p}}+(\overset{1}{x}-\overset{3}{x})\delta S(\overset{0}{x},\overset{4}{x})/h]}. \end{aligned}$$

Using the obvious relation

$$e^{it\overset{2}{\hat{p}}}\varphi(\overset{1}{x}) = e^{it\overset{1}{\hat{p}}}\varphi(\overset{2}{x} + th)$$

we can permute the operators $\overset{2}{\hat{p}}$ and $\hat{x}$ and obtain

$$e^{it[\![\hat{p}+\partial S/\partial x]\!]} = e^{i[t\overset{2}{\hat{p}}+(\overset{1}{x}+th-\overset{3}{x})\delta S(\overset{0}{x},\overset{4}{x})/h]} = e^{(it\overset{2}{\hat{p}}+\delta S(\overset{1}{x},\overset{3}{x}))}.$$

Representing now both sides of the formula (III.13) via the Fourier transform of $f(z)$ and using the latter relation we obtain the desired result.

This is indeed a remarkable formula. It gives a closed form expression for

$$f\left([\![\hat{p} + \frac{\partial S}{\partial t}]\!]\right),$$

and the expansion in powers of h can be obtained in a very simple way, just by expanding into the ordinary Taylor series in powers of $(-ih\partial/\partial x)$.

1.7 Summary: the General Scheme

Here we try to summarize the observations made above.

Suppose that we intend to construct an asymptotic expansion of some type for the equation

$$P\left(x, -i\frac{\partial}{\partial x}\right) u(x) = f(x),$$

or for the Cauchy problem

$$\begin{cases} \left(-i\frac{\partial}{\partial t}\right)^m u + P\left(x, t, -i\frac{\partial}{\partial x}\right) u = f, \\ \frac{\partial^j u}{\partial t^j}(x, 0) = 0, \quad j = 1, \dots, m-1, \end{cases}$$

or for any other well-posed problem for a differential equation.

The first stage of analysis consists in introducing operators $A_1, \dots, A_N$ (probably not commuting with each other), which would define the type of asymptotic expansion.

This means that one constructs an asymptotic expansion of the solution to the original problem with respect to the scale $H^s(A)$ of function spaces with norms determined by the equation

$$\|u\|_s = \left\| (1 + A_1^2 + \cdots + A_N^2) u \right\|_H .$$

Next, it is necessary to represent the main problem in operator form. As we will see in concrete examples in the subsequent sections, this might require introducing some additional operators $B_1, \dots, B_m$ that do not occur among $A_1, \dots, A_n$. It may even happen that one and the same operator occurs twice in the sequence $A_1, \dots, A_n$, $B_1, \dots, B_m$, once among the A_j's, defining the type of asymptotic expansions, and once more among $B_1, \dots, B_m$; this looks as if such an operator splits into two on passing to the operator interpretation.

At this stage we represent the problem under investigation in the form of an operator equation

$$[\![H(\overset{i_1}{A_1}, \dots, \overset{i_n}{A_n}, \overset{j_1}{B_1}, \dots, \overset{j_m}{B_m})]\!] \, [\![\Phi(\overset{i_1}{A_1}, \dots, \overset{i_n}{A_n}, \overset{j_1}{B_1}, \dots, \overset{j_m}{B_m})]\!] = 1$$

(for stationary equations such as the Helmholtz equation) or of a Cauchy problem

$$\begin{cases} [\![\left(-i\frac{\partial}{\partial t}\right)^m + H(\overset{i_1}{A_1}, \dots, \overset{i_n}{A_n}, \overset{j_1}{B_1}, \dots, \overset{j_m}{B_m})]\!] \, [\![\Phi(\overset{i_1}{A_1}, \dots, \overset{i_n}{A_n}, \overset{j_1}{B_1}, \dots, \overset{j_m}{B_m})]\!] = 0, \\ \left.\dfrac{\partial^j \Phi}{\partial t^j}\right|_{t=0} = 0, \quad j = 0, \dots, m-2; \quad \left.\dfrac{\partial^j \Phi}{\partial t^j}\right|_{t=0} = 1, \end{cases}$$

or in some other operatorial form (as usual, 1 denotes the identity operator in the function space considered). The operators $A_1, \dots, A_n$ and $B_1, \dots, B_m$ occurring in these representations should be chosen in such a way that

1) the class of operators

$$\Phi(\overset{i_1}{A_1}, \dots, \overset{i_n}{A_n}, \overset{j_1}{B_1}, \dots, \overset{j_m}{B_m})$$

for the chosen symbol class $\Phi \in \mathcal{F}$ forms an operator algebra;

2) the symbol class $\mathcal{F}$ is large enough as to contain asymptotic solutions of equations for the symbols given below.

Using the first requirement, we pass from the operator equations to equations for the symbols of these operators by using the technique of left ordered representations. To this end, one computes the operators $\hat{l}_{A_j}$, $j = 1, \dots, N$, and $\hat{l}_{B_j}$, $j = 1, \dots, M$, of the left ordered representation associated with the tuple $(\overset{i_1}{A_1}, \dots, \overset{i_n}{A_n}, \overset{j_1}{B_1}, \dots, \overset{j_m}{B_m})$ and writes out the equation for the symbol:

$$H(\overset{i_1}{l}_{A_1}, \dots, \overset{i_n}{l}_{A_n}, \overset{j_1}{l}_{B_1}, \dots, \overset{j_m}{l}_{B_m}) \, \Phi(z_1, \dots, z_n, w_1, \dots, w_m) = 1$$

(or the corresponding symbolic Cauchy problem for the operator Cauchy problem).

One should now construct the asymptotic solution of the resulting symbolic problem with respect to decay as $(z_1, \ldots, z_N) \to \infty$ (if desired, one can pass to smoothness asymptotic expansions by considering the Fourier transform with respect to z). This stage of constructing asymptotic expansions has nothing to do with noncommuting operators except for the fact that it is at this stage that the minimal extent of the symbol space $\mathcal{F}$ to be considered becomes clear. Of course, most appropriate would be the variant in which $\mathcal{F}$ is very large and so automatically contains the solutions of problems of this type. However, it should be taken into account that if $\mathcal{F}$ is too large, then unavoidable function-analytic difficulties may well occur. Thus one should be very careful.

If we succeed, then it is quite easy to construct the asymptotic solutions. For example, the asymptotic expansion of the solution to the equation $P(x, -i\partial/\partial x)\,u = f$ has the form

$$u(x) = \tilde{\Phi}(\overset{i_1}{A_1}, \ldots, \overset{i_n}{A_n}, \overset{j_1}{B_1}, \ldots, \overset{j_m}{B_m})\, f(x),$$

where $\Phi(z, w)$ is an asymptotic solution of the corresponding symbolic problem with respect to negative powers of z, and the asymptotic solution of the Cauchy problem has the form

$$u(x,t) = \int_0^t \tilde{\Phi}((t-\tau), \overset{i_1}{A_1}, \ldots, \overset{i_n}{A_n}, \overset{j_1}{B_1}, \ldots, \overset{j_m}{B_m})\, f(\tau, x)\, d\tau,$$

where $\Phi(t, z, w)$ is the asymptotic solution of the corresponding symbolic Cauchy problem.

Thus, the operator calculus provides a method of reducing the problem of asymptotic expansions of arbitrary type for (differential) equations to a problem of constructing asymptotic solutions for large z for pseudodifferential equations.

2 Difference and Difference-Differential Equations

Noncommutative analysis offers convenient techniques to deal with difference and difference-differential approximations to differential equations. Such approximations are obtained by replacing the derivatives occurring in a differential equation by their finite-difference counterparts. For example, $\partial f/\partial x$ can be substituted on a uniform grid by either

$$\delta_x^+ f(x) = \frac{f(x+h) - f(x)}{h}$$

(the forward difference), or

$$\delta_x^- f(x) = \frac{f(x) - f(x-h)}{h}$$

(the backward difference), or, more rarely,

$$\delta_x^c f(x) = \frac{f(x+h/2) - f(x-h/2)}{h}$$

(the central difference); here h is the mesh width. The second derivative $\partial^2 f/\partial x^2$ is usually replaced by its central-difference approximation

$$\delta_x^+ \delta_x^- f(x) = \frac{f(x+h) - f(x-h) - 2f(x)}{h^2},$$

and the mixed derivatives $\partial^2 f/\partial x_i \partial x_j$ turn into something like $\delta_{x_i}^{\pm} \delta_{x_j}^{\pm} f$ with various combinations of signs. Similar expressions can be written out for higher-order derivatives.

The resulting equation can be viewed as a finite linear algebraic system (or a system of ordinary differential equations if, say, the t-derivatives are retained in the equation) for the finite set of values of the unknown function(s) at the mesh points. It is to be supplemented with the related boundary and (or) initial conditions and then solved numerically. There is an advanced theory concerning the choice of grids and approximations, the convergence and stability of solutions as $h \to 0$, etc.

How can noncommutative calculus be useful in these topics? Without going into detail, let us briefly outline three different approaches that can be applied here.

2.1 Difference Approximations as Pseudodifferential Equations

Since the translation operator

$$T^h f(x) = f(x+h)$$

can be presented as a function of $-i\partial/\partial x$,

$$T^h = e^{h\partial/\partial x} = e^{ih(-i\partial/\partial x)},$$

the same is clearly true of difference approximations to derivatives. Specifically, we have

$$\begin{aligned}
\delta_x^+ &= \frac{e^{h\partial/\partial x} - 1}{h} = \frac{e^{ih(-i\partial/\partial x)} - 1}{h}, \\
\delta_x^- &= \frac{e^{-ih(-i\partial/\partial x)} - 1}{h}, \\
\delta_x^+ \delta_x^- &= \frac{e^{ih(-i\partial/\partial x)} + e^{-ih(-i\partial/\partial x)} - 2}{h^2} = \frac{2\cos(-ih\partial/\partial x) - 2}{h^2} \\
&= -\frac{4}{h^2} \sin^2\left(-\frac{ih}{2}\frac{\partial}{\partial x}\right),
\end{aligned}$$

etc. With these expressions substituted for derivations, a differential operator turns into a pseudodifferential operator.

For example, consider the wave equation in an inhomogeneous medium

$$\frac{\partial^2 u}{\partial t^2} - c^2(x)\frac{\partial^2 u}{\partial x^2} = 0, \quad t \in \mathbb{R}^1, \quad x \in \mathbb{R}^1,$$

where $c(x)$ is the speed of sound depending on the position x. By replacing $\partial^2/\partial x^2$ with its central-difference approximation, we obtain the difference-differential equation

$$\frac{\partial^2 u}{\partial t^2} + \frac{4c^2(x)}{h^2}\sin^2\left(-\frac{ih}{2}\frac{\partial}{\partial x}\right)u = 0.$$

Denote by

$$u_n(t) = u(nh, t)$$

the values of u at the mesh points. Then the equation takes the form

$$\ddot{u}_n = \frac{4c_n^2(u_{n+1} - 2u_n + u_{n-1})}{h^2}, \quad n = 0, \pm 1, \pm 2, \dots,$$

where $c_n = c(nh)$. The latter system describes vibrations of a one-dimensional crystal lattice in which the atoms only interact with their nearest neighbours and the interaction is Hookeian.

It is interesting to study the relationship between the solutions to the two equations (with appropriately related initial data) as $h \to 0$. For u sufficiently smooth, the difference-differential operator is a good approximation to the wave operator (one can expand

$$\sin^2\left(-\frac{ih}{2}\frac{\partial}{\partial x}\right)$$

into the MacLaurin series, and the remainder will decay as $h \to 0$ since u is smooth); but what can be said of arbitrary (say discontinuous) u? The advantage of the operator approach is that the difference-differential equation is considered as a pseudodifferential equation, and so the standard WKB and canonical operator machinery can be used to find its asymptotic solutions with respect to either smoothness, $h \to 0$, or both.

Precise analysis (not reproduced here) shows that, for smooth initial data, the solutions of the difference-differential equation tend to the corresponding solutions of the wave equation, whereas the situation is a bit more complicated for discontinuous initial data. Specifically, the "leading" discontinuity of the solution to the difference-differential equation propagates along the characteristics of the wave equation, but it is accompanied by a "tail of vibrations" that occupies a certain region in the space $\mathbb{R}^2_{x,t}$. This region can be described in terms of the characteristics (we refer the reader to [129] for a detailed exposition).

Thus, it is possible to study the behaviour as the mesh width tends to zero of solutions to difference and difference-differential equations with ill-behaved initial data or right-hand sides.

2.2 Difference Approximations as Functions of x and $\delta_x^{\pm}$

Another approach, which can also prove valuable in some applications, consists in representing difference operators as functions of x, δ_x^+, and δ_x^-. Then difference equations can be written as something like

$$F(\overset{3}{x}, \overset{1}{\delta_x^+}, \overset{1}{\delta_x^-})u = v,$$

and the problem is easily reduced to finding the symbol $G(y_1, y_2, y_3)$ of the (asymptotic) inverse $G(\overset{3}{x}, \overset{2}{\delta_x^+}, \overset{1}{\delta_x^-})$ of $F(\overset{3}{x}, \overset{1}{\delta_x^+}, \overset{1}{\delta_x^-})$, where the equation for $G(y_1, y_2, y_3)$ is obtained via the ordered representation.

Therefore, let us calculate the left ordered representation for the Feynman tuple $(\overset{3}{x}, \overset{2}{\delta_x^+}, \overset{1}{\delta_x^-})$; the corresponding variables will be denoted by

$$y_1 \leftrightarrow \overset{3}{x}, \quad y_2 \leftrightarrow \overset{2}{\delta_x^+}, \quad y_3 \leftrightarrow \overset{1}{\delta_x^-}$$

(for clarity, we assume that $x \in \mathbb{R}^1$, so that no additional indices appear).

First of all, let us find the commutation relations satisfied by these operators. Denote $\delta_x^+ = \delta^+$ and $\delta_x^- = \delta^-$ (i.e., omit the index x). We have

$$\delta^+(xf(x)) = \frac{\delta}{\delta x}(xf(x))(x, x+h)$$

and, by the Leibniz rule for difference derivatives,

$$\delta^+(xf(x)) = f(x)\,\delta^+(x) + (x+h)\delta^+(f(x)) = f(x) + (x+h)\delta^+(f(x)),$$

that is,

$$\delta^+ x = (x+h)\,\delta^+ + 1.$$

Similarly,

$$\delta^- x = (x-h)\,\delta^- + 1.$$

Finally, both δ^+ and δ^- are linear combinations of translations and scalar operators. Since the translations commute, so do δ^+ and δ^-,

$$[\delta^+, \delta^-] = 0.$$

We can now evaluate the ordered representation operators. We have

$$x[\![f(\overset{3}{x}, \overset{2}{\delta^+}, \overset{1}{\delta^-})]\!] = \overset{3}{x} f(\overset{3}{x}, \overset{2}{\delta^+}, \overset{1}{\delta^-}),$$

so that

$$l_x = y_1.$$

Next,

$$\begin{aligned}
\delta^+[\![f(\overset{3}{x},\overset{2}{\delta^+},\overset{1}{\delta^-})]\!] &= \overset{4}{\delta^+} f(\overset{3}{x},\overset{2}{\delta^+},\overset{1}{\delta^-}) \\
&= \overset{4}{\delta^+} f(\overset{5}{\overline{x+h}},\overset{2}{\delta^+},\overset{1}{a^-}) + \frac{\delta f}{\delta y_1}(\overset{5}{\overline{x+h}},\overset{3}{x},\overset{2}{\delta^+},\overset{1}{\delta^-})\overset{4}{1} \\
&= \overset{2}{\delta^+} f(\overset{3}{\overline{x+h}},\overset{2}{\delta^+},\overset{1}{\delta^-}) + \frac{\delta f}{\delta y_1}(\overset{3}{\overline{x+h}},\overset{3}{x},\overset{2}{\delta^+},\overset{1}{\delta^-}),
\end{aligned}$$

and we obtain

$$l_{\delta^+} = y_2 T^h_{y_1} + \delta^+_{y_1}$$

where the operator $T^h_{y^1}$ is the shift along the axis y^1:

$$T^h_{y^1} + f(y') = f(y'+h).$$

Similarly,

$$\begin{aligned}
\delta^-[\![f(\overset{3}{x},\overset{2}{\delta^+},\overset{1}{\delta^-})]\!] &= \overset{4}{\delta^-} f(\overset{3}{x},\overset{2}{\delta^+},\overset{1}{\delta^-}) \\
&= \overset{4}{\delta^+} f(\overset{5}{\overline{x-h}},\overset{2}{\delta^+},\overset{1}{\delta^-}) + \frac{\delta f}{\delta y_1}(\overset{5}{\overline{x-h}},\overset{3}{x},\overset{2}{\delta^+},\overset{1}{\delta^-})\overset{4}{1} \\
&= \overset{2}{\delta^-} f(\overset{3}{\overline{x-h}},\overset{2}{\delta^+},\overset{1}{\delta^-}) + \frac{\delta f}{\delta y_1}(\overset{3}{\overline{x-h}},\overset{3}{x},\overset{2}{\delta^+},\overset{1}{\delta^-}),
\end{aligned}$$

so that

$$l_{\delta^-} = y_3 T^{-h}_{y_1} + \delta^-_{y_1}.$$

The tuple $(l_x, l_{\delta^+}, l_{\delta^-})$ satisfies the generalized Jacobi condition. Indeed, we have

$$[l_{\delta^+}, l_{\delta^-}] = 0$$

since both operators involve translations with respect to y_1 and multiplications by y_2 or y_3. Furthermore,

$$l_{\delta^+} l_x = (y_2 T^h_{y_1} + \delta^+_{y_1}) y_1 = y_2(y_1+h)T^h_{y_1} + (y_1+h)\delta^+_{y_1} + 1 = (l_x + h) l_{\delta^+} + 1,$$

and similarly

$$l_{\delta^-} l_x = (l_x - h)\, l_{\delta^-} + 1.$$

Now, to invert the difference operator $F(\overset{3}{x}, \overset{2}{\delta^+_x}, \overset{1}{\delta^+_x})$, one should solve the equation

$$f(\overset{3}{y_1}, \overset{2}{\overline{y_2 T^h_{y_1} + \delta^+_{y_1}}}, \overset{1}{\overline{y_3 T^h_{y_1} + \delta^-_{y_1}}})(G(y_1, y_2, y_3)) = 1,$$

which is again a difference equation but with a special right-hand side.

2.3 Another Approach to Difference Approximations

The commutation relations between δ^+ (or δ^-) and x in the preceding subsection can be considered as small perturbations of those between $\partial/\partial x$ and x. So far, the logical sequence of reasoning was as follows: we replace derivatives by difference operators and study the operators arising, commutation relations, etc. Putting a different emphasis on it, we might wish to perturb the commutation relations directly without being bound to any particular replacements in the equation itself. We have also considered an example of this sort in Subsection 2.2; let us present another one.

Consider the Schrödinger equation

$$\left[-\frac{h^2}{2}\frac{\partial^2}{\partial x^2} + V(x)\right]\psi(x) - E\psi(x) = 0,$$

where E is a constant and $V(x)$ a smooth potential. Set

$$B = ihe^{-x}\frac{\partial}{\partial x}, \quad A = e^x.$$

Then

$$[A, B] = -ih,$$

and the Schrödinger equation takes the form

$$\left[-\frac{1}{2}(AB)^2 + V(\ln A)\right]\psi(x) - E\psi(x) = 0.$$

We fix the operator $A = e^x$ and assume that the commutation relation undergoes a small perturbation,

$$e^a AB - BA = -ih,$$

where a is small. What form will the equation take? To answer this question, we should find the perturbed operator B. We have

$$A^{-1}BA = e^a\, B + ihA^{-1},$$

or

$$e^{-x}\, B\, e^x = e^a \cdot B + ihe^{-x}.$$

Let us seek B in the form $B = B\left(\overset{2}{x}, \overset{1}{\partial/\partial x}\right)$. We have

$$e^{-x}[\![B\left(\overset{2}{x}, \overset{1}{\frac{\partial}{\partial x}}\right)]\!]e^x = B\left(\overset{2}{\overline{e^{-x}\cdot x\cdot e^x}}, \overset{1}{\overline{e^{-x}\cdot\frac{\partial}{\partial x}\cdot e^x}}\right) = B\left(\overset{2}{x}, \overset{1}{\frac{\partial}{\partial x}+1}\right),$$

so that the equation for B becomes

$$B\left(\overset{2}{x}, \frac{\overset{1}{\partial}}{\partial x} + 1\right) = e^a B\left(\overset{2}{x}, \frac{\overset{1}{\partial}}{\partial x}\right) + ihe^{-x},$$

or, passing to the symbols on both sides,

$$B(x, p+1) = e^a B(x, p) + ihe^{-x}.$$

Obviously the last equation has quite a few solutions. We take the particular solution

$$B(x, p) = ihe^{-x} \frac{1 - e^{ap}}{1 - e^a}.$$

Indeed, we have

$$\begin{aligned} B(x, p+1) &= ihe^{-x} \frac{1 - e^{a(p+1)}}{1 - e^a} = ihe^{-x} \frac{(1 - e^a) + (e^a - e^a e^{ap})}{1 - e^a} \\ &= ihe^{-x} + e^a B(x, p). \end{aligned}$$

The corresponding operator has the form

$$B\left(\overset{2}{x}, \frac{\overset{1}{\partial}}{\partial x}\right) = ihe^{-x} \frac{1 - e^{a\partial/\partial x}}{1 - e^a}$$

and

$$B\left(\overset{2}{x}, \frac{\overset{1}{\partial}}{\partial x}\right) \to ihe^{-x} \frac{\partial}{\partial x} \quad \text{as} \quad a \to 0.$$

With this choice of B, the perturbed equation can be rewritten in the form

$$\frac{h^2}{2(1 - e^a)^2} \{\psi_{n-1} + \psi_{n+1} - 2\psi_n\} V((n-1)a)\psi_{n-1} - E\psi_{n-1} = 0,$$

where

$$\psi_n = \psi(na).$$

Hence, we have arrived at a difference approximation of the Schrödinger equation on the uniform grid $\{x = na\}$; the example is of course trivial, but in less evident situations the idea may be helpful.

3 Propagation of Electromagnetic Waves in Plasma

Let us consider the construction of high-frequency asymptotic expansions for a wave generated in plasma by a point source. This problem is a classical one (see [120]) and is characterized by an interesting physical phenomenon. Aside from the usual rays predicted by geometric optics, there occur so-called *transient* rays; in the illuminated region their contribution is of an order less than that of geometric optics rays, but they still give the leading term of the asymptotic expansion in the umbral region.

Here we show that the appearance of transient rays is caused by turning on the point source at the initial moment of time ($t = 0$). Concurrently, we compute the so-called *diffraction coefficient*, which is in fact the ratio of two amplitudes corresponding, respectively, to the rays of geometric optics and to the transient rays.

The cited phenomenon (the appearance of transient rays) was discovered very long ago, but until very recently the diffraction coefficient was computed by a semiheuristic method based on model problems. The uniform asymptotic expansion with respect to smoothness and parameter provides rigorous justification for both the diffraction coefficient and the appearance of the transient part of the asymptotic expansion.

Since it it just *simultaneous asymptotic expansion* (with respect to parameter and smoothness) that is needed, we are led to the use of noncommuting operators. Their usage provides uniform asymptotic expansions admitting analysis in physical terms.

Our exposition is arranged as follows. In the first subsection we describe the statement of the problem and choose a family of noncommuting operators adequate to the type of asymptotic expansion required. The second subsection deals with the construction of the asymptotic expansion for the considered problem from the viewpoint of noncommutative analysis. The third (and the last) subsection is devoted to the analysis of the obtained asymptotic expansion in different zones of the physical space (the illuminated region, the umbral region, etc.)

Of necessity, the exposition in this section is more technical than the authors would like. However, we tried to avoid clumsy calculations by considering the asymptotic solution of the Cauchy problem only for small values of t, which permits us to use solely the WKB method and to sidestep the much more complicated language of Maslov's canonical operator. Of course, with the help of the latter all constructions of this section can be carried out for arbitrary values of t. We refer the reader to [145] and [146] for further details.

3.1 Statement of the Problem

We consider the Cauchy problem

$$\begin{cases} -\dfrac{\partial^2 u}{\partial t^2} + c^2 - \dfrac{\partial^2 u}{\partial x^2} - \lambda^2 b^2(x) u = \lambda \delta(x) r(t) e^{-i\lambda q(t)}, \\ u|_{t=0} = \frac{\partial u}{\partial t}\big|_{t=0} = 0 \end{cases}$$

describing propagation of electromagnetic waves in plasma (see, e.g., Lewis [120]). Here $\lambda b(x)$ is the plasma frequency and λ the average plasma frequency considered as a large parameter. The function $b(x)$ hence describes spatial inhomogeneity of the plasma and is dimensionless; it is assumed to be everywhere positive.

The right-hand side of the equation describes a point source at the origin with amplitude $r(t)$ and instantaneous frequency $\lambda q'(t)$. Our intention is to study the asymptotic behaviour of the solutions as $\lambda \to \infty$.

We begin with several remarks. First, we need to take smoothness into account when constructing asymptotic expansions with respect to λ, since the right-hand side of the equation contains a distribution. Otherwise each subsequent term of the asymptotic expansion would be less smooth than the preceding one, which makes the expansion physically unjustified. Second, it will be more convenient to consider the Cauchy problem for a homogeneous equation. This can be achieved with the help of the classical Duhamel principle. Specifically, it is easy to see that the solution can be represented in the form

$$u(x,t) = -\int_0^t v(x,t,\tau)\,d\tau, \tag{III.14}$$

where $v(x,t,\tau)$ is the solution of the following problem:

$$\begin{cases} -\dfrac{\partial^2 v}{\partial t^2} + \hat{H}v = 0, \\ v|_{t=\tau} = 0, \ \dfrac{\partial v}{\partial t}\Big|_{t=\tau} = F(x,\lambda,\tau). \end{cases} \tag{III.15}$$

Here we use the following notation. By $\hat{H}$ we denote the operator

$$\hat{H} = c^2\frac{\partial^2}{\partial x^2} - \lambda^2 b^2(x),$$

and

$$F = \lambda\delta(x)r(t)e^{-i\lambda q(t)}$$

is the right-hand side of the original equation. Having in mind that the solutions of (III.15) and the original problem are related, we see that it suffices to construct the simultaneous asymptotic expansion of a solution to (III.15) with respect to smoothness and the large parameter λ. This being done, the Duhamel formula yields the asymptotic expansion of the same type for the solution to the original problem.

We consider the following set of operators:

$$\overset{1}{A_1} = -i\frac{\partial}{\partial x}, \quad \overset{1}{A_2} = \lambda, \quad \overset{2}{B} = x,$$

acting on the Hilbert space $H_0 = L_2(\mathbb{R}^2_x \times [1,\infty)_\lambda)$. Only the operators A_1 and A_2 are "large parameters" of the asymptotic expansion; consequently, the scale $H^1(A)$ is defined by the following sequence of norms:

$$\|u\|_1 = \left\| (1 + A_1^2 + A_2^2)^{s/2} u \right\|_{L_2} = \left\| (1 - \Delta + \lambda^2)^{s/2} u \right\|_{L_2}.$$

In Subsection 3.2 we construct the asymptotic solution of problem (III.15) in the scale $H^s(A)$. The asymptotic behaviour of the solution $u(x, t)$ to the original problem is carried out in Subsection 3.3.

3.2 The Construction of the Asymptotic Expansion

We begin with an operator interpretation of the Cauchy problem under investigation. First of all, we represent $\hat{H}$ as a function of the operator tuple (A_1, A_2, B),

$$\hat{H} = -c^2 \overset{1}{A_1^2} - \overset{1}{A_2^2}\, b^2(\overset{2}{B}).$$

Next, we seek the solution to (III.15) in the form

$$v(x, t, \tau) = \Phi(\overset{1}{A_1}, \overset{1}{A_2}, \overset{2}{B}, t, \tau)\, F = \hat{\Phi} F, \tag{III.16}$$

where $\hat{\Phi}$ is the operator with unknown symbol $\Phi(p, \lambda, x, t, \tau)$. Here we use the correspondence

$$p \leftrightarrow A_1, \quad \lambda \leftrightarrow A_2, \quad \text{and} \quad x \leftrightarrow B$$

between the variables and the operators; it will always be clear whether, say, x denotes the variable or the corresponding multiplication operator.

The operator $\hat{\Phi}$ should satisfy the following operator Cauchy problem:

$$\begin{cases} -\dfrac{\partial^2 \Phi}{\partial t^2} + \hat{H}\hat{\Phi} = 0, \\ \Phi|_{t=\tau} = 0, \quad \hat{\Phi}\Big|_{t=\tau} = 1. \end{cases}$$

since $\hat{H}$ is independent of t, it suffices to obtain the solution for $\tau = 0$ and then substitute $t - \tau$ for t (in fact, we here use the homogeneity in time of our system). The next step is to transform the operator Cauchy problem into a usual Cauchy problem for the symbol $\Phi(p, \lambda, x, t, \tau)$ of the operator $\hat{\Phi}$. As was explained in Chapter II, to this end we should use the operators of the left ordered representation for the tuple $(\overset{1}{A_1}, \overset{1}{A_2}, \overset{2}{B})$. These easy-to-compute operators have the following form:

$$l_{A_1} = p - i\frac{\partial}{\partial x}, \quad l_{A_2} = \lambda, \quad l_B = x.$$

We substitute these operators into the symbol of $\hat{H}$ and obtain a Cauchy problem for the symbol Φ:

$$\begin{cases} \dfrac{\partial^2 \Phi}{\partial t^2} - c^2 \left(p - i\dfrac{\partial}{\partial x}\right)^2 \Phi - \lambda^2 b^2(x)\Phi = 0, \\ \Phi|_{t=0} = 0, \quad \dfrac{\partial \Phi}{\partial t}\Big|_{t=0} = 1. \end{cases} \tag{III.17}$$

Of course, we are interested in an *asymptotic* rather than a precise solution. Taking into account the expression for the norms in the spaces $H^s(A)$, where our asymptotic expansions live, we see that it suffices to solve the latter problem to within functions decaying sufficiently rapidly as $(p, \lambda) \to \infty$. Operators corresponding to such symbols are of large negative order in the scale $H^s(A)$. It is therefore natural to consider $\mu = \sqrt{\lambda^2 + p^2}$ as a large parameter and to seek the asymptotic expansion of the solution as $\mu \to \infty$. We set $\omega_1 = p/\mu$ and $\omega_2 = \lambda/\mu$; the "angular parameters" ω_1 and ω_2 range over the unit sphere $\omega_1^2 + \omega_2^2 = 1$.

To obtain the asymptotic solution of (III.17) as $\mu \to \infty$ for large (but finite) values of t one should use Maslov's canonical operator method. We have no intention of describing this theory here (see, e.g. [137]) and limit ourselves to obtaining an asymptotic expansion of the solution for small t. For these t, the WKB approximation can be used:

$$\Phi = e^{i\mu S(\omega,x,t)}(a_0(\omega, x, t) + \mu^{-1} a_1(\omega, x, t) + \cdots).$$

Moreover, here we only find the *leading* term of the asymptotic expansion, via the functions $S(\omega, x, t)$ and $a_0(\omega, x, t)$. Following the standard procedure of the WKB method, we insert Φ into (III.17), collect similar terms, and equate the coefficients of powers of μ to zero. This procedure yields the Hamilton–Jacobi equation

$$-\left(\frac{\partial S}{\partial t}\right)^2 + c^2\left(\omega_1 + \frac{\partial S}{\partial x}\right)^2 + \omega_2^2 b^2(x) = 0$$

for $S(\omega, x, t)$ and a transport equation for $a_0(\omega, x, t)$.

The Cauchy data in (III.17) induce the initial data

$$S|_{t=0} = 0$$

for the Hamilton–Jacobi equation. Clearly, the latter splits into two equations:

$$\frac{\partial S_\pm}{\partial t} \pm \sqrt{c^2\left(\omega_1 + \frac{\partial S_\pm}{\partial x}\right)^2 + \omega_2^2 b^2(x)} = 0,$$

and hence we obtain *two* solutions $S_\pm = S_\pm(\omega, x, t)$. These solutions can be evaluated explicitly by the standard method of characteristics. Specifically, denote by $\mathcal{H}(\lambda, x, q)$ the Hamiltonian function

$$\mathcal{H}(\omega, x, q) = \sqrt{c^2(\omega_1 + p)^2 + \omega_2^2 b^2(x)}$$

(we consider the "+" sign). The corresponding Hamiltonian system for the characteristics has the form

$$\begin{cases} \dot{x} = \mathcal{H}_q(\omega, x, q) = c^r(\omega_1 + q)/\mathcal{H}, \\ \dot{q} = -\mathcal{H}_x(\omega, x, q) = -2\omega_2^2 b(x) b'(x)/\mathcal{H}, \end{cases}$$

with the initial data

$$\begin{cases} x(0) = x_0, \\ q(0) = 0. \end{cases}$$

Let

$$x = x^+(x_0, t), \quad q = q^+(x_0, t)$$

be its solution. Then, for small t, the function $S_1(\omega, x, t)$ can be calculated according to the formula

$$S_+(\omega, x, t) = \left[\int_0^t \frac{c^2\omega_1(\omega_1 + q) + b^2(x)\omega_2^2}{\mathcal{H}} \Bigg|_{x=x^+(x_0,t), q=q^+(x_0,t)} dt \right]_{x_0=x_0^+(x,t)},$$

where the integral is in fact taken along the trajectories of the Hamiltonian system and $x_0 = x_0^+(x, t)$ is the solution of the equation $x = x^+(x_0, t)$ for x_0. Similarly, the function $S(\omega, x, t)$ is given by the integral

$$S_-(\omega, x, t) = \left[-\int_0^t \frac{c^2\omega_1(\omega_1 + q) + b^2(x)\omega_2^2}{\mathcal{H}} \Bigg|_{x=x^-(x_0,t), q=q^-(x_0,t)} dt \right]_{x_0=x_0^-(x,t)},$$

where the functions x^-, q^-, and x_0^- are defined by analogy with x^+, q^+, and x_0^+ from the Hamiltonian system corresponding to the "−" sign in the Hamilton–Jacobi equation.

Since, on the one hand, there are two functions $S_\pm$, the solutions of the Hamilton–Jacobi equation, and, on the other hand, there are two initial conditions in (III.17), the asymptotics of the solution to problem (III.17) is represented as the sum

$$\Phi = e^{i\mu S_+(\lambda,x,t)} a_0^+(\omega, x, t) + e^{i\mu S_-(\omega,x,t)} a_0^-(\omega, x, t)$$

(recall that we only construct the leading term of the asymptotic expansion). The initial conditions in (III.17) lead to the initial data for the functions $a_0^\pm$; each of these functions is found from the transport equation associated with the corresponding function $S_\pm(\omega, x, t)$. We shall not dwell upon the computational aspect and merely write out the leading term of the asymptotic expansion of the symbol $\Phi(x, p, \lambda, t)$ with respect to the parameter $\mu = \sqrt{\lambda^2 + p^2}$. It has the form

$$\Phi(x, p, \lambda, t) = \frac{i}{2} \sum_{j\in(+,-)} j e^{i\mu S_j(\omega,x,t)} \left[\left(c^2 \left(p + \frac{\partial S_j}{\partial x} \right)^2 + b^2(x)\lambda^2 \right) J_j \right]^{-1/2},$$

where

$$J_j = \det\left[\frac{\partial x_j}{\partial x_0}(x_0, t)\right]_{x_0 = x_0^{\pm}(x,t)}$$

is the corresponding Jacobian. Now the simultaneous asymptotic expansion of the solution to the original problem with respect to smoothness and parameter is given by (III.14) and (III.16). On substituting the expression obtained for Φ into (III.16) we obtain an explicit expression

$$u(x, t, \lambda) \cong u_+(x, t, \lambda) + u_-(x, t, \lambda)$$

for the solution $u(x, t, \lambda)$ of the original problem, where

$$u_\pm(x, t, \lambda) = \pm\frac{i\lambda}{2(2\pi)^n}\int_0^t\int_{\mathbb{R}_p^n} e^{i\varphi_\pm(p,x,\lambda,t,\tau)} a_\pm(p, x, \lambda, t, \tau)\, dp\, d\tau, \qquad \text{(III.18)}$$

and the phase and the amplitude of the integrand are given by the formulas

$$\varphi_\pm = px + S_\pm(p, x, \lambda, t-\tau) - \lambda q(t),$$
$$a_\pm = \left[J_\pm\left(c^2\left(\frac{\partial\varphi_\pm}{\partial x}\right)^2 + \lambda^2 b^2(x)\right)\right]^{-1/2} r(\tau).$$

Of course, the integral over $\mathbb{R}_p^n$ is in the sense of the theory of distributions.

3.3 Analysis of the Asymptotic Solution

First, consider the simplest case in which $b(x) = b = \text{const}$. Then

$$J_+ = J_- = 1,$$
$$S_\pm = \pm\sqrt{c^2p^2 + \lambda^2 b^2}\, t.$$

Straightforward computations show that

$$\Phi_\pm = px \pm \sqrt{c^2p^2 + \lambda^2b^2}(t-\tau) - \lambda q(\tau) = \lambda(kx \pm \sqrt{c^2k^2 + b^2}(t-\tau) - q(\tau)),$$

where $k = p/\lambda$, and

$$u_\pm = \pm\frac{i\lambda^{2-n}}{2(2\pi)^n}\int_0^t\int_{\mathbb{R}_k^n}\frac{r(t)e^{i\lambda\Phi_\pm}}{\sqrt{c^2k^2 + b^2}}\, d\tau\, dk$$

(see R. Lewis [120], p. 848). Note in particular that the method described gives a precise solution for constant coefficients since the right-hand sides vanish for higher transport equations.

Let us now study the general case $b = b(x)$. The asymptotic expansion of the integral (III.18) can be computed with the help of the stationary phase method. The stationary point equations for the phase function have the form

$$\begin{cases} x + \dfrac{\partial S}{\partial p}(p, \lambda, x, t - \tau) = 0, \\ \dfrac{\partial S}{\partial t}(p, \lambda, x, t - \tau) + \lambda q'(\tau) = 0 \end{cases} \tag{III.19}$$

(here we consider only the function u_+; the function u_- can be considered in a similar way).

However, the integration domain in (III.18) has boundary points, and henceforth we must allow for the contribution of boundary stationary points at $\tau = 0$ and $\tau = t$. The equations of these stationary points are as follows:

$$\begin{cases} x + \dfrac{\partial S}{\partial p}(p, \lambda, x, t) = 0, \\ \tau = 0 \end{cases} \tag{III.20}$$

for the boundary point $\tau = 0$ and

$$x = 0, \quad \tau = t \tag{III.21}$$

for the boundary point $\tau = t$.

We shall not analyze asymptotic expansions in the neighborhood of the point $x = 0$ (i.e., of the source).

Hence we do not have to take the stationary points (III.21) into consideration. By the change of variables $p = \lambda k$ we reduce (III.18) to the form

$$u \sim \frac{i\lambda^{2-n}}{2(2\pi)^n} \int_0^t \int_{\mathbb{R}^n_k} \frac{e^{i\lambda\Phi(k,x,1,t,\tau)} r(\tau)\, dk\, d\tau}{\sqrt{J\left(c^2\left(\frac{\partial\Phi}{\partial x}\right)^2 + b^2(x)\right)}}.$$

The stationary point equations can be obtained from (III.19) and (III.20) by the substitution $p = k$, $\lambda = 1$. In order to apply the stationary phase method correctly, we should require that the stationary points $\{k(x,t), \tau(x,t)\}$ of system (III.19) and $k'(x,t)$ of (III.20) be located in a bounded domain $|k| \le R < \infty$ of the space $\mathbb{R}^n_k$. In that case we can use a partition of unity $\{e_1(|k|), e_2(|k|)\}$ such that $e_1(|k|) + e_2(|k|) = 1$ and $e_2(z) = 0$ for $z < R$. The integral in question reduces to a sum of two terms, of which the second one can be shown to be $O(\lambda^{-\infty})$ uniformly with respect to smoothness. This can be performed by integration by parts with respect to k. The first term is an integral over a compact domain in $\mathbb{R}^n_k$, so one can apply the stationary phase method to obtain its asymptotic expansion.

The boundedness of the set of stationary points is equivalent to the nontrivial solvability of systems (III.19) and (III.20) for p at $\lambda = 0$.

The relations

$$\begin{cases} S(p,0,x,t-\tau) = cp(t-\tau), \\ \frac{\partial S}{\partial p}(p,0,c,t-\tau) = c(t-\tau), \\ \frac{\partial S}{\partial p}(p,0,x,t-\tau) = cp \end{cases}$$

imply that for $\lambda = 0$ these systems can be written as

$$x + c(t-\tau) = 0, \quad cp = 0,$$

and

$$x + ct = 0, \tag{III.22}$$

respectively. Since the first system implies that $p = 0$, we see that system (III.19) does not have any "bad" solutions, and (III.22) shows that neither does system (III.20), unless the point (x,t) lies on the characteristic cone. Since we wish to obtain a uniform asymptotic expansion, we have to use uniform asymptotic expansions given by the stationary phase method for the case in which the boundary stationary point (i.e., the solution of equation (III.20)) coincides with the interior stationary point (III.19).

This case corresponds to the points (x,t) lying on the so-called *shadow boundary*. We obtain the following results by using the stationary phase method:

A. If the integral has interior stationary points (the illuminated region), the leading term of the asymptotic expansion has the form

$$u \sim \lambda^{2-3n/2} e^{i\lambda S(x,t)} \varphi(x,t),$$

which coincides with the approximation given for the original problem by geometric optics.

B. If the integral has no interior stationary points, but rather has stationary points at the boundary, then the asymptotic expansion of the solution has the form

$$u \sim \frac{e^{i\lambda S(x,t)} \varphi(x,t)}{\frac{\partial \Phi}{\partial t}(k,x,1,t,0)}$$

(see [48], p. 141). This formula corresponds to the so-called *transient rays* of geometric optics. The factor

$$\frac{1}{\frac{\partial \Phi}{\partial t}(k,x,1,t,0)} = \left.\frac{1}{\varkappa - q'(\tau)}\right|_{\tau=0} = \frac{1}{\sqrt{c^2k^2 + b^2(0)} - q'(0)} = \frac{1}{\omega_s - \omega_0}$$

is known as the *diffraction coefficient*. Here ω_0 is the frequency of the source, and ω_s is the instantaneous frequency on a given transient ray.

We do not compute the asymptotic expansion for the case in which an interior stationary point coincides with a boundary stationary point. This can be done with the help of an appropriate version of the stationary phase method (see, e.g., [48]).

4 Equations with Symbols Growing at Infinity

In this section we consider the technique of noncommuting operators on the example of constructing asymptotic solutions to differential equations with respect to smoothness and decay at infinity. We only consider a model example that displays all the main features of the theory. We refer the reader to [134] for more detailed information.

4.1 Statement of the Problem and its Operator Interpretation

We consider the Cauchy problem

$$\begin{cases} \dfrac{\partial^2 u}{\partial t^2} = \dfrac{\partial^2 u}{\partial x^2} - x^{2l} c(x) u + f(x,t), \\ u|_{t=0} = u_0(x), \quad \left.\dfrac{\partial u}{\partial t}\right|_{t=0} = u_1(x), \end{cases} \tag{III.23}$$

where $t \in \mathbb{R}$, $x \in \mathbb{R}$, and the function $c(x)$ is bounded together with all its derivatives and satisfies the condition

$$c(x) \geq c_0 > 0.$$

In order to construct the simultaneous asymptotic solution with respect to smoothness and growth as $x \to \infty$ we define the scale $H^s(A, B)$, where

$$A = -i\frac{\partial}{\partial x} \quad \text{and} \quad B = x.$$

The norm in $H^s(A, B)$ is given by the expression

$$\| u(x) \|_s^2 = \left\| (1 + A^2 + B^2)^{s/2} u(x) \right\|_{L_2(\mathbb{R}^1)}.$$

However, even at first glance it is clear that the role of the variable x is quite different in the factor x^{2l} than in the coefficient $c(x)$. Whereas x^{2l} increases at infinity, $c(x)$ is uniformly bounded. Recall that a necessary condition for success of our method is that the operator in question be homogeneous with respect to the operators defining the scale $H^s(A, B)$. To this end, we introduce three rather than two operators, forming a Feynman tuple:

$$\overset{1}{A} = -i\frac{\partial}{\partial x}, \quad \overset{2}{B_1} = x, \quad \overset{2}{B_2} = x$$

and use $\overset{2}{B_1}$ for "homogeneous" factors and $\overset{2}{B_2}$ for "well-behaved" bounded factors.

We can rewrite the Cauchy problem in the form

$$\begin{cases} \dfrac{\partial^2 u}{\partial t^2} = -[A^2 + B_1^{2l} c(B_2)] u + f(x,t), \\ u|_{t=0} = u_0(x), \quad \left.\dfrac{\partial u}{\partial t}\right|_{t=0} = u_1(x). \end{cases} \tag{III.24}$$

Here only the operators A and B_1 occur in the definition of the scale H^s (i.e., one should replace B by B_1, and the operator B_2 is, in a sense, an "operator parameter".

We can now present an operator setting for the Cauchy problem (III.24). Clearly, in order to find the solution of (III.24), it suffices to construct the *Green operator* of this problem, i.e., an operator $\hat{R}(t)$ depending on the parameter t and satisfying the following operator Cauchy problem:

$$\begin{cases} \dfrac{\partial^2 \hat{R}(t)}{\partial t^2} = -[A^2 + B_1^{2l} c(B_2)]\,\hat{R}(t), \\ \hat{R}(0) = 0, \quad \dfrac{\partial \hat{R}}{\partial t}(0) = 1. \end{cases} \tag{III.25}$$

Indeed, by Duhamel's principle, the solution of (III.24) can be expressed via $\hat{R}(t)$ by the formula

$$u(x,t) = \frac{\partial \hat{R}(t)}{\partial t} u_0 + \hat{R}(t) u_1 + \int_0^t \hat{R}(t-\tau)\, f(x,\tau)\, d\tau$$

(we take into account that the coefficients of the operator do not depend on t). Moreover, it is clear that for constructing the asymptotic expansion of the solution to (III.24) in the scale of spaces $H^s(A, B)$ it suffices to construct the asymptotic solution of (III.25) modulo operators of low negative order in this scale.

Taking all this into account, we seek the operator $\hat{R}(t)$ (or, more precisely, its asymptotic expansion) in the form

$$\hat{R}(t) = \Phi(t, \overset{1}{A}, \overset{2}{B_1}, \overset{2}{B_2})$$

(the choice of Feynman indices reflects the fact that B_1 and B_2 commute with each other).

Let us denote by z, w_1, and w_2 the arguments of the function Φ corresponding to the operators A, B_1, and B_2, respectively. The left ordered representation operators for the tuple $(\overset{1}{A}, \overset{2}{B_1}, \overset{2}{B_2})$ can be computed easily. They are

$$l_A + z - i\frac{\partial}{\partial w_1} - i\frac{\partial}{\partial w_2}, \quad l_{B_1} = w_1, \quad \text{and} \quad l_{B_2} = w_2.$$

(Numerous examples of similar computations are given in Chapter II; we leave detailed computations to the reader.) On substituting $\hat{R}(t)$ into the operator equation (III.25), we obtain the following Cauchy problem for the symbol $\Phi(t, z, w_1, w_2)$:

$$\begin{cases} \dfrac{\partial^2 \Phi}{\partial t^2} + \left[\left(z - i\dfrac{\partial}{\partial w_1} - i\dfrac{\partial}{\partial w_2}\right)^2 + w_1^{2l} c(w_2)\right] \Phi = 0, \\ \Phi|_{t=0} = 0, \quad \left.\dfrac{\partial \Phi}{\partial t}\right|_{t=0} = 1. \end{cases} \tag{III.26}$$

We should construct the asymptotic expansion of the solution Φ to the Cauchy problem (III.26) as $(z, w_1) \to \infty$. This expansion should be uniform with respect to w_2. This difference between the roles played by w_1 and w_2 clarifies the reasons for "duplicating" the operators B_1 and B_2 in the expression for the Green operator $\hat{R}(t)$.

4.2 Asymptotic Solution of the Symbolic Equation

We shall now construct the required asymptotic expansion of the solution to (III.26). As was mentioned above, the problem of constructing such asymptotic expansions is not an intrinsic problem of noncommutative analysis. However, the difficulties encountered in constructing asymptotic solutions to (III.26) are quite characteristic of the problems arising in the applications of the theory of noncommuting operators in the asymptotic theory of differential equations. Therefore, we devote some space to the construction of the asymptotic solution to problem (III.26).

In order to spare the reader superfluous technical details we restrict ourselves to constructing the asymptotic expansion of the solution for small values of t. The solution of (III.26) will be sought in the form

$$\Phi = \Phi_+ + \Phi_-,$$

where

$$\Phi_\pm = e^{iS_\pm(t,z,w)} a_\pm(t, z, w).$$

By substituting the components $\Phi_\pm$ of Φ into equation (III.26) we obtain the following equation (the $\pm$ sign is omitted):

$$\left[-\left(\frac{\partial S}{\partial t}\right)^2 + \left(z + \frac{\partial S}{\partial w_1} + \frac{\partial S}{\partial w_2}\right)^2 + w_1^{2l} c(w_2) \right] a(t, z, w) - i\hat{P}_1 a(t, z, w)$$
$$+\hat{P}_2 a(t, z, w) = 0, \tag{III.27}$$

where $\hat{P}_j$ are operators of order j with coefficients depending on derivatives of S; the explicit form of these coefficients is inessential to us as yet.

Let us focus our attention on the first term on the left-hand side of this equation; in the following we shall see that this term gives the Hamilton–Jacobi equation.

The construction of asymptotic expansions of such a form usually employs homogeneous functions (thus, the action S is usually a homogeneous function of order 1 of the variables with respect to which the asymptotic expansion is constructed; in our case, these are z and w_1). However, since different powers of z and w_1 occur there, it is natural to use *generalized homogeneity*. Specifically, we assume that S is a homogeneous function of z and w_1 of the form

$$S(t, \lambda^l z, \lambda w_1, w_2) = \lambda^l S(t, z, w_1, w_2),$$

and we seek the amplitude a in the form of a sum of homogeneous functions of the same type and decreasing orders of homogeneity.

The orders of homogeneity of the functions occurring in the first term on the left-hand side are as follows:

$$\operatorname{ord} z = l; \quad \operatorname{ord} w_1 = 1.$$

Hence, the terms of the highest homogeneity degree $2l$ give

$$\left[-\left(\frac{\partial S}{\partial t}\right)^2 + \left(z + \frac{\partial S}{\partial w_2}\right)^2 + w_1^{2l} c(w_2) \right] a_0,$$

where a_0 is the zero-order homogeneous component of a. The next order of homogeneity present is $(2l - 1)$. The corresponding terms have the following forms:

$$2\left(z + \frac{\partial S}{\partial w_2}\right) \frac{\partial S}{\partial w_1} a_0. \tag{III.28}$$

Evidently, following the standard method, we should equate both quantities to zero, thus obtaining *two* different Hamilton–Jacobi equations for S. This is because the functions occurring in the second summand in expression (III.27),

$$\begin{aligned} \hat{P}_1 a = & -2\frac{\partial S}{\partial t}\frac{\partial a}{\partial t} + 2\left(z + \frac{\partial S}{\partial w_1} + \frac{\partial S}{\partial w_2}\right)\frac{\partial a}{\partial w_1} + \left(z + \frac{\partial S}{\partial w_1} + \frac{\partial S}{\partial w_2}\right)\frac{\partial a}{\partial w_2} \\ & - \left[\frac{\partial^2 S}{\partial t^2} + \frac{\partial^2 S}{\partial w_1^2} + 2\frac{\partial^2 S}{\partial w_1 \partial w_2} + \frac{\partial^2 S}{\partial w_2^2}\right] a, \end{aligned}$$

all have homogeneity degrees not exceeding l, and therefore do not occur in the second term in (III.28) provided that $l > 1$.

The remedy is to include all terms homogeneous of degrees from $l + 1$ to $2l$ into the leading term. This gives the Hamilton–Jacobi equation

$$\left(\frac{\partial S}{\partial t}\right)^2 = \left(z + \frac{\partial S}{\partial w_1} + \frac{\partial S}{\partial w_2}\right)^2 + w_1^{2l} c(w_2), \tag{III.29}$$

with the initial condition

$$S|_{t=0} = 0.$$

Note that the function $S(t, z, w)$ thus determined is no longer a generalized homogeneous function of z and w_1. However, it can be shown that S is an *asymptotically* generalized homogeneous function of these variables, which is more than sufficient for proving appropriate estimates. The same can be said of the components $a_0, a_{-1}, \dots$ of the amplitude $a(t, z, w)$. For these components transport equations can be obtained in a standard way, and solving these equations leads to *asymptotically* generalized homogeneous functions of z and w_1.

Thus, the Hamiltonian of problem (III.23) corresponding to the asymptotic expansion with respect to smoothness and growth at infinity is the function

$$(z + p_1 + p_2)^2 + w_1^{2l} c(w_2). \tag{III.30}$$

However, it should be noted that the principal (generalized homogeneous) component of $S(t, z, w)$ is not defined by the Hamiltonian function, but rather by the Hamiltonian

$$(z + p_2)^2 + w_1^{2l} c(w_2). \tag{III.31}$$

This is because the term $\partial S/\partial w_1$ is of degree lower than that of any other term.

It is now clear that all conditions on the behaviour of the trajectories of the Hamiltonian system (usually occurring in stationary rather than Cauchy problems) should be imposed on the Hamiltonian (III.31), whereas S is a solution of problem (III.29) with the Hamiltonian (III.30). Thus two Hamiltonian functions have arisen in the problem! The second one (III.31) is referred to merely as the Hamiltonian, whereas the first one is called the *essential* Hamiltonian of problem (III.30). Such a situation is characteristic of problems related to generalized homogeneity (see [129]).

The subsequent steps are not difficult and can be carried out by analogy with their counterparts in the preceding section. An interested reader can either do this himself or herself or look through the literature recommended above.

4.3 Equations with Fractional Powers of x in the Coefficients

We have considered an example of an equation with coefficients growing at infinity as an integral power of x, so that the operator $\mathcal{L}$ occurring in the equation could be written in the form

$$\mathcal{L} = f(A_1, B_1, B_2),$$

where $f(y_1, y_2, y_3)$ is a quasihomogeneous function of (y_1, y_2) and is smooth for $y_1^2 + y_2^2 \neq 0$.

Let us now consider a more complicated example in which the coefficients behave as a fractional power of x. For simplicity, we restrict ourselves to the first-order equation

$$\mathcal{L}u \equiv -i\frac{\partial u}{\partial t} - i\frac{\partial u}{\partial x} + c(x)|x|^\alpha u + b\left(\overset{2}{x}, -i\overset{1}{\frac{\partial}{\partial x}}\right) u = 0 \tag{III.32}$$

with the initial condition

$$u|_{t=0} = u_0(x), \tag{III.33}$$

where $\alpha > 0$, $c(x)$ is bounded together with all its derivatives, $c(x) \geq c_0 > 0$ for $|x| \geq 1$, $c(x)|x|^\alpha$ is everywhere smooth, and $b(x, \xi) = b_1(x, x, \xi)$, where $b_1(x, y, z)$ is an $(\alpha^{-1}, 1)$-quasihomogeneous function of order 0 of (y, z) for $|y|^\alpha + |z| \geq 1$, that is,

$$b_1(x, \lambda^{1/\alpha} y, \lambda z) = b_1(x, y, z)$$

for $\lambda \geq 1$ and $|y|^\alpha + |z| \geq 1$.

Introduce the operators

$$A_1 = -i\frac{\partial}{\partial x}, \quad A_2 = \psi(x), \quad B = x, \tag{III.34}$$

where $\psi(x)$ is a real-valued function such that $\psi(x) \in C^\infty(\mathbb{R})$, $\psi(x) = |x|$ for $|x| \geq 2$ and $\psi(x) \geq 1$ everywhere.

The operators (III.34) satisfy the commutation relations

$$[A_1, A_2] = -i\chi(B), \quad [A_1, B] = -i, \quad [A_2, B] = 0, \tag{III.35}$$

where $\chi(x)$ is a smooth real-valued function, $\chi(x) = \operatorname{sign} x$ for $|x| \geq 2$. It is easy to compute the left ordered representation operators for the tuple $(\overset{1}{A_1}, \overset{2}{A_2}, \overset{2}{B})$. They have the form

$$l_{A_1} = y_1 - i\frac{\partial}{\partial y_3} - i\chi(y_3)\frac{\partial}{\partial y_2},$$
$$l_{A_2} = y_2, \quad l_B = y_3$$

(we assume the correspondence $y_1 \leftrightarrow A_1$, $y_2 \leftrightarrow A_2$, $y_3 \leftrightarrow A_3$).

Equation (III.32) takes the form

$$-i\frac{\partial u}{\partial t} + A_1 u + c_1(B)A_2^\alpha u + b_1(\overset{2}{B}, \overset{2}{A_2}, \overset{1}{A_1})u + b_2(\overset{2}{B}, \overset{1}{A_1})u = 0,$$

where the function

$$c_1(x) = c(x)\left[\frac{|x|}{\psi(x)}\right]^\alpha$$

is smooth and bounded together with all its derivatives, $c_1(x) \geq c_0 > 0$, and $b_2(x, \xi)$ is homogeneous of degree 0 in ξ for large $|\xi|$ and is compactly supported with respect to x.

Let us seek the solution of problem (III.32) – (III.33) in the form

$$u = u(t) = \hat{G}(t)u_0 = g(\overset{1}{A_1}, \overset{2}{A_2}, \overset{2}{B}, t)u_0,$$

then for $g(y_1, y_2, y_3, t)$ we obtain the equation

$$\Big[-i\frac{\partial}{\partial t} + y_1 - i\frac{\partial}{\partial y_3} - i\chi(y_3)\frac{\partial}{\partial y_2} + c_1(y_3)y_2^\alpha$$
$$+b_1\left(\overset{2}{y_3}, \overset{2}{y_2}, \overset{1}{[\![} y_1 - i\frac{\partial}{\partial y_3} - i\chi(y_3)\frac{\partial}{\partial y_2}]\!]\right)$$
$$+b_2\left(\overset{2}{y_3}, \overset{1}{[\![} y_1 - i\frac{\partial}{\partial y_3} - i\chi(y_3)\frac{\partial}{\partial y_2}]\!]\right)\Big] g(y_1, y_2, y_3, t) = 0 \tag{III.36}$$

with the initial conditions

$$g(y_1, y_2, y_3, 0) = 1. \tag{III.37}$$

The following assertion is valid.

Theorem III.1 *For any N there exists an $N_1 = N_1(N)$ such that the bounds*

$$\left| \frac{\partial^\beta T}{\partial y^\beta}(y_1, y_2, y_3) \right| \leq C(1 + |y_1| + |y_2|)^{-N_1+|\beta|}, \quad |\beta| = 1, \ldots, N_1, \quad y_2 \geq 1/2,$$

imply that the operator

$$x^\gamma \left(-i\frac{\partial}{\partial x}\right)^\delta T(\overset{1}{A_1}, \overset{2}{A_2}, \overset{2}{B})$$

is bounded in L_2 for $\gamma + \delta \leq N$.

Proof. The spectrum of the self-adjoint operator A_2 lies in the domain $y_2 \geq 1$, and hence the operator $T(\overset{1}{A_1}, \overset{2}{A_2}, \overset{2}{B})$ does not depend on the values of its symbol $T(y)$ for $y_2 < 1$. Let $\varphi(y_2) \in C^\infty(\mathbb{R})$, $\varphi(y_2) \equiv 1$ for $y_2 \geq 1$, $\varphi(y_2) = 0$ for $y_2 < 1/2$. If T satisfies the estimates in the theorem for $y_2 \geq 1/2$, then so does $\varphi(y_2)$, and $T(y)$ for all y. On the other hand, as was mentioned above,

$$T(\overset{1}{A_1}, \overset{2}{A_2}, \overset{2}{B}) = (\varphi T)(\overset{1}{A_1}, \overset{2}{A_2}, \overset{2}{B}),$$

and the required assertion follows from the results of Section 3. The theorem is proved. □

Let us now make the change of variables

$$y_1 = \lambda x_1, \quad y_2 = \lambda^{1/\alpha} x_2, \quad y_3 = x_3,$$

where λ is a large parameter, in equation (III.36).

We obtain the equation

$$\begin{aligned} \Bigg[-i\lambda^{-1}\frac{\partial}{\partial t} + x_1 - i\lambda^{-1}\frac{\partial}{\partial x_3} - i\varepsilon\chi(x_3)\lambda^{-1}\frac{\partial}{\partial x_2} + c_1(x_3)x_2^\alpha \\ +\lambda^{-1}b_1\left(\overset{2}{x_3}, \overset{2}{x_2}, \overset{1}{[\![} x_1 - i\lambda^{-1}\frac{\partial}{\partial x_3} - i\varepsilon\chi(x_3)\lambda^{-1}\frac{\partial}{\partial x_2}]\!]\right) \\ +\lambda^{-1}b_2\left(\overset{2}{x_3}, \overset{1}{[\![} x_1 - i\lambda^{-1}\frac{\partial}{\partial x_3} - i\varepsilon\chi(x_3)\lambda^{-1}\frac{\partial}{\partial x_2}]\!]\right) \Bigg] G(x, t, \varepsilon, \lambda) = 0, \end{aligned} \tag{III.38}$$

where $\varepsilon = \lambda^{-1/\alpha}$ and

$$G(x, t, \varepsilon, \lambda) = g(y, t)|_{y_1 = \lambda x_1, y_2 = \varepsilon^{-1}x_2, y_3 = x_3}.$$

It suffices to construct an asymptotic solution to equation (III.38) in the domain $x_2 \geq \varepsilon$, i.e., to find a function $G(x, t, \varepsilon, \lambda)$ such that the left-hand side in (III.38) is $O(\lambda^{-s})$ in $x_2 \geq \varepsilon$ uniformly with respect to ε. We seek $G(x, t, \varepsilon, \lambda)$ in the form

$$G(x, t, \varepsilon, \lambda) = e^{i\lambda S(x,t,\varepsilon)} \sum_{k=0}^{s} \lambda^{-k}\varphi_k(x, t, \varepsilon, \lambda^{-1}). \tag{III.39}$$

We insert (III.39) into (III.38) and obtain the following equations for S and φ_k:

$$\frac{\partial S}{\partial t} + x_1 + \frac{\partial S}{\partial x_3} + \varepsilon\chi(x_3)\frac{\partial S}{\partial x_2} + c_1(x_3)x_2^{\alpha} = 0 \tag{III.40}$$

(the Hamilton–Jacobi equation) and the transport equations

$$\begin{aligned} &\frac{\partial \varphi_k}{\partial t} + \frac{\partial \varphi_k}{\partial x_3} + \varepsilon\chi(x_3)\frac{\partial \varphi_k}{\partial x_2} \\ &+ \; i\left[b_1\left(x_3, x_2, x_1 + \frac{\partial S}{\partial x_3} + \varepsilon\chi(x_3)\frac{\partial S}{\partial x_2}\right)\right. \\ &\left. + b_2\left(x_3, x_1 + \frac{\partial S}{\partial x_3} + \varepsilon\chi(x_3)\frac{\partial S}{\partial x_2}\right)\right]\varphi_k = \mathcal{B}\varphi_{k-1} \end{aligned} \tag{III.41}$$

where the right-hand side of the latter equations is equal to

$$\begin{aligned} \mathcal{B}\varphi_{k-1} \;=\; & -i\lambda\left\{b_1\left(\overset{2}{x_3}, \overset{3}{x_2}, \overset{1}{[\![} x_1 + \frac{\partial S}{\partial x_3} + \varepsilon\chi(x_3)\frac{\partial S}{\partial x_2} - i\lambda^{-1}\frac{\partial}{\partial x_3}\right.\right. \\ & \left. - i\varepsilon\chi(x_3)\lambda^{-1}\frac{\partial}{\partial x^2}]\!]\right) \\ & - b_1\left(x_3, x_2, x_1 + \frac{\partial S}{\partial x_3} + \varepsilon\chi(x_3)\frac{\partial S}{\partial x_2}\right) \\ & + b_2\left(\overset{2}{x_3}, \overset{1}{[\![} x_1 + \frac{\partial S}{\partial x_3} + \varepsilon\chi(x_3)\frac{\partial S}{\partial x_2} - i\lambda^{-1}\frac{\partial}{\partial x_3} - i\varepsilon\chi(x_3)\lambda^{-1}\frac{\partial}{\partial x_2}]\!]\right) \\ & \left. - b_2\left(x_3, x_1 + \frac{\partial S}{\partial x_3} + \varepsilon\chi(x_3)\frac{\partial S}{\partial x_2}\right)\right\}\varphi_{k-1} \end{aligned}$$

(here $\varphi_{-1} \equiv 0$ by convention). To be able to use the equation (III.41) for constructing the asymptotic solution, one must verify that the right-hand side $\mathcal{B}\varphi_{k-1}$ of this equation does not, increase as $\lambda \to +\infty$. This fact can be verified directly from the explicit formula for $\mathcal{B}\varphi_{k-1}$ with the help of the Newton series expansion. The corresponding simple calculation is left to the reader.

It is now rather simple to write down the explicit formulas for the action $S(x, t)$ and the amplitudes $\varphi_k(x, t)$ via integrals along trajectories of the vector field

$$V = \frac{\partial}{\partial t} + \frac{\partial}{\partial x_3} + \varepsilon\chi(x_3)\frac{\partial}{\partial x_2}$$

occurring on in the left-hand side in (III.41). These formulas give an asymptotic solution of equation (III.38) with respect to powers of λ^{-1} in the domain

$$\left\{x_2 - \varepsilon\int_0^t \chi(x_3 - \tau)d\tau \geq \varepsilon/2\right\},$$

i.e., on substituting this solution into the equation we obtain the remainder $R_s(x, t, \varepsilon, \lambda)$ on the right-hand side such that the following estimates are valid:

$$\left\| \frac{\partial^{|\alpha|} R_s(x, t, \varepsilon, \lambda)}{\partial x^\alpha} \right\| \leq C\lambda^{-s-1+|\alpha|}$$

for $(x, t) \in K$, $x_2 - \varepsilon \int_0^t \chi(x_3 - \tau)d\tau \geq \varepsilon/2$, and it can be shown that the constant C depends only on K and does not depend on ε. Hence, we obtain a simultaneous asymptotic expansion of the solution to the equation considered with respect to smoothness and growth at infinity.

The condition

$$x_2 \geq \varepsilon \left(\frac{1}{2} + \int_0^t \chi(x_3 - \tau)d\tau \right)$$

imposes restrictions on the size of the interval $[0, T]$. Namely, since we seek the solution in the domain $\{x_2 \geq \varepsilon\}$, T can be defined from the condition

$$\sup_{x_3} \int_0^T |\chi(x_3 - \tau)|d\tau > 1/2,$$

e.g., one can set

$$T = (2 \sup \chi(x_3))^{-1}.$$

To increase T, it suffices to choose $\psi(x)$ so that $\inf \psi(x) \geq M$; then one can take

$$T = (M - 1/2)/ \sup_{x_3} |\chi(x_3)|.$$

5 Geostrophic Wind Equations

Let us consider an example of an asymptotic solution with simultaneous smoothness-growth at infinity for a model but still physically meaningful problem.

We will consider the so-called geostrophic wind and study the evolution of small deviations of velocities and pressure from their geostrophic values in the equatorial zone.

According to [153], the geostrophic wind is described as follows. The equations of the gradient wind are derived from the equations of atmospheric dynamics under the assumption that the motion is stationary and the trajectories of particles are isobars. In considering large-scale atmospheric phenomena it is often possible to neglect the nonlinear terms describing the acceleration of particles due to the curvature of the trajectories in the equations of motion. The expression for the components of the flow velocity obtained under this assumption in conjunction with the hypothesis that the flow is stationary is known as the *geostrophic wind*.

Obukhov showed that a deviation from the geostrophic state results in the appearance of rapidly propagating waves which "dissolve" the perturbation in a short period of time. Namely, the pressure field adapts to the velocity field. His conclusions were based on the following argument. He considered the approximate system

$$\begin{cases} \dfrac{\partial u}{\partial t} + \dfrac{\partial u}{\partial x} + v\dfrac{\partial u}{\partial y} + w\dfrac{\partial u}{\partial z} - 2\omega_z v = -\dfrac{1}{8}\dfrac{\partial p}{\partial x}, \\ \dfrac{\partial v}{\partial t} + u\dfrac{\partial u}{\partial x} + v\dfrac{\partial v}{\partial y} + w\dfrac{\partial v}{\partial z} + 2\omega_z u = -\dfrac{1}{\rho}\dfrac{\partial p}{\partial y}, \\ 0 = -\dfrac{1}{\rho}\dfrac{\partial p}{\partial z} - g \end{cases}$$

for the flow velocity $\vec{v} = (u, v, w)$ and pressure p, where ρ is the density and ω_z the vertical component of the Earth's angular velocity, $\omega_z = \omega_0 \sin\theta$ (here $\omega_0 = 7.29 \times 10^{-5}\ \text{sec}^{-1}$ and θ is the latitude). Hence, $\partial w/\partial t$, the component F_z of the Coriolis force, and quantities such as $\omega_x w$, $\omega_y w$, etc. are neglected.

Further, additional simplifying assumptions are made.

1) The surface of the Earth is assumed to be flat (i.e., a Cartesian coordinate system (x, y) is used);
2) ω_z is assumed to be independent of (x, y);
3) the terms quadratic in $\vec{v}$ and derivatives of $\vec{v}$ are discarded;
4) averaging over z is performed;
5) the barotropic scheme for the dependence of P on ρ is used.

This results in the following averaged system:

$$\begin{cases} \dfrac{\partial U}{\partial t} - 2\omega_z V = -gH_0\dfrac{\partial \chi}{\partial x}, \\ \dfrac{\partial V}{\partial t} + 2\omega_z U = -gH_0\dfrac{\partial \chi}{\partial y}, \\ \dfrac{\partial \chi}{\partial t} = -\left(\dfrac{\partial U}{\partial x} + \dfrac{\partial V}{\partial y}\right), \end{cases}$$

where

$$U = \frac{z}{p_0}\int_0^\infty \rho u\,dz, \quad V = \frac{z}{p_0}\int_0^\infty \rho v\,dz, \quad \text{and} \quad \chi = \frac{p - p_0}{p_0}$$

(p_0 is the pressure averaged over the surface and $H_0 \sim 8$ km).

The geostrophic wind takes place for

$$\frac{\partial U}{\partial t} = \frac{\partial V}{\partial t} = \frac{\partial \chi}{\partial t} = 0.$$

The adaptation of perturbations of the stationary solution goes as follows. There occurs a wave with velocity $c = \sqrt{gH_0} \cong 280$ m/sec, which "carries the perturbation away" in time $t \sim 2R/C$, where R is the radius of the perturbed region.

Here we study perturbations of the geostrophic state under different assumptions. We do not assume that ω_z is constant. Quite the opposite, we consider the flow in the equatorial zone. Since $\omega_z = \omega_0 \sin\theta$, we see that $\omega_z = 0$ on the equator. We assume that the zone in small enough to take $\sin\theta \cong \theta$, but the characteristic length of the perturbation is relatively small with respect to the size of the zone, so that the distance y from the equator can be considered as a large parameter. With these assumptions in mind, we will state the problem of construction of simultaneous asymptotic expansions with respect to smoothness and large y.

Hence, if we denote the longitude by x and the latitude by y, the asymptotic expansion of the solution will be sought with respect to the tuple of operators

$$A_1 = -i\frac{\partial}{\partial x}, \quad A_2 = -i\frac{\partial}{\partial y}, \quad \text{and} \quad A_3 = y.$$

Here we derive the geostrophic wind equations in the situation described, pass to the ordered representation, and construct the leading term of the asymptotic expansion.

We assume that the surface of the Earth is the unit sphere

$$X^2 + Y^2 + Z^2 = 1$$

in space $\mathbb{R}^3$ with coordinates (X, Y, Z) (not to be confused with the lower case letters x and y, whose meaning is quite different). The equator is given by the equation $Z = 0$ and the angular velocity of the Earth by the vector

$$\Omega = (0, 0, 1).$$

The Coriolis force is

$$F = 2m[V, \Omega],$$

where

$$V = (\dot{X}, \dot{Y}, \dot{Z});$$

in what follows we assume that $2m = 1$. Let φ be the longitude and θ the latitude. Then we obtain

$$X = \sin\varphi\cos\theta, \quad Y = \cos\varphi\cos\theta, \quad Z = \sin\theta,$$

In the derivation of the equations, it suffices to consider an arbitrary value of φ, say $\varphi = 0$; then

$$\dot{X} = \dot{\varphi}\cos\theta, \quad \dot{Y} = -\dot{\theta}\sin\theta, \quad \dot{Z} = \dot{\theta}\cos\theta,$$

and

$$[V, \Omega] = (-\dot{\theta}\sin\theta, -\dot{\varphi}\cos\theta, 0).$$

The tangent plane at any point (φ, θ), $\varphi = 0$, is determined by the outer normal $\hat{n} = (0, \cos\theta, \sin\theta)$. The projection of F on this plane has the form

$$\begin{aligned} P_r(F) &= [V, \Omega] - \hat{n}(\hat{n}, [V, \Omega]) \\ &= (-\dot{\theta}\sin\theta, -\dot{\varphi}\cos\theta\sin^2\theta, \dot{\varphi}\cos^2\theta\sin\theta). \end{aligned}$$

By neglecting second- and higher-order infinitesimals and denoting $\varphi = x$, $\theta = y$, $\dot{\varphi} = v$, and $\dot{\theta} = w$, we obtain

$$F_x = -yw, \quad F_y = yv.$$

The hydrodynamic equations for $\vec{v} = (v, w)$ have the form

$$\begin{cases} \rho\left[\frac{\partial \vec{v}}{\partial t} + (\vec{v}\Delta)\vec{v}\right] = -\Delta p + F, \\ \frac{\partial \rho}{\partial t} + \operatorname{div}(\rho\vec{v}) = 0, \\ p = p(\rho). \end{cases}$$

We linearize this system assuming that

$$\rho = 1 + \rho', \quad \frac{\partial p}{\partial \rho}(1) = 1,$$

and the quantities ρ' and v are small.

The linearized system can be written in the form

$$\begin{pmatrix} -i\frac{\partial}{\partial t} & -iy & -i\frac{\partial}{\partial x} \\ iy & -i\frac{\partial}{\partial t} & -i\frac{\partial}{\partial y} \\ -i\frac{\partial}{\partial x} & -i\frac{\partial}{\partial y} & -i\frac{\partial}{\partial t} \end{pmatrix} \begin{pmatrix} v \\ w \\ \rho \end{pmatrix} = 0. \tag{III.42}$$

We assume the correspondence

$$p_0 \leftrightarrow -i\frac{\partial}{\partial t}, \quad p_1 \leftrightarrow -i\frac{\partial}{\partial x}, \quad p_2 \leftrightarrow -i\frac{\partial}{\partial y}$$

between the derivation operators and the corresponding arguments of symbols; we then obtain the following symbol matrix for this system:

$$\begin{pmatrix} p_0 & -iy & p_1 \\ iy & p_0 & p_2 \\ p_1 & p_2 & p_0 \end{pmatrix}.$$

Its determinant is equal to

$$\det\begin{pmatrix} p_0 & -iy & p_1 \\ iy & p_0 & p_2 \\ p_1 & p_2 & p_0 \end{pmatrix} = p_0(p_0^2 - p_1^2 - p_2^2 - y^2),$$

and it is clear intuitively that the system is "hyperbolic" with respect to the desired type of asymptotic expansions.

We equip equation (III.42) with the initial data

$$\left.\begin{pmatrix} v \\ w \\ \rho \end{pmatrix}\right|_{t=0} = \begin{pmatrix} v_0 \\ w_0 \\ \rho_0 \end{pmatrix} \tag{III.43}$$

and seek for the solution of the Cauchy problem (III.42) – (III.43) in the form

$$\begin{pmatrix} v \\ w \\ \rho \end{pmatrix} = G\left(\overset{2}{x}, \overset{2}{y}, -i\overset{1}{\frac{\partial}{\partial x}}, -i\overset{1}{\frac{\partial}{\partial y}}, t\right)\begin{pmatrix} v_0 \\ w_0 \\ \rho_0 \end{pmatrix}, \tag{III.44}$$

G is a 2π-periodic matrix function with respect to the variable x and

$$G|_{t=0} = 1.$$

The equation for G is readily obtained with the help of the left ordered representation of the tuple $\left(\overset{1}{-i\partial/\partial x}, \overset{1}{-i\partial/\partial y}, \overset{2}{x}, \overset{2}{y}\right)$. Recall that this representation has the form

$$l_{-i\partial/\partial x} = p_1 - i\frac{\partial}{\partial x}, \quad l_{-i\partial/\partial y} = p_2 - i\frac{\partial}{\partial y}, \quad l_x = x, \quad l_y = y,$$

so that for $G(x, y, p_1, p_2, t)$ we obtain the system

$$-i\frac{\partial G}{\partial t} + \begin{pmatrix} 0 & -iy & p_1 - i\frac{\partial}{\partial x} \\ iy & 0 & p_2 - i\frac{\partial}{\partial y} \\ p_1 - i\frac{\partial}{\partial x} & p_2 - i\frac{\partial}{\partial y} & 0 \end{pmatrix} G = 0$$

with the initial conditions $G|_{t=0} = 1$. To visualize the orders of the terms in the expansion, let us make the change of variables

$$x = x_1, \quad y = \lambda x_2, \quad p_1 = \lambda\eta_1, \quad p_2 = \lambda\eta_2,$$

where λ is a large parameter (note that we have assigned λ to exactly those variables that correspond to the operators $-i\partial/\partial x$, $-i\partial/\partial y$, and y, which define the type of the asymptotic expansion. We denote

$$g(x, y, t, \lambda) = G(x_1, \lambda x_2, \lambda\eta_1, \lambda\eta_2, t).$$

The problem for g reads

$$-i\lambda^{-1}\frac{\partial g}{\partial t} + \begin{pmatrix} 0 & -ix_2 & \eta_1 - i\lambda^{-1}\frac{\partial}{\partial x_1} \\ ix_2 & 0 & \eta_2 - i\lambda^{-1}\frac{\partial}{\partial x_2} \\ \eta_1 - i\lambda^{-1}\frac{\partial}{\partial x_1} & \eta_2 - i\lambda^{-1}\frac{\partial}{\partial x_2} & 0 \end{pmatrix} g = 0,$$

$$g|_{t=0} = 1. \tag{III.45}$$

Clearly, g is independent of x_1. We seek for a particular solution to the equation in this problem of the form

$$g = e^{i\lambda S(x,\eta,t)}\varphi(x,\eta,\lambda^{-1},t),$$

where S and φ are smooth functions of their arguments and S is real-valued. Substituting g into the system and collecting the coefficients of the powers of λ^{-1}, we obtain the following system of equations:

$$\left[\left(\frac{\partial S}{\partial t} + \begin{pmatrix} 0 & -ix_2 & \eta_1 \\ ix_2 & 0 & \eta_2 \\ \eta_1 & \eta_2 & 0 \end{pmatrix}\right) - i\lambda^{-1}\left(\frac{\partial}{\partial t} + \begin{pmatrix} 0 & 0 & 0 \\ 0 & 0 & i\frac{\partial S}{\partial x_2} \\ 0 & i\frac{\partial S}{\partial x_2} & 0 \end{pmatrix}\right)\right.$$

$$\left. -\lambda^2 \begin{pmatrix} 0 & 0 & 0 \\ 0 & 0 & i\frac{\partial}{\partial x_2} \\ 0 & i\frac{\partial}{\partial x_2} & 0 \end{pmatrix}\right] \varphi(x,\eta,t,\lambda^{-1}) = 0$$

(we omit the derivatives $\partial/\partial x_1$ since φ and S are independent of x_1, $\partial S/\partial x_1 = \partial\varphi/\partial x_1 = 0$). Next, we equate to zero the coefficients of powers of λ^{-1} in the equation obtained. First of all, we should have

$$\left[\frac{\partial S}{\partial t} + \begin{pmatrix} 0 & -ix_2 & \eta_1 \\ ix_2 & 0 & \eta_2 \\ \eta_1 & \eta_2 & 0 \end{pmatrix}\right] \varphi(x,\eta,t,0) = 0,$$

but this equation can be satisfied for $\varphi(x,\eta,t,0) \not\equiv 0$ if and only if

$$\det\begin{pmatrix} \frac{\partial S}{\partial t} & -ix_2 & \eta_1 \\ ix_2 & \frac{\partial S}{\partial t} & \eta_2 \\ \eta_1 & \eta_2 & \frac{\partial S}{\partial t} \end{pmatrix} = 0,$$

that is,

$$\frac{\partial S}{\partial t}\left[\left(\frac{\partial S}{\partial t}\right)^2 - \eta_1^2 - \eta_2^2 - x_2^2\right] = 0.$$

Also, we should have $S|_{t=0} = 0$ (this follows from the initial data). There are three solutions satisfying these conditions:

$$S_0(x,\eta,t) \equiv 0$$

and

$$S_\pm(x,\eta,t) = \pm t\sqrt{\eta_1^2 + \eta_2^2 + x_2^2}.$$

Let us find the corresponding eigenvectors of the symbol matrix, of which $\varphi(x,\eta,t,0)$ is a multiple. Let[1] ${}^t(a,b,c)$ be an eigenvector with eigenvalue zero. We have

$$0=\begin{pmatrix} 0 & -ix_2 & \eta_1 \\ ix_2 & 0 & \eta_2 \\ \eta_1 & \eta_2 & 0 \end{pmatrix}\begin{pmatrix} a \\ b \\ c \end{pmatrix}=\begin{pmatrix} c\eta_1-ibx_2 \\ iax_2+c\eta_2 \\ a\eta_1+b\eta_2 \end{pmatrix}.$$

We normalize the eigenvector by the condition $c=1$; then we obtain

$$b=-i\eta_1/x_2, \quad a=i\eta_2/x_2,$$

and the eigenvector normalized to unity has the form

$$\chi_0=\frac{{}^t(i\eta_2,-i\eta_1,x_2)}{\sqrt{\eta_1^2+\eta_2^2+x_2^2}}.$$

The eigenvectors corresponding to $S_{\pm n}(x,\eta,t)$ can be obtained in a similar way. Specifically, the system of equations for these vectors has the form

$$\begin{pmatrix} \pm\sqrt{\eta_1^2+\eta_2^2+x_2^2} & -ix_2 & \eta_1 \\ ix_2 & \pm\sqrt{\eta_1^2+\eta_2^2+x_2^2} & \eta_2 \\ \eta_1 & \eta_2 & \pm\sqrt{\eta_1^2+\eta_2^2+x_2^2} \end{pmatrix}\begin{pmatrix} a \\ b \\ c \end{pmatrix}=0$$

(one takes the lower or the upper sign in all three rows simultaneously). The normalized eigenvectors have the form

$$\chi_\pm=\frac{{}^t(\pm\eta_1\sqrt{\eta_1^2+\eta_2^2+x_2^2}+i\eta_1x_2,\pm\eta_2\sqrt{\eta_1^2+\eta_2^2+x_2^2}-i\eta_1x_2,\eta_1^2+\eta_2^2)}{\sqrt{2}\sqrt{\eta_1^2+\eta_2^2}\sqrt{\eta_1^2+\eta_2^2+x_2^2}}.$$

The orthogonal projections onto the one-dimensional eigenspaces have the form

$$P_0 = \frac{1}{\eta_1^2+\eta_2^2+x_2^2}\begin{pmatrix} \eta_2^2 & -\eta_1\eta_2 & ix_2\eta_2 \\ -i\eta_1\eta_2 & \eta_1^2 & -i\eta_1x_2 \\ -ix_2\eta_1 & i\eta_1x_2 & x_2^2 \end{pmatrix},$$

$$P_\pm = \frac{1}{2(\eta_1^2+\eta_2^2+x_2^2)}\begin{pmatrix} A_{11} & A_{12} & A_{13} \\ A_{21} & A_{22} & A_{23} \\ A_{31} & A_{32} & A_{33} \end{pmatrix}$$

where

$$A_{11}=\eta_1^2+x_2^2, \quad A_{22}=\eta_2^2+x_2^2, \quad A_{33}=\eta_1^2+\eta_2^2,$$

$$A_{21}=\eta_1\eta_2\mp ix_2\sqrt{\eta_1^2+\eta_2^2+x_2^2}, \quad A_{31}=i\eta_2x_2\mp i\eta_1\sqrt{\eta_1^2+\eta_2^2+x_2^2},$$

$$A_{32}=-i\eta_1x_2\mp i\eta_2\sqrt{\eta_1^2+\eta_2^2+x_2^2}$$

[1] Here the superscript t stands for the transposed matrix.

and the matrix $||A_{ij}||$ is Hermitian, that is, $A_{ij} = \bar{A}_{ij}$ (the bar stands for complex conjugation).

Clearly, we have

$$P_0 + P_+ + P_- \equiv 1.$$

We now seek the asymptotic solution to problem (III.45) (more precisely, we seek its leading term) in the form

$$g(x, \eta, \lambda, t) = e^{i\lambda S_0}\varphi_0 + e^{i\lambda S_+}\varphi_+ + e^{i\lambda S_-}\varphi_-,$$

with

$$\varphi_j = c_j(x, \eta, t)P_j,$$
$$\varphi_j\big|_{t=0} = P_j, \quad j = 0, +, -.$$

Substituting this representation of the solution into the equation, we obtain

$$\left[\frac{\partial}{\partial t} + \begin{pmatrix} 0 & 0 & 0 \\ 0 & 0 & i\dfrac{\partial S}{\partial x_2} \\ 0 & i\dfrac{\partial S}{\partial x_2} & 0 \end{pmatrix}\right]\varphi_j = \Delta(S)\varphi_{j,1}, \quad j = 0, +, -,$$

where $\varphi_{j,1}$ is the next term of the asymptotic expansion and

$$\Delta(S) = \begin{pmatrix} \dfrac{\partial S}{\partial t} & -ix_2 & \eta_1 \\ ix_2 & \dfrac{\partial S}{\partial t} & \eta_2 \\ \eta_1 & \eta_2 & \dfrac{\partial S}{\partial t} \end{pmatrix}.$$

A necessary and sufficient condition for the solvability of this equation is that

$$P_j\left[\frac{\partial}{\partial t} + \begin{pmatrix} 0 & 0 & 0 \\ 0 & 0 & i\dfrac{\partial S}{\partial x_2} \\ 0 & \dfrac{\partial S}{\partial x_2} & 0 \end{pmatrix}\right]c_j P_j = 0,$$

or

$$P_j\frac{\partial C_j}{\partial t} + i\frac{\partial S}{\partial x_2}c_j P_j\begin{pmatrix} 0 & 0 & 0 \\ 0 & 0 & 1 \\ 0 & 1 & 0 \end{pmatrix}P_j = 0.$$

However, it is easy to check that

$$P_j\begin{pmatrix} 0 & 0 & 0 \\ 0 & 0 & 1 \\ 0 & 1 & 0 \end{pmatrix}P_j = 0, \quad j = 0, +, -.$$

Henceforth, $c_j \equiv 1$ and we obtain, in the first approximation, the following expression for the symbol of the resolvent operator:

$$G(y, p_x p_y, t) = \frac{1}{p_x^2 + p_y^2 + y^2} \begin{pmatrix} p_y^2 & -p_x p_y & iyp_y \\ -p_x p_y & p_x^2 & -iyp_x \\ -iyp_y & iy & y^2 \end{pmatrix}$$

$$+ \frac{\cos[t\sqrt{p_x^2 + p_y^2 + y^2}]}{p_x^2 + p_y^2 + y^2} \begin{pmatrix} p_x^2 + y^2 & p_x p_y & -iyp_y \\ p_x p_y & p_y^2 + y^2 & iyp_x \\ ixp_y & -iyp_x & p_x^2 + p_y^2 \end{pmatrix}$$

$$+ \frac{\sin[t\sqrt{p_x^2 + p_y^2 + y^2}]}{p_x^2 + p_y^2 + y^2} \begin{pmatrix} 0 & -y & -p_x \\ y & 0 & -ip_y \\ -ip_x & -ip_y & 0 \end{pmatrix} + \text{lower-order terms.}$$

This expression provides the leading term of the asymptotic solution to the model geostrophic-wind problem (III.42) – (III.43) by formula (III.44).

6 Degenerate Equations

In this section we consider a simple model example showing the application of noncommutative analysis to a problem of constructing the asymptotic expansion of the solution to a degenerate equation. The problem itself is probably of little interest mainly because it can be solved by applying an appropriate "quantized canonical transformation" that takes the operator in the equation into the Hamiltonian operator of the harmonic oscillator. However, it is the technique of noncommutative analysis that we intend to illustrate by considering this example; specifically, we first obtain a "weakened" asymptotic solution for localized right-hand sides. Then, using the known exact solution for some "standard" operator (in this particular example, this is the energy operator of the harmonic oscillator), we obtain asymptotic solutions for arbitrary right-hand sides.

6.1 Statement of the Problem

Consider the ordinary differential operator

$$\hat{H} = -a(x)\frac{\partial^2}{\partial x^2} + b(x)x^2\Omega^2, \quad x \in \mathbb{R},$$

where $\Omega \in \mathbb{R}$ is a large parameter. We assume that $a(x)$ and $b(x)$ are smooth real-valued functions bounded from below by a positive constant. Moreover, we assume that $a(x)$ and $b(x)$ are bounded together with all their derivatives. Without loss of generality it can be assumed that $a(0) = b(0) = 1$, which can always be achieved by a linear change of variables followed by the multiplication of $\hat{H}$ by an appropriate constant. Under these conditions $\hat{H}$ can be defined by closure from $C_0^\infty(\mathbb{R})$ as a closed operator on $L_2(\mathbb{R})$.

Consider the following stationary problem for $\hat{H}$: find the solution $u \in L_2(\mathbb{R})$ of the equation

$$\hat{H}u(x) \equiv -a(x)\frac{\partial^2 u(x)}{\partial x^2} + b(x)x^2\Omega^2 u(x) = v(x), \tag{III.46}$$

where $v \in L_2(\mathbb{R})$ is a given element.

Definition III.3 An asymptotic solution as $\Omega \to \infty$ of problem (III.46) is a sequence of operators $\hat{G}_N$, $N = 1, 2, \ldots$, on $L_2(\mathbb{R})$ such that

$$||\hat{H}\hat{G}_N - 1||_{L_2(\mathbb{R}) \to H^{k(N)}(\mathbb{R})} \leq C_N \cdot \Omega^{-k(N)}, \quad \Omega \geq 1,$$

where $k(N) \to \infty$ as $N \to \infty$ and $H^k(\mathbb{R}) \equiv W_2^k(\mathbb{R})$ is the Sobolev space of order k.

The construction of the asymptotic solution to (III.46) is carried out in three stages. Namely, at the first stage we reduce the problem to a similar one with localized right-hand side. At the second stage we present the solution of this problem. Finally, at the third stage we construct the asymptotic solution for a general case.

6.2 Localization of the Right-Hand Side

Near the degeneracy point $x = 0$ the operator $\hat{H}$ is close to the operator

$$\hat{H}_0 = -\frac{\partial^2}{\partial x^2} + \Omega^2 x^2,$$

which coincides with the energy operator of the quantum mechanical harmonic oscillator up to a constant factor. In this subsection we construct an operator $\hat{G^1}_N$ such that the substitution $u = \hat{G^1}_N v$ yields an equation whose right-hand side is sufficiently smooth and, in some sense, is localized near the point $x = 0$. This enables us to solve the resulting equation with the help of perturbation theory, using the known inverse $(\hat{H}_0)^{-1}$.

We seek $\hat{G^1}_N$ in the form of a function of the following self-adjoint operators in $L_2(\mathbb{R})$:

$$A_1 = -i\frac{\partial}{\partial x}, \quad A_2 = \Omega x, \quad A_3 = \Omega, \quad B = x;$$

$$\hat{G^1}_N = G_N^1(\overset{1}{A_1}, \overset{2}{A_2}, \overset{2}{A_3}, \overset{2}{B}).$$

The operators A_1, A_2, A_3 and B generate the Lie algebra with the commutation relations

$$[A_1, A_2] = -iA_3, \quad [A_1, B] = -i$$

(all the other commutators are zero). According to the general theory presented in Chapter II, the product of two operators $\hat{f_i} \equiv f_i(\overset{1}{A_1}, \overset{2}{A_2}, \overset{2}{A_3}, \overset{2}{B})$, $i = 1, 2$, can be written as

$$\hat{f_1} \circ \hat{f_2} = f_3(\overset{1}{A_1}, \overset{2}{A_2}, \overset{2}{A_3}, \overset{2}{B}),$$

where the symbol $f_3(y, \alpha)$, $y \in \mathbb{R}^3, \alpha \in \mathbb{R}$, of the product $\hat{f_1}\hat{f_2}$ is given by the formula

$$f_3(y, \alpha) = f_1(\overset{1}{l_{A_1}}, \overset{2}{l_{A_2}}, \overset{2}{l_{A_3}}, \overset{2}{l_B})(f_2(y, \alpha)),$$

where $l_{A_1}, l_{A_2}, l_{A_3}$ and l_B are the left ordered representation operators,

$$l_{A_1} = y_1 - iy_3\frac{\partial}{\partial y_2} - i\frac{\partial}{\partial \alpha}, \quad l_{A_2} = y_2, \quad l_{A_3} = y_3, \quad l_3 = \alpha.$$

In particular, since

$$\hat{H} = a(\overset{2}{B})\overset{1}{A_1}{}^2 + b(\overset{2}{B})\overset{1}{A_2}{}^2,$$

it follows that

$$\hat{H}\hat{G}^1{}_N = 1 + \hat{R}^1{}_N \equiv 1 + R^1_N(\overset{1}{A_1}, \overset{2}{A_2}, \overset{2}{A_3}, \overset{2}{B}),$$

where

$$R^1_N(y, \alpha) = \left(a(\alpha)\left(y_1 - iy_3\frac{\partial}{\partial y_2} - i\frac{\partial}{\partial \alpha}\right)^2 + b(\alpha)y_2^2\right) G^1_N(y, \alpha) - 1.$$

Being solved exactly, the equation $R^1_N(y, \alpha) = 0$ for $G^1_N(y, \alpha)$ provides the exact solution of the original problem after the substitution of A_1, A_2, A_3 and B for y_1, y_2, y_3, and α.

We will construct an asymptotic solution of that equation assuming that $1/\sqrt{y_1^2 + y_2^2}$ and $y_3/(y_1^2 + y_2^2)$ are small.

Clearly,

$$a(\alpha)\left(y_1 - iy_3\frac{\partial}{\partial y_2} - i\frac{\partial}{\partial \alpha}\right)^2 + b(\alpha)y_2^2 = P_0 + P_1 + P_2,$$

where P_j, $j = 1, 2, 3$, are differential operators of order j; more precisely,

$$P_0 = a(\alpha)y_1^2 + b(\alpha)y_2^2,$$

$$P_1 = -2ia(\alpha)y_1\left(y_3\frac{\partial}{\partial y_2} + \frac{\partial}{\partial \alpha}\right),$$

and

$$P_2 = -a(\alpha)\left(y_3\frac{\partial}{\partial y_2} + \frac{\partial}{\partial \alpha}\right)^2.$$

Set

$$g^1_{-1}(y,\alpha) \equiv 0, \quad g^1_0(y,\alpha) = \frac{1}{\{a(\alpha)y_1^2 + b(\alpha)y_2^2\}}$$

and define the functions $g^1_k(y,\alpha)$, $k = 1, 2, \ldots$, by the iterative formulas

$$g^1_k(y,\alpha) = -\frac{1}{a(\alpha)y_1^2 + b(\alpha)y_2^2}(P_1 g^1_{k-1}(y,\alpha) + P_2 g^1_{k-2}(y,\alpha))$$

(note that $a(\alpha)y_1^2 + b(\alpha)y_2^2 \neq 0$ for $y_1^2 + y_2^2 \neq 0$ by virtue of the conditions imposed on $a(\alpha)$ and $b(\alpha)$).

Next, choose a function $\psi \in C_0^\infty(\mathbb{R})$ such that $\psi(z) \equiv 1$ in the neighborhood of zero. Set[1]

$$G^1_N(y,\alpha) = \psi\left(\frac{y_3}{y_1^2 + y_2^2}\right)\sum_{k=0}^{N} g^1_k(y,\alpha).$$

Proposition III.2 *For this choice of the symbol $G^1_N(y,\alpha)$ the remainder $R^1_N(y,\alpha)$ is bounded along with all its derivatives and satisfies the estimates*

$$\left|\frac{\partial^{|\beta|+\gamma}}{\partial y_1^{\beta_1}\partial y_2^{\beta_2}\partial\alpha^\gamma} R^1_N(y,\alpha)\right| \leq C_{\beta\gamma} \circ \left(\frac{y_3}{y_1^2+y_2^2}\right)^{(N+1)/2}\left(\frac{1}{\sqrt{y_1^2+y_2^2}}\right)^{|\beta|},$$
$$|\beta|, \gamma = 0, 1, \ldots.$$

Proof. First let us show that the following assertion is valid.

Lemma III.1 *The functions $g^1_k(y,\alpha)$, $k = -1, 0, 1, 2, \ldots$, have the form*

$$g^1_k(y,\alpha) = \sum_{j=1}^{k} \sigma_{jk}(y_1, y_2, \alpha) y_3^j,$$

where the functions $\sigma_{jk}(y_1, y_2, \alpha)$ are smooth for $y_1^2 + y_2^2 \neq 0$ and homogeneous with respect to (y_1, y_2) of degree $(k + j + 2)$. On the sphere $y_1^2 + y_2^2 = 1$ all derivatives of these functions are bounded uniformly in $\alpha \in \mathbb{R}$.

This lemma can easily be proved by induction on k.

---

[1]The function G^1_N thus defined has a singularity. However, we are only interested in the domain $y_3 \geq 1$, where no singularities occur (y_3 is replaced by Ω)). Since in this case the function ψ vanishes for $y_1^2 + y_2^2$ sufficiently small.

Let us now proceed to the proof of Proposition III.2. The formula for $R_N^1(y,\alpha)$ can be rewritten in the form

$$R_N^1(y,\alpha) = (P_0 + P_1 + P_2)\psi\left(\frac{y_3}{y_1^2+y_2^2}\right)\sum_{k=0}^{N} g_k^1(y,\alpha) - 1$$

$$= \left[P_1 + P_2, \psi\left(\frac{y_3}{y_1^2+y_2^2}\right)\right]\sum_{k=0}^{N} g_N^1(y,\alpha)$$

$$+\psi\left(\frac{y_3}{y_1^2+y_2^2}\right)\left\{(P_0 g_0^1(y,\alpha) - 1) + \sum_{k=1}^{N}(P_0 g_k^1(y,\alpha) + P_1 g_{k-1}^1(y,\alpha)P_2 g_{k-2}^1(y,\alpha))\right\}$$

$$+\psi\left(\frac{y_3}{y_1^2+y_2^2}\right)(P_1 g_N^1(y,\alpha) + P_2 g_{N-1}^1(y,\alpha) + P_2 g_N^1(y,\alpha)).$$

The expression in curly brackets is equal to zero, which follows from the construction of $g_k^1(y,\alpha)$. Let us estimate the remaining terms.

(a) The derivatives of $\psi\left(y_3/(y_1^2+y_2^2)\right)$ of order $|\beta| \geq 1$ satisfy the bound

$$\left|\frac{\partial^{|\beta|}}{\partial y_1^{\beta_1}\partial y_2^{\beta_2}}\psi\left(\frac{y_3}{y_1^2+y_2^2}\right)\right| \leq C_{m\beta}\left(\frac{y_3}{y_1^2+y_2^2}\right)^m\left(\frac{1}{y_1^2+y_2^2}\right)^{|\beta|/2},$$

for any m, which can be checked by straightforward derivation (one should take into account that the supports of all derivatives of $\psi(z)$ lie in the domain $1/C \leq z \leq C$ for some $C > 0$).

(b) The commutator $[P_1, \psi]$ has the form

$$\left[P_1, \psi\left(\frac{y_3}{y_1^2+y_2^2}\right)\right] = ia(\alpha)y_1y_2\left(\frac{y_3}{y_1^2+y_2^2}\right)^2\psi'\left(\frac{y_3}{y_1^2+y_2^2}\right).$$

It follows immediately from Lemma III.1 on g_k^1 and from the bounds for ψ that $[P_1,\psi]\sum_{k=0}^{N} g_k^1$ satisfy the needed estimates. Similarly we can estimate the term containing $[P_2,\psi]$.

(c) Let us estimate $P_1 g_N^1(y,\alpha)$:

$$P_1 g_N^1(y,\alpha) = -2ia(\alpha)\sum_{j=0}^{N}\left(y_3^{j+1}y_1\frac{\partial\sigma_{jN}}{\partial y_2} + y_3^j y_1\frac{\partial\sigma_{jN}}{\partial\alpha}\right).$$

The derivative $\partial^{|\beta|+\gamma}/\partial y_1^{\beta_1}\partial y_2^{\beta_2}\partial\alpha^\gamma$ of this expression is a sum of terms of the form

$$y_3^3\sigma_a(y_1,y_2,\alpha)$$

where $0 \leq s \leq N+1$ and $\sigma(y_1,y_2,\alpha)$ is a function homogeneous of order $r \leq -(N+|\beta|+s+1)$. We have, for $y_1^2+y_2^2 \geq c > 0$, $y_3 \geq 1$,

$$y_3^s\sigma(y_1,y_2,\alpha) \leq \text{const}\, y_3^s(y_1^2+y_2^2)^{-|\beta|/2}(y_1^2+y_2^2)^{-(N+s+1)/2}$$

$$\leq \left(\frac{y_3}{y_1^2+y_2^2}\right)^{(N+1)/2}(y_1^2+y_2^2)^{-|\beta|/2}\left(\frac{y_3}{y_1^2+y_2^2}\right)^{s/2},$$

In analogy with the above, we can estimate $P_2 g_{N_1}^1(y,\alpha)$ and $P_2 g_{N_1}^1(y,\alpha)$. In conjunction with the estimates (a), these estimates imply Proposition III.2, since $y_3/(y_1^2+y_2^2)$ and $1/\sqrt{y_1^2+y_2^2}$ are bounded on supp ψ (owing to the condition $y_3 \geq 1$). Proposition III.2 is proved. □

Now let $v(x, \Omega) \in L_2(\mathbb{R})$ be a function such that the norm $||v||_{L_2(\mathbb{R})}$ is bounded uniformly in Ω. In the equation $\hat{H}u = v$ we replace u by $u_N = \hat{G}^1_N v$ and obtain the following equation for the difference $(u - u_N)$:

$$\hat{H}(u - u_N) = -\hat{R}^1_N v \equiv r_N(x, \Omega).$$

As was already mentioned at the beginning of this section, the right-hand side $r_N(x, \Omega)$ of this equation is localized in a neighbourhood of the degeneracy point $x = 0$. We will now give a precise meaning to this assertion. Introduce the new variable ξ by setting $x = \xi\Omega^{-1/2}$, $\tilde{f}(\xi, \Omega) \stackrel{\text{def}}{=} f(\xi\Omega^{-1/2}, \Omega) \equiv f(x, \Omega)$ for any function $f \in L_2(\mathbb{R}_x)$ depending on the parameter Ω.

Proposition III.3 *The function $\tilde{r}_N(\xi, \Omega)$ lies in the domain of $\left(\xi^2 - \partial^2/\partial\xi^2\right)^{(N+1)/2}$ and*

$$\left\| \left(\xi^2 - \frac{\partial^2}{\partial\xi^2}\right)^{(N+1)/2} \tilde{r}_N(\xi, \Omega) \right\|_{L_2(\mathbb{R}_\xi)} \leq C\,\Omega^{1/4} ||v(x, \Omega)||_{L_2(\mathbb{R}_x)},$$

where the constant C is independent of Ω.

This proposition can be proved by the standard technique of L_2-estimates. We omit this proof here since such estimates are not the topic of the present book. The reader can carry them out for himself or herself.

Let us now proceed to the second stage of the solution.

6.3 Solving the Equation with Localized Right-Hand Side

Let ε denote the small parameter $\varepsilon = \Omega^{-1/2}$. We again use the change of variables $x = \varepsilon\xi$.

For brevity, we denote the difference $u - u_N$ by the same letter u. The equation for u reads

$$\hat{H}u \equiv a(\varepsilon\xi)\frac{\partial^2 u(\xi, \varepsilon)}{\partial\xi^2} + b(\varepsilon\xi)\xi^2 u(\xi, \varepsilon) = v(\xi, \varepsilon),$$

where $v(\xi, \varepsilon) = \varepsilon^2 \tilde{r}_N(\xi, \varepsilon^{-2})$. Thus, the right-hand side $v(\xi, \varepsilon)$ satisfies the estimate

$$\left\| \left(\xi^2 - \frac{\partial^2}{\partial\xi^2}\right)^{(N+1)/2} v(\xi, \varepsilon) \right\|_{L_2(\mathbb{R}_\xi)} \leq C \cdot \varepsilon^{3/2}, \qquad \text{(III.47)}$$

where the constant is independent of ε.

The function $u(\xi, \varepsilon)$ will be calculated with the help of perturbation theory, by expanding in ε. To this end, let us choose a positive integer l (the precise value of l will be fixed later) and rewrite the equation for u in the following form using the Taylor expansions of a and b:

$$\left(\hat{H}_{\text{osc}} + \sum_{k=1}^{l} \varepsilon^k \hat{H}_k + \varepsilon^{l+1}\hat{R}_{l+1}(\varepsilon)\right) u = v,$$

where

$$\hat{H}_{\text{osc}} = \xi^2 - \frac{\partial^2}{\partial \xi^2}$$

is the energy operator of the quantum oscillator,

$$\hat{H}_k = \frac{1}{k!}\left(b_k \xi^{k+2} - a_k \xi^k \frac{\partial^2}{\partial \xi^2}\right),$$

$$\hat{R}_{l+1}(\varepsilon) = \frac{1}{l!}\left(b_{l+1}(\varepsilon\xi)\xi^{l+3} - a_{l+1}(\varepsilon\xi)\xi^{l+1}\frac{\partial^2}{\partial \xi^2}\right),$$

a_k, b_k are the MacLaurin coefficients of $a(x)$ and $b(x)$, and

$$a_{l+1}(x) = \int_0^1 \frac{\partial^{l+1} a}{\partial x^{l+1}}(\tau x)(1-\tau)^l \, d\tau, \quad b_{l+1}(x) = \int_0^1 \frac{\partial^{l+1} b}{\partial x^{l+1}}(\tau x)(1-\tau)^l \, d\tau.$$

Let us seek $u(x, \varepsilon)$ in the form

$$u(x, \varepsilon) = \sum_{k=0}^{l} \varepsilon^k u_k(x, \varepsilon).$$

For $u_k(x, \varepsilon)$ we obtain the system of equations

$$\hat{H}_{\text{osc}} u_0(x, \varepsilon) = v(x, \varepsilon),$$

$$\hat{H}_{\text{osc}} u_k(x, \varepsilon) = -\sum_{i=1}^{k} \hat{H}_i u_{k-i}(x, \varepsilon),$$

Assuming this system to be satisfied we obtain the following remainders on the right-hand side:

$$[\hat{H}u(x, \varepsilon) - v(x, \varepsilon)] = \varepsilon^{l+1}\left\{\sum_{k=1}^{l} \hat{H}_k \sum_{j=0}^{k-1} \varepsilon^j u_{l-k+j+1}(x, \varepsilon) + \hat{R}_{l+1}(\varepsilon)\sum_{j=0}^{l} \varepsilon^j u_j(x, \varepsilon)\right\}.$$

Let us solve the system for $u_k(x, \varepsilon)$. The operator $\hat{H}_{\text{osc}}$ is boundedly invertible in $L_2(\mathbb{R})$. The inverse has the form

$$\hat{H}_{\text{osc}}^{-1} = \sum_{n=0}^{\infty} \lambda_n^{-1} P_n,$$

where $\lambda_n = 2n + 1$ is the nth eigenvalue of $\hat{H}_{\text{osc}}$ and P_n the orthogonal projector on the corresponding (one-dimensional) eigenspace. Thus, formally we obtain

$$u_0 = \hat{H}_{\text{osc}}^{-1} v, \quad u_k = -\hat{H}_{\text{osc}}^{-1} \sum_{i=1}^{k} \hat{H}_i u_{k-i}, \quad k = 1, 2, \ldots, l.$$

In order to prove that the above formulas indeed define an asymptotic solution of the problem $\hat{H}u = v$ as $\varepsilon \to 0$, we should show that the functions u_k successively obtained lie in the domains of $\hat{H}_{\rm osc}$, $\hat{H}_i$, $i = 1, \dots, l$ and $\hat{R}_{l+1}(\varepsilon)$. Also, it is necessary to estimate the remainder. We shall prove the following assertion.

Proposition III.4 *Under the above assumptions on $v(\xi, \varepsilon)$ the formulas constructed define correctly for each $l \leq N+1$ an element u of $L_2(\mathbb{R})$ such that u lies in the domain of $\hat{H}$. For each $k = 0, \dots, l$ the element u lies in the domain of $(\hat{H}_{\rm osc})^{(N+3-k)/2}$; and the remainder satisfies the following estimates for $l \leq \pi/2$:*

$$||\hat{H}_{\rm osc}^{N/2-1}(\hat{H}u - v)||_{L_2(\mathbb{R}_\xi)} \leq C\,\varepsilon^{l+1}||\hat{H}_{\rm osc}^{(N+1)/2}v||_{L_2(\mathbb{R}_\xi)}.$$

(Recall that v satisfies (III.47)*, and, consequently, the right-hand side of the last equation does not exceed $c\,\varepsilon^{l+3/2}$).*

Thus, for $l \leq N/2$ we obtain a true asymptotic solution modulo $O(\varepsilon^{l+1})$, and the greater N, the more derivations and multiplications by ξ can be applied to this asymptotic expansion.

We omit the proof of this proposition due to its purely technical nature.

The constructed solution of the equation

$$\hat{H}u = v$$

has the form

$$u = \hat{G^2}_l v,$$

where $\hat{G^2}_l$ is the unbounded operator in $L_2(\mathbb{R})$ determined by the formulas

$$\hat{G^2}_l = \sum_{k=0}^{l} \varepsilon^k \hat{G^2}_{lk};$$
$$\hat{G^2}_{l0} = \hat{H}_{\rm osc}^{-1};$$
$$\hat{G^2}_{lk} = -\hat{H}_{\rm osc}^{-1} \sum_{i=1}^{k} \hat{H}_i \hat{G^2}_{l,k-i}, \quad k = 1, \dots, l.$$

The operators $\hat{G^2}_{lk}$ are $(\hat{H}_{\rm osc})^{l+1/2}$-bounded in $L_2(\mathbb{R})$ uniformly in ε (recall that an operator A is said to be B-bounded if $||Au|| \leq \text{const}\,||Bu||$ for all u in the domain of the operator B). The operator $\hat{G}_l^2$ written in the original coordinates (x, Ω) will also be denoted by $\hat{G^2}_l$.

These operators together with the operators G_N^1 determined in Subsection 6.2 will be used in construction of an asymptotic solution in the next subsection.

6.4 The Asymptotic Solution in the General Case

The following theorem describes the asymptotic solution in the general case.

Theorem III.2 *The sequence of operators*

$$\hat{G}_N = \hat{G^1}_{4N} - \hat{G^2}_N \hat{R^1}_{4N}$$

is an asymptotic solution as $\Omega \to \infty$ *to* (III.46) *in the sense of Definition* III.3.

Proof. We have

$$\hat{H}\hat{G}_N = \hat{H}\hat{G^1}_{4N} - \hat{H}\hat{G^2}_N \hat{R^1}_{4N} = 1 + (1 - \hat{H}\hat{G^2}_N)\hat{R^1}_{4N}.$$

Thus, we should prove that there exists a function $K(N)$ such that

$$\lim_{N\to+\infty} K(N) = +\infty$$

and

$$||(1 - \hat{H}\hat{G^2}_N)\hat{R^1}_{4N}||_{L_2(\mathbb{R}_x)\to H^{K(N)}(\mathbb{R}_x)} \le C_N \Omega^{-K(N)}, \quad \Omega \ge 1.$$

By Proposition III.4,

$$||(\hat{H}_{\text{osc}})^k (1 - \hat{H}\hat{G^2}_N)u||_{L_2(\mathbb{R}_\xi)} \le C\,\varepsilon^{N+1}||(\hat{H}_{\text{osc}})^{2N+1/2}u||_{L_2(\mathbb{R}_\xi)}$$

for $k \le N$.

By Proposition III.3,

$$||(\hat{H}_{\text{osc}})^{2N+1/2}\hat{R}^1_{4N}v||_{L_2(\mathbb{R}_\xi)} \le C\,\Omega^{1/4}||v||_{L_2(\mathbb{R}_x)}.$$

Returning to the variables x, we obtain

$$\left\| \Omega^{-k}\left(\Omega^2 x^2 - \frac{\partial^2}{\partial x^2}\right)^k ((1 - \hat{H}\circ\hat{G}^2_N)\hat{R}_{4N})v \right\|_{L_2(\mathbb{R}_x)} \le C\,\Omega^{-(N+1)/2}||v||_{L_2(\mathbb{R}_x)}.$$

To continue the proof we need the following statement.

Lemma III.2 *For any natural k there exists a constant* C_k *independent of* $\Omega \ge 1$ *and such that for any* $v \in S(\mathbb{R}_x)$ *we have*

$$\left\| \left(1 - \frac{\partial^2}{\partial x^2}\right)^k v \right\|_{L_2(\mathbb{R}_x)} \le C_k \left\| \left(\Omega^2 x^2 - \frac{\partial^2}{\partial x^2}\right)^k v \right\|_{L_2(\mathbb{R}_x)}.$$

Proof. Let us make the change of variable $x = \xi/\sqrt{\Omega}$, then the inequality in question becomes

$$\left\| \left(\frac{1}{\Omega} - \frac{\partial^2}{\partial \xi^2} \right)^k v \right\|_{L_2(\mathbb{R}_\xi)} \leq C_k \left\| \left(\xi^2 - \frac{\partial^2}{\partial \xi^2} \right)^k v \right\|_{L_2(\mathbb{R}_\xi)} .$$

Since $1/\Omega \leq 1$, it suffices to prove that

$$\left\| \frac{\partial^l v}{\partial \xi^l} \right\|_{L_2(\mathbb{R}_\xi)} \leq C_k \left\| \left(\xi^2 - \frac{\partial^2}{\partial \xi^2} \right)^k v \right\|_{L_2(\mathbb{R}_\xi)}$$

for $l \leq k$, which follows from the fact that the operator

$$\xi^r \left(\frac{\partial}{\partial \xi} \right)^s (\hat{H}_{\mathrm{osc}})^{-(r+1)/2}$$

is bounded for any r, s. The last assertion can be verified with the help of the standard technique of L_2-estimates. The lemma is proved. □

Hence, with the help of Lemma III.2 we obtain

$$\|(1 - \hat{H}\hat{G}_N^2)\hat{R}_{4N}\|_{L_2(\mathbb{R}_x) \to H^k(\mathbb{R}_x)} \leq C\, \Omega^{k-(N+1)/2}.$$

Putting $K(N) = (N+1)/4$, we get the desired estimate. The theorem is proved. □

7 Microlocal Asymptotic Solutions for an Operator with Double Characteristics

In this section we consider (pseudo)differential equations with degeneration of a type somewhat similar to that discussed in the preceding example. The equations in question are a quite special case of *equations with double characteristics*. The latter have been considered in a number of papers (e.g., see Boutet de Monvel [15], Kohn [108], Boutet de Monvel and Treves [16], Hörmander [80], Radkevich [157], Grushin [74], and Taylor [175]). We do not obtain any new results; our only purpose is to show how the operator calculus machinery can be used in combination with conventional techniques in order to solve problems in the theory of differential equations. Accordingly, we put the main emphasis on the operator calculus and are rather brief on the other topics arising in this example.

We consider the equation

$$\hat{L}u = f, \tag{III.48}$$

where $\hat{L} = L(\overset{2}{x}, -i\overset{1}{\partial/\partial x})$ is a (pseudo)differential operator satisfying several conditions to be given below. In what follows we deal with *microlocal asymptotic solutions* of (III.48). One says that a function w is an asymptotic of a function v microlocally at a point (x_0, p_0) of the phase space $T^*\mathbb{R}^n$ if the wavefront $WF(w - v)$ of their difference does not contain the point (x_0, p_0)[1]. An operator $\hat{R}$ is called a *right microlocal regularizer* of $\hat{L}$ if for any right-hand side f the function $\hat{L}\hat{R}f$ is microlocally an asymptotic of f at (x_0, p_0). Equivalently, $\hat{R}$ is a right microlocal regularizer of $\hat{L}$ if

$$\hat{L}\hat{R} = \hat{S} + \hat{Q}, \tag{III.49}$$

where $\hat{S}$ is a pseudodifferential operator whose symbol is equal to 1 in some neighbourhood of the point (x_0, p_0) and $\hat{Q}$ is an operator of order $-\infty$ in the Sobolev scale $\{H^S(\mathbb{R}^n)\}$. We shall only construct microlocal regularizers of finite order; that is, the operator $\hat{Q}$ in (III.49) will be of arbitrarily large but finite negative order. The passage to infinitely smoothing $\hat{Q}$ is standard and can be accomplished by using the Borel lemma.

Let us state the conditions that we impose on $\hat{L}$.

We assume that L is an operator of order m with total symbol

$$L(x, p) = \sum_{k=m}^{-\infty} (-i)^{m-k} L_k(x, p),$$

where $L_k(x, p)$ is kth-order homogeneous in p. It is assumed that (x_0, p_0) is a characteristic point of $\hat{L}$ (that is, $L_m(x_0, p_0) = 0$) and that the following conditions are satisfied in a homogeneous neighbourhood of the point (x_0, p_0):

(i) $L_m(x, p) \geq 0$ for $p \neq 0$;

(ii) the characteristic set $S = \{(x, p) \mid L_m(x, p) = 0\}$ of the operator $\hat{L}$ is a smooth manifold of dimension $2n - 2$;

(iii) for any $(x, p) \in S$ and any vector ξ at the point (x, p) the inequality

$$\langle d^2 L_m(x, p)\xi, \xi\rangle > 0$$

holds provided that ξ is not tangent to S;

(iv) we have

$$\operatorname{rank} i^*\omega = 2n - 2 = \dim S,$$

where $i : S \longrightarrow \mathbb{R}^n \oplus \mathbb{R}_n$ is the natural embedding, $\omega = \sum_{i=1}^{n} dp_i \wedge dx_i$ is the standard symplectic form in the symplectic space $\mathbb{R}^n \oplus \mathbb{R}_n$ with the coordinates (x, p). Thus, $\hat{L}$ is (microlocally, in a neighbourhood of (x_0, p_0)) an elliptic operator degenerating on the surface S.

[1] See Mishchenko, Sternin, and Shatalov [137] for a comprehensive discussion of wavefronts and microlocalization.

Note that in view of condition (i), we have $dL_m(x, p) = 0$ for $(x, p) \in S$. Hence condition (iii) is invariant with respect to changes of variables.

Before solving equation (III.48) we bring it into a simpler form by transformations that clearly do not affect the possibility of constructing microlocal asymptotic solutions. The transformations are as follows:

1. Right and left multiplications by a microlocally elliptic pseudodifferential operator $\hat{A}$:

$$\hat{L} \mapsto \hat{L}\hat{A}, \quad \hat{L} \mapsto \hat{A}\hat{L}.$$

2. Conjugation

$$\hat{L} \longmapsto \hat{\Phi}_g^* \hat{L} \Phi_g,$$

by a microlocally unitary Fourier operator associated with a homogeneous symplectic transformation g (here $\hat{\Phi}_g^*$ is the adjoint of $\hat{\Phi}_g$ in L_2, and microlocal *unitarity* means that the total symbol of $\hat{\Phi}_g^* \hat{\Phi}_g$ is equal to 1 in a neighbourhood of (x_0, p_0)).

Remark III.1 Without loss of generality it can be assumed (and *is* assumed throughout the remaining part of this section) that $m = 2$. Indeed, otherwise, one can always apply transformation 1 with ord $\hat{A} = 2 - m$.

The following theorem shows that the operator $\hat{L}$ can be reduced to a quite simple "normal form."

Theorem III.3 *Let $\hat{L}$ be an operator satisfying the above conditions. Then there exist elliptic pseudodifferential operators $\hat{A}$ and $\hat{B}$ and a symplectic transformation g such that:*

(a) *g^{-1} takes the surface S to the surface $\{x_1 = p_1 = 0\}$ and the point (x_0, p_0) to a point $(0, x_0'; 0, p_0')$, $p_0' \neq 0$, where $x' = (x_2, \dots, x_n)$, $p' = (p_2, \dots, p_n)$;*

(b) *the composition of the corresponding transformations 1 and 2 takes $\hat{L}$ to the operator*

$$\hat{H} = H\left(\overset{2}{x}, -\frac{\overset{1}{\partial}}{\partial x}\right) = \hat{A}\hat{\Phi}_g^* \hat{L} \hat{\Phi}_g \hat{B}$$

with total symbol

$$H(x, p) = p_1^2 + x_1^2 |p'|^2 - iH_1(x', p') + \sum_{j=0}^{-\infty} (-i)^j H_j(x', p').$$

Here ord $H_j(x', p') = j$, $j = 1, 0, -1, \dots$, and

$$H_1(x', p') = g^* \left\{ 2L_{\text{sub}} \left\{ \det[\omega^{-1} d^2 L_2] \right\}^{1/2} \right\} \Bigg|_{x_1 = p_1 = 0} |p'|.$$

In the last formula

$$L_{\mathrm{sub}}(x, p) = L_1(x, p) - \frac{1}{2}\sum_{i=1}^{n} \frac{\partial^2 L_2(x, p)}{\partial x_i \partial p_i}$$

is the subprincipal symbol of L,

$$\omega : T\mathbb{R}^{2n} \to T^*\mathbb{R}^{2n}$$

is the mapping induced by the standard symplectic form $\omega = dp \wedge dx$ (and denoted by the same letter);

$$d^2 L_2 : T\mathbb{R}^{2n} \to T^*\mathbb{R}^{2n}$$

is the mapping taking any vector ξ to the form $d^2L_2(\xi, \cdot)$; $\omega^{-1} \circ d^2 L_2$ is restricted to the skew-orthogonal complement of TS prior to evaluating the determinant.

Note that the function $H_1(x', p')$ is determined uniquely up to a symplectic transformation in the space $\mathbb{R}^{2n}_{x',p'}$.

The proof is based on the technique of microlocal reduction to a normal form (e.g., see [172], [124], [143], [144]). This technique has nothing to do with noncommutative analysis, and so we omit the proof and deal directly with the reduced operator in what follows.

Hence we consider the equation

$$\hat{H} u = f,$$

where $\hat{H}$ is the pseudodifferential operator with symbol

$$H(x, p) = p_1^2 + x_1^2 |p'|^2 - i H_1(x', p') + \sum_{j=0}^{-\infty} (-i)^j H_j(x', p').$$

Suppose for the moment that x' does not occur in the equation, that is, the operator $\hat{H}$ does not depend on x'. Then, by applying the Fourier transform with respect to x', we arrive at the equation

$$\left[-\frac{\partial^2}{\partial x_1^2} + |p'|^2 x_1^2 - \sum_{j=1}^{-\infty} (-i)^j H_j(p') \right] \tilde{u} = \tilde{f}$$

where the tilde stands for the Fourier transform. The task is to obtain the asymptotic expansion of the solution as $|p'| \to \infty$. This is pretty easy. Denoting

$$|p'| = \lambda, \quad \sqrt{\lambda} x_1 = y, \quad \omega = \frac{p'}{|p'|}$$

we rewrite the equation as

$$\left[-\frac{\partial^2}{\partial y^2} + y^2 - H_1(\omega) + \sum_{j=0}^{-\infty} \lambda^{j-1}(-i)^j H_j(\omega)\right] \tilde{u} = \lambda^{-1} \tilde{f}.$$

The asymptotics as $\lambda \to \infty$ of the solution to the last equation can be obtained by regular perturbation theory provided that $H_1(\omega)$ is not an eigenvalue of the operator $-\partial^2/\partial y^2 + y^2$ for any $\omega \in S^{n-2}$.

Hence, the idea was to consider p' as parameter thus reducing the problem to the familiar quantum-mechanical oscillator.

Let us now return to the general case, in which x' does occur in the equation. Then the Fourier transform does not help us make the primed variables into parameters. However, still there is an appropriate technique, namely, that of operator-valued symbols[2].

Let us say a few words as to what an operator-valued symbol is.

Let $\mathcal{A}$ be an operator algebra and $\mathcal{F}$ a proper symbol space. The usual definition of the spaces of n-ary symbols reads

$$\mathcal{F}_n = \underbrace{\mathcal{F}\hat{\otimes}\cdots\hat{\otimes}\mathcal{F}}_{n\text{ copies}}.$$

We consider the space

$$\hat{\mathcal{F}}_n = \mathcal{A}\hat{\otimes}\mathcal{F}\hat{\otimes}\cdots\hat{\otimes}\mathcal{F} = \mathcal{A}\otimes\mathcal{F}_n.$$

The elements of $\hat{\mathcal{F}}_n$ are called *operator-valued symbols*. If $A_1, \dots, A_n$ are $\mathcal{F}$-generators in $\mathcal{A}$ and $f \in \hat{\mathcal{F}}_n$ an operator-valued symbol, then we can substitute $A_1, \dots, A_n$ into f. However, in this case not only $A_1, \dots, A_n$, but also f itself should be equipped with Feynman indices. If $f = a \otimes g$, $a \in \mathcal{A}$, $a \in \mathcal{F}_n$, then we set

$$\overset{s}{f}(\overset{j_1}{A_1}, \dots, \overset{j_n}{A_n}) = \overset{s}{a}g(\overset{j_1}{A_1}, \dots, \overset{j_n}{A_n}).$$

The definition is extended to the entire $\hat{\mathcal{F}}_0$ by linearity and continuity.

The calculus of operator-valued symbols includes many useful counterparts of formulas valid for ordinary symbols; they are usually easy to prove by considering factorable elements first. Let us give a few examples:

(a) The conjugation formula

$$U^{-1}[\![\overset{s}{f}(\overset{j_1}{A_1}, \dots, \overset{j_n}{A_n})]\!]U = \overset{3}{\overline{U^{-1}fU}}(\overset{j_1}{\overline{U^{-1}A_1U}}, \dots, \overset{j_n}{\overline{U^{-1}A_nU}});$$

[2]This technique, which originally arose in context of differential equations, is now well-integrated into noncommutative analysis.

(b) The left multiplication formula

$$L_{\overset{s}{f}(\overset{j_1}{A_1},\dots,\overset{j_n}{A_n})} = \overset{s}{L_f}(\overset{j_1}{L_{A_1}},\dots,\overset{j_n}{L_{A_n}}).$$

Returning to our example, the idea we have in mind is to represent $\hat{H}$ as a function with operator-valued symbol that can be considered as a perturbation of the quantum oscillator.

Namely, we write

$$\hat{H} = O\left(\overset{2}{x'}, -i\overset{1}{\frac{\partial}{\partial x'}}\right),$$

where

$$O(x', p') = x_1^2|p'|^2 - \frac{\partial}{\partial x_1^2} + \left[-iH_1(x', p') + \sum_{j=0}^{\infty}(-i)^{-j}H_j(x', p')\right] I,$$

(I is the identity operator). Note that the symbol $O(x', p')$ satisfies the inequalities

$$\left\|\frac{\partial^{\alpha+\beta}}{(\partial x')^{\alpha}(\partial p')^{\beta}}O(x', p')u\right\|_{L_2} \leq C_{\alpha\beta}(1+|p'|^{2-|\beta|})\left\|\left(1+x_1^2-\frac{\partial^2}{(\partial x')^2}\right)u\right\|_{L_2}.$$

Naturally, the regularizer will also be sought in the form

$$\hat{R} = R\left(\overset{2}{x'}, -i\overset{1}{\frac{\partial}{\partial x'}}\right)$$

with operator-valued symbol $R(x', p')$.

We should have

$$\hat{H}\hat{R} = 1 + \hat{Q},$$

where $\hat{Q}$ is a smoothing operator in the Sobolev scale; if so, then the operator

$$\hat{R}_1 = \hat{\Phi}_g\hat{B}\hat{R}\hat{A}\hat{\Phi}_g^*,$$

where $\hat{\Phi}_g$, $\hat{A}$, and $\hat{B}$ are the operators constructed in Theorem III.3, is a microlocal right regularizer of $\hat{L}$.

In order to compute the product $\hat{H}\hat{R}$, we use the left ordered representation of the tuple $(\overset{2}{x}, -i\overset{1}{\partial/\partial x'})$; this representation has the form (cf. Chapter II)

$$l_{x'} = x', \quad l_{-i\partial/\partial x'} = p' - i\partial/\partial x'.$$

since both symbols $\hat{H}$ and $\hat{R}$ are operator-valued, care should be taken so as to guarantee that the usual product formula be valid. In our case, the symbol $O(x', p')$ commutes with each of the operators x' and $-i\partial/\partial x'$, and we shall require that the same be true of the symbol $R(x', p')$. Then we have

$$\mathrm{smbl}(\hat{H}\hat{R}) = O\left(\overset{2}{x'}, p\, \overset{1}{-i\frac{\partial}{\partial x'}}\right) R(x', p') = \sum_{\alpha\geq 0} \frac{(-i)^{|\alpha|}}{\alpha!} \frac{\partial O(x', p')}{(\partial p')^{\alpha}} \frac{\partial R(x', p')}{(\partial x')^{\alpha}}. \tag{III.50}$$

We intend to solve the equation

$$\mathrm{smbl}(\hat{H}\hat{R}) = 1$$

asymptotically. For convenience, we make a certain transformation of the symbols O and R of the operators $\hat{H}$ and $\hat{R}$. Our aim is to transform the principal part of $O(x', p')$ into the operator of the quantum oscillator. Let

$$U_{\lambda}: \quad L_2(\mathbb{R}^1) \longrightarrow L_2(\mathbb{R}^1),$$

be the operator that acts according to the formula

$$U_{\lambda} f(y) = f(y/\lambda).$$

Let

$$U \equiv U(p') = U_{\sqrt{|p'|}}.$$

Clearly, we have

$$O(x', p') = U^{-1}\left[|p'|\left\{\hat{H}_{\mathrm{osc}} - iG(x', p')\right\} + N(x', p')\right] U,$$

where

$$G(x', p') = H_1(x', p')/|p'|, \quad N(x', p') = \sum_{j=0}^{-\infty} (-i)^{-j} H_j(x', p'),$$

and

$$\hat{H}_{\mathrm{osc}} = x_1^2 - \left(-i\frac{\partial}{\partial x_1}\right)^2$$

is the operator of the quantum oscillator.

We seek R in the form

$$R = U^{-1} C\, U.$$

We equate the right-hand side of (III.50) to 1 term-by-term and solve the resulting infinite system of equations recursively. The difficulty is that U does not commute with $\partial/\partial p'$, and this leads to additional terms in the transformed equation (III.50).

To this end, we should be able to compute the derivatives $-i\partial/\partial p_i(U^{-1}AU)$ and $-i\partial/\partial x_i(U^{-1}AU)$ for a given symbol A. Clearly, we have

$$\left[\frac{\partial U}{\partial p_i} f\right](y) = \frac{\partial}{\partial p_i} f\left(\frac{y}{\sqrt{|p'|}}\right) = -\frac{1}{2}|p'|^{-3/2}\frac{p_i}{|p'|}\left(y\frac{\partial f}{\partial y}\right)\left(\frac{y}{\sqrt{|p'|}}\right)$$
$$= -\frac{1}{2}\frac{p_i}{|p'|^2}\frac{y}{\sqrt{|p'|}}\frac{\partial f}{\partial y}\left(\frac{y}{\sqrt{|p'|}}\right) = -\frac{1}{2}\frac{p_i}{|p'|^2}\left(y\frac{\partial}{\partial y}\right) f\left(\frac{y}{\sqrt{|p'|}}\right).$$

Hence

$$\frac{\partial U}{\partial p_i} = -\frac{1}{2}\frac{p_i}{|p'|^2} y\frac{\partial}{\partial y} U.$$

On the other hand,

$$\frac{\partial U^{-1}}{\partial p_i} = -U^{-1}\frac{\partial U}{\partial p_i}U^{-1} = \frac{1}{2}\frac{p_i}{|p'|^2}U^{-1}y\frac{\partial}{\partial y}.$$

These formulas yield

$$\frac{\partial}{\partial p_i}(U^{-1}AU) = U^{-1}\left\{\frac{\partial A}{\partial p_i} + \frac{1}{2}\frac{p_i}{|p'|^2}\left[y\frac{\partial}{\partial y}, A\right]\right\}U.$$

The formula for $(\partial/\partial x_i)(U^{-1}AU)$ is evident,

$$\frac{\partial}{\partial x_i}(U^{-1}AU) = U^{-1}\frac{\partial A}{\partial x_i}U.$$

We obtain the following equation for the symbol C:

$$\operatorname{smbl}(\hat H \hat R) = U^{-1}\left[|p'|\left\{\hat H_{\mathrm{osc}} - iG(x', p')\right\} + N(x', p')\right]CU \qquad \text{(III.51)}$$
$$+ \sum_{|\alpha|\ge 1}\frac{(-i)^\alpha}{\alpha!}U^{-1}\left(\frac{\partial}{\partial p'} + \frac{1}{2}\frac{p'}{|p'|^2}\operatorname{ad}_{y\partial/\partial y}\right)^\alpha\left(\frac{\partial}{\partial x}\right)^\alpha CU = 1.$$

Let

$$C = C_0 - iC_1 - C_2 + \cdots$$

be the expansion of C by homogeneity degree. Equation (III.51) results in the following system of equations:

$$|p'|\left(\hat H_{\mathrm{osc}} - iG(x', p')\right)C_0 = 1,$$
$$|p'|\left(\hat H_{\mathrm{osc}} - iG(x', p')\right)C_1 = \mathcal{F}_1(C_0),$$
$$|p'|\left(\hat H_{\mathrm{osc}} - iG(x', p')\right)C_2 = \mathcal{F}_2(C_0, C_1),$$
$$\vdots$$

where

$$\mathcal{F}(C_0),\ \mathcal{F}(C_0,\ C_1), \ldots$$

are expressions that depend only on the symbols determinable from the preceding equations. The symbols $C_j(x', p')$ can be computed from this system if the operator

$$\hat{H}_{\text{osc}} - iG(x',\ p')$$

is invertible, that is, if $iG(x', p')$ does not assume values in the spectrum of $\hat{H}_{\text{osc}}$. Let us write down the expressions for the C_0, C_1: C_j:

$$\begin{aligned}
C_0(x',\ p') &= \frac{1}{|p'|}\left\{\hat{H}_{\text{osc}} - iG(x',\ p')\right\}^{-1}, \\
C_1(x',\ p') &= -\frac{1}{|p'|}\left\{\hat{H}_{\text{osc}} - iG(x',\ p')\right\}^{-1}\left\{\frac{N(x',\ p')}{|p'|}\left\{\hat{H}_{\text{osc}} - iG(x',\ p')\right\}^{-1}\right. \\
&\quad + \sum_{i=2}^{n}\left\{|p'|\frac{\partial G(x',\ p')}{\partial p_i} + \frac{1}{2}\frac{p_i}{|p'|}\hat{H}^{-}_{\text{osc}} + \frac{p_i}{|p'|}\hat{H}_{\text{osc}}\right\} \\
&\quad \left.\times \frac{1}{|p'|}\left\{\hat{H}_{\text{osc}} - iG(x',\ p')\right\}^{-2}\frac{\partial G}{\partial x_i}(x',\ p')\right\},
\end{aligned}$$

where

$$\hat{H}^{-}_{\text{osc}} = y^2 + \partial^2/\partial y^2.$$

We have taken into account the fact that

$$\left[y\frac{\partial}{\partial y},\ y^2 \pm \frac{\partial^2}{\partial y^2}\right] = 2\left(y^2 \mp \frac{\partial^2}{\partial y^2}\right).$$

Theorem III.4 *If the function $iG(x', p')$ does not assume values belonging to the spectrum of the operator $\hat{H}_{\text{osc}}$, the microlocal regularizer of the operator $\hat{H}$ exists and has the form*

$$\hat{R} = (U^{-1}CU)\left(\overset{2}{x'},\ -i\overset{1}{\frac{\partial}{\partial x'}}\right),$$

where the symbol C is the solution of equation (III.51). *This equation possesses an (asymptotically) unique solution, whose first two components are given by the above formulas.*

In closing, let us point out that this is the simplest example, which indicates two main features of the approach:

1) The operator-valued symbol, in conjunction with operator transformations of the symbol, is used in such a manner that reduces the principal part of the equation to a simple (though infinite-dimensional) operator.

2) Ordered representations are used to compute products of operators with operator-valued symbols. In that case certain commutativity conditions should be imposed in order that the usual product formulas be valid.

Chapter IV
Functional-Analytic Background of Noncommutative Analysis

1 Topics on Convergence

Functions of several Feynman-ordered operators were introduced in Chapter I. Their definition and the proofs of theorems concerning their properties in fact rely heavily upon the notions of continuity and convergence, completion of tensor products, etc. However, we were not too rigorous about all these concepts and abandoned precise statements in favour of brief and clear exposition based primarily on common sense. The present chapter is intended to justify the definitions and assertions made earlier in this book; we begin with locating the bottlenecks of our argument.

1.1 What Is Actually Needed?

Let us briefly look through Section 2 of Chapter I and find out exactly in which places the convergence and related notions are used and in what context (that section is a crucial one, since the sections following it are based on its results rather on the properties of convergence itself). There are quite a few such places where some mapping is required or proved to be continuous, but the most important are as follows:

(a) In Definition I.2 the Feynman quantization mapping

$$\begin{aligned} \mu_A : \mathcal{F}_n &\to \mathcal{A} \\ f(y_1, \dots, y_n) &\mapsto f(A) \end{aligned}$$

for some Feynman tuple of operators $A = (\overset{1}{A_1}, \dots, \overset{n}{A_n})$ is described as the unique extension by continuity from the subspace $\mathcal{F} \otimes \cdots \otimes \mathcal{F}$, where $\mathcal{F}$ is the space of unary symbols; moreover, the space $\mathcal{F}_n$ of n-ary symbols is defined as the completion $\mathcal{F}_n = \mathcal{F} \hat{\otimes} \cdots \hat{\otimes} \mathcal{F}$ of the tensor product $\mathcal{F} \otimes \cdots \otimes \mathcal{F}$. However, nothing was said about the topology chosen on $\mathcal{F} \otimes \cdots \otimes \mathcal{F}$, so that this point needs clarification.

(b) The main hypothesis in the uniqueness theorem (Theorem I.1) is that the difference derivative is a continuous mapping

$$\frac{\delta}{\delta y} : \mathcal{F} \to \mathcal{F} \hat{\otimes} \mathcal{F};$$

the same type of objection applies here.

(c) In the statements and proofs of the main theorems in Section 2 the passage from an operator algebra $\mathcal{A}$ to the algebra $\mathcal{L}(\mathcal{A})$ of continuous linear operators on $\mathcal{A}$ and to (at least 2×2) matrix algebras over $\mathcal{A}$ is freely used, and, for a linear space E, algebras of continuous operators on E are considered. Hence, we should equip the newly arising spaces and algebras with a natural notion of convergence.

Let us examine the question about tensor products more closely. Given a space $\mathcal{F}$ of unary symbols with some topology, what topology is actually needed on $\mathcal{F} \otimes \mathcal{F}$? The answer becomes clear if we look at *how this tensor product is used* in our definitions. Suppose we are given two generators $A, B \in \mathcal{A}$ and, hence, two continuous mappings

$$\mu_A : \mathcal{F} \to \mathcal{A} \quad \text{and} \quad \mu_B : \mathcal{F} \to \mathcal{A}.$$

We define the mapping

$$\mu_{\substack{1\ 2\\ A,B}} : \mathcal{F} \otimes \mathcal{F} \to \mathcal{A}$$

by setting

$$\mu_{\substack{1\ 2\\ A,B}}(f \otimes g) = \mu_B(g)\mu_A(f) = g(B)f(A)$$

the mapping is extended to the whole $\mathcal{F} \otimes \mathcal{F}$ by linearity and to $\mathcal{F} \hat{\otimes} \mathcal{F}$, the completion of $\mathcal{F} \otimes \mathcal{F}$, by continuity. We see that our definitions must ensure the continuity of $\mu_{\substack{1\ 2\\ A,B}}$ on $\mathcal{F} \otimes \mathcal{F}$, otherwise the construction fails.

Suppose momentarily that $\mathcal{F}$ and $\mathcal{A}$ are Banach algebras; we can define a mapping

$$\nu : \mathcal{F} \otimes \mathcal{F} \to \mathcal{A}$$

by setting

$$\nu(f, g) = g(B)f(A);$$

clearly, ν is a *continuous* bilinear mapping, that is,

$$\|\nu(f, g)\| \le C\|f\|\,\|g\|,$$

where $C = \|\mu_A\| \cdot \|\mu_B\|$. Consider the following *projective norm* on $\mathcal{F} \otimes \mathcal{F}$:

$$\|\varphi\| = \inf \sum_{i=1}^{N} \|f_i\|\,\|g_i\|,$$

where the infimum is taken over all representations

$$\varphi = \sum_{i=1}^{N} f_i \otimes g_i$$

(where N finite is not fixed). We claim that ν extends by linearity to a continuous mapping $\mathcal{F} \otimes \mathcal{F} \to \mathcal{A}$. Indeed, denote the extension by $\bar{\nu}$ and let $\varphi \in \mathcal{F} \otimes \mathcal{F}$. For each $\varepsilon > 0$ there exists a positive integer N and elements $f_i, g_i \in \mathcal{F}$ such that $\varphi = \sum_{i=1}^{n} f_i \otimes g_i$ and

$$\sum_{i=1}^{N} \|f_i\| \, \|g_i\| \leq \|\varphi\| + \varepsilon.$$

Then

$$\|\bar{\nu}(\varphi)\| = \Big\| \sum_{i=1}^{N} g_i(B) f_i(A) \Big\| \leq C \sum_{i=1}^{N} \|f_i\| \, \|g_i\| \leq C(\|\varphi\| + \varepsilon).$$

Since ε is arbitrary, we obtain

$$\|\bar{\nu}(\varphi)\| \leq C(\|\varphi\|).$$

The completion $\mathcal{F}\hat{\otimes}\mathcal{F}$ of $\mathcal{F}\otimes\mathcal{F}$ with respect to the chosen norm is called the *projective tensor product* of $\mathcal{F}$ and $\mathcal{F}$ and is characterized by the following property: for any continuous bilinear mapping $\nu : \mathcal{F} \otimes \mathcal{F} \to \mathcal{A}$ into a Banach space $\mathcal{A}$ there exists a unique continuous mapping $\tilde{\nu} : \mathcal{F}\hat{\otimes}\mathcal{F} \to \mathcal{A}$ such that the diagram

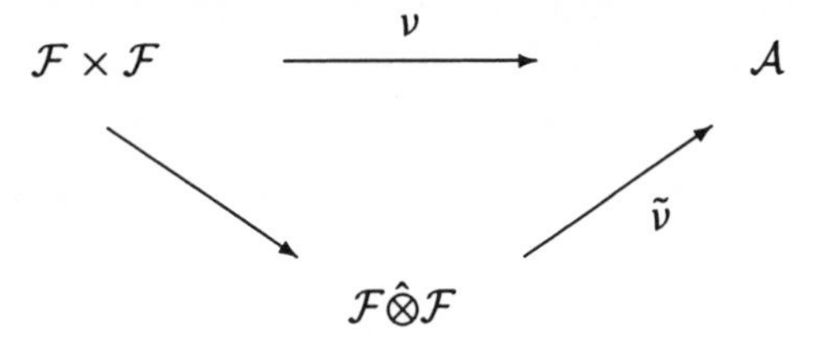

commutes, where the left arrow is the natural embedding $(a, b) \mapsto a \otimes b$.

Let us consider a simple example. Suppose that $\mathcal{F}$ is the Banach space of continuous functions on the unit circle, $\mathcal{F} = C(S^1)$. Then the space $\mathcal{F}_2 = \mathcal{F}\hat{\otimes}\mathcal{F}$ of binary symbols has a rather odd-looking structure. It is continuously embedded[1] into $C(S^1 \times S^1) = C(T^2)$ since if $\varphi = \sum f_i \otimes g_i \in C(S^1) \otimes C(S^1)$, then

$$\begin{aligned} \|\varphi\|_{C(T^2)} &= \max_{x,y \in T^2} \Big| \sum f_i(x) g_i(y) \Big| \leq \sum \max_x f_i(x) \max_y g_i(y) \\ &= \sum \|f_i\|_{C(S^1)} \|g_i\|_{C(S^1)}, \end{aligned}$$

[1] We omit the proof of injectivity.

and we can pass to the infimum over all possible representations of φ. However, the norm on $\mathcal{F}\hat{\otimes}\mathcal{F}$ is *not* the one induced from $C(T^2)$. Indeed, let $\Psi(x)$ be a continuous function with C-norm 1 and support in $[0, 2\pi]$. Set

$$\varphi_N(x, y) = \sum_{k=1}^{N} \Psi(Nx - 2\pi k)\Psi(Ny - 2\pi k);$$

then

$$\|\varphi_N(x, y)\|_{C(T^2)} = 1 \quad \text{and} \quad \|\varphi_N(x, y)\|_{C(S^1)\hat{\otimes}C(S^1)} \geq N$$

(the latter inequality is easy to prove since the supports of the terms in the sum defining $\varphi_N(x, y)$ do not intersect one another). Hence,

$$C(S^1 \times S^1) \neq C(S^1)\hat{\otimes}C(S^1).$$

Fortunately, the situation is quite different for infinitely differentiable symbols, which, by Remark I.6, are of primary interest to us.

Let us take $\mathcal{F} = C^\infty(S^1)$ and prove that

$$\mathcal{F}\hat{\otimes}\mathcal{F} = C^\infty(T^2)$$

(the symbol space $C^\infty(S^1)$ is not used anywhere in this book; however, we think it is useful to consider this example, because the idea of the proof, which is quite obvious here, remains essentially the same for the symbol spaces $S^\infty(\mathbb{R})$, but the proof itself becomes rather cumbersome).

We should first equip $C^\infty(S^1)$ and $C^\infty(T^2)$ with some topology. We consider $C^\infty(S^1)$ as the Fréchet space with topology defined by the system of seminorms

$$\|f\|_k = \max_{x\in S^1} \sum_{j=0}^{k} |f^{(j)}(x)|, \quad k = 0, 1, 2, \ldots$$

Similarly, the system of seminorms

$$|||f|||_k = \max_{(x,y)\in T^2} \sum_{i+j\leq k} |f^{(i,j)}(x, y)|, \quad k = 0, 1, 2, \ldots$$

makes $C^\infty(T^2)$ into a Fréchet space. Consider the diagram

$$\begin{array}{ccc} C^\infty(S^1)\hat{\otimes}C^\infty(S^1) & \overset{\bar{\omega}}{\dashrightarrow} & C^\infty(T^2) \\ \uparrow & & \uparrow \\ C^\infty(S^1) \otimes C^\infty(S^1) & \xrightarrow{\ \omega\ } & C^\infty(S^1) \otimes C^\infty(S^1), \end{array}$$

where the vertical arrows are embeddings and ω is the identity mapping on

$$C^\infty(S^1) \otimes C^\infty(S^1)$$

considered with two different topologies: on the left, this is the topology of the projective tensor product determined by the system of seminorms

$$\|\varphi\|_{\hat{\otimes},j} = \inf \sum_i \|f_i\|_j \|g_i\|_j,$$

where the infimum is taken over all representations $\varphi = \sum_i f_i \otimes g_i$, whereas on the right, this is the topology induced from $C^\infty(T^2)$.

We claim that the dashed arrow $\bar{\omega}$ is well-defined and is a continuous isomorphism. The proof is in several steps.

1) Let us prove that ω is continuous. In fact, this is obvious, since if $\varphi = \sum_i f_i \otimes g_i$, then

$$\begin{aligned} |||\varphi|||_k &= \max_{(x,y)\in T^2} \sum_{s+j\le k} \Big|\sum_i f_i^{(s)}(x) g_i^{(j)}(y)\Big| \\ &\le \text{const} \cdot \sum_i \|f_i\|_k \|g_i\|_k, \end{aligned}$$

where const is independent of the number of summands, and, by passing to the infimum over all possible representations, we obtain

$$|||\varphi|||_k \le \text{const}\, \|\varphi\|_{\hat{\otimes},k}$$

for any k. Thus, the continuity of ω is proved, and we can uniquely complete the dashed arrow by continuity.

2) Next, let us prove that $C^\infty(S^1) \otimes C^\infty(S^1)$ is dense in $C^\infty(T^2)$. This is however trivial, too, since for any $\varphi \in C^\infty(T^2)$ its Fourier series

$$\sum_{k,l=-\infty}^{\infty} \varphi_{kl} e^{ikx} e^{ily},$$

where

$$\varphi_{kl} = \frac{1}{4\pi^2} \int_{T^2} e^{-ikx-ily} \varphi(x,y)\, dx\, dy,$$

converges to φ uniformly with all derivatives, and its partial sums

$$\varphi^{[N]}(x,y) = \sum_{k,l=-N}^{N} \varphi_{kl} e^{ikx} e^{ily}$$

belong to $C^\infty(S^1) \otimes C^\infty(S^1)$.

3) Finally, let us prove that ω^{-1} is continuous. In conjunction with the above, this clearly implies the existence and continuity of $\bar{\omega}^{-1}$.

We will show that there exists a constant C and a positive integer m such that

$$\|\varphi\|_{\hat{\otimes},0} \leq C|||\varphi|||_m \tag{IV.1}$$

for any $\varphi \in C^\infty(S^1) \otimes C^\infty(S^1)$; then, by induction over s, it can easily be shown that

$$\|\varphi\|_{\hat{\otimes},s} \leq C_s|||\varphi|||_{m+s}$$

(we omit the induction step).

Thus, let

$$\varphi = \sum_{i=1}^{N} f_i \otimes g_i \in C^\infty(S^1) \otimes C^\infty(S^1).$$

No matter how large m is, we can represent f_i and g_i as

$$f_i = F_i + \tilde{f}_i, \quad g_i = G_i + \tilde{g}_i,$$

where F_i and G_i are trigonometric polynomials and the remainder terms $\tilde{f}_i$ and $\tilde{g}_i$ satisfy the estimates

$$\|\tilde{f}_i\|_0 \leq \|\tilde{f}_i\|_m \leq \varepsilon,$$
$$\|\tilde{g}_i\|_0 \leq \|\tilde{g}_i\|_m \leq \varepsilon,$$

where $\varepsilon > 0$ is arbitrarily small. Set

$$\overset{0}{\varphi} = \sum_{i=1}^{N} F_i \otimes G_i.$$

Then

$$\varphi - \overset{0}{\varphi} = \sum_{i=1}^{N} (\tilde{f}_i \otimes g_i + f_i \otimes \tilde{g}_i - \tilde{f}_i \otimes \tilde{g}_i),$$

and we have

$$|||\varphi - \overset{0}{\varphi}|||_m \leq M_1\varepsilon,$$
$$\|\varphi - \overset{0}{\varphi}\|_{\hat{\otimes},0} \leq M_2\varepsilon,$$

where M_1 and M_2 depend only on N, $\max_i \|f_i\|_m$, and $\max_i \|g_i\|_m$. Since $\varepsilon > 0$ is arbitrary, we see that it suffices to prove (IV.1) for trigonometric polynomials. Thus, we can assume that

$$\varphi(x, y) = \sum_{k,l=-R}^{R} \varphi_{kl} e^{ikx} e^{ily},$$

and, consequently,

$$\|\varphi\|_{\hat{\otimes},0} \leq \sum_{k,l=-R}^{R} |\varphi_{kl}|.$$

Take $m = 2$. We have

$$\begin{aligned}\sum_{k,l=-R}^{R} |\varphi_{kl}| &= \sum_{k,l=-R}^{R} |\varphi_{kl}|\sqrt{(1+k^2)(1+l^2)}\frac{1}{\sqrt{(1+k^2)(1+l^2)}} \\ &\leq \left\{\sum_{k,l=-R}^{R} |\varphi_{kl}|^2(1+k^2)(1+l^2) \sum_{k,l=-R}^{R} \frac{1}{(1+k^2)(1+l^2)}\right\}^{1/2}\end{aligned}$$

by the Cauchy–Schwarz–Bunyakovskii inequality. The second factor in the braces is bounded uniformly with respect to R, since the series

$$\sum_{k=-\infty}^{\infty} \frac{1}{1+k^2}$$

is convergent. The first factor can be rewritten as

$$\begin{aligned}&\sum_{k,l=-R}^{R} |\varphi_{kl}|^2(1+k^2)(1+l^2) \\ &= \frac{1}{4\pi^2}\int\!\!\int_{T^2}\left(1-\frac{\partial^2}{\partial x^2}\right)\varphi(x,y)\left(1-\frac{\partial^2}{\partial y^2}\right)\varphi(x,y)\,dx\,dy \leq |||\varphi|||_2^2,\end{aligned}$$

for the same reason. Hence, we arrive at inequality (IV.1), and the desired assertion is proved.

Let us now outline the requirements on the class of spaces and algebras to be used in the constructions of noncommutative analysis. Partially, these requirements follow from the preceding considerations, and partially they are implied by general mathematical principles.

The spaces and algebras should be endowed with the notion of convergence, and notions such as " Cauchy sequence", "completion", "dense subspace" should be well-defined. If E_1 and E_2 belong to the considered category, then the same should be true of $E_1 \oplus E_2$ and of the space $\mathcal{L}(E_1, E_2)$ of continuous linear mappings from E_1 to E_2. Moreover, the projective tensor product should be well-defined (as the universal object for polylinear mappings).

From the practical viewpoint, we should have the possibility to deal with unbounded operators (such as differential operators), either by involving the technique of scales of spaces, or otherwise. However, it seems impossible to devise a consistent approach to the construction of symbol spaces and operator algebras within the framework of *topological* linear spaces. The most adequate apparatus is probably supplied by the theory of polynormed spaces and algebras, some elements of which are presented below.

1.2 Polynormed Spaces and Algebras

The exposition in this and the following section is neither complete nor self-contained owing to space limitations. We only give the necessary definitions and provide examples making these definitions understandable; most of the proofs are omitted and can be found elsewhere.

Following Moore and Smith, we describe convergence in terms of generalized sequences, and so it is appropriate to say a few words on that topic.

Let I be a poset (a partially ordered set). The set I is said to be *directed* if for any $\alpha \in I$ and $\beta \in I$ there exists a $\gamma \in I$ such that $\gamma \geq \alpha$ and $\gamma \geq \beta$.

Definition IV.1 A *generalized sequence* in a set X is a mapping $I \to X$, $\alpha \mapsto x_\alpha$, where I is a directed set. One says that $\{x_\alpha\}$ is a (generalized) sequence indexed by (elements of) I; the words in parentheses may be omitted.

Suppose that X is a topological space. A generalized sequence $\{x_\alpha\}_{\alpha \in I}$ in X is said to be *convergent* to an element $x \in X$ if for any neighborhood $U \subset X$ of the point x there exists an $\alpha_0 \in I$ such that $x_\alpha \in U$ for all $\alpha \geq \alpha_0$. A topology on X is uniquely determined by the class of convergent generalized sequences in X.

Next, we need the notion of a filter.

Definition IV.2 Let X be a nonempty set. A family F of subsets of X is called a *filter in* X if the following conditions are satisfied:

i) $\emptyset \notin F$;
ii) if $A \in F$ and $B \in F$, then $A \cap B \in F$;
iii) if $A \in F$, then for any subset $B \supset A$ of X we have $B \in F$.

Let F be a filter in X. A subset $F_0 \subset F$ is called a *filter base* if each $A \in F$ includes some $A_0 \in F_0$. Note that if F_0 is a filter base, then it necessarily has the following property: for any $A, B \in F_0$ there exists a $C \in F_0$ such that $C \subset A \cap B$. Indeed, $A \cap B \in F$ since F is a filter, and the existence of C follows since F_0 is a base of F.

Definition IV.3 A *bidirected set* is a poset I such that both I and I' are directed sets (here I' is the poset obtained from I by inverting the ordering relation).

Hence, a poset I is a bidirected set if for any $\alpha, \beta \in I$ there exist $\gamma, \delta \in I$ such that $\gamma \geq \alpha$, $\gamma \geq \beta$ and $\delta \leq \alpha$, $\delta \leq \beta$.

Definition IV.4 Let I be a (bi)directed set. A *section* of I is a nonempty subset $S \subset I$ such that if $\alpha \in S$, then each $\delta \in I$ satisfying $\delta \geq \alpha$ belongs to S.

We can now give the definition of a polynormed space.

Let H be a linear space over the field $\mathbb{C}$ (spaces over $\mathbb{R}$ are considered in the same way).

Definition IV.5 A *polynorm* on H is a mapping

$$p : H \times I \to \overline{\mathbb{R}}_+ = [0, \infty) \cup \{\infty\},$$

where I is a bidirected set, such that the following properties hold:

1) $p(\lambda x, \alpha) = |\lambda| p(x, \alpha)$ for any $\lambda \in \mathbb{C}$, $x \in H$, and $\alpha \in I$;
2) $p(x + y, \alpha) \leq p(x, \alpha) + p(y, \alpha)$ for any $x, y \in H$ and $\alpha \in I$;
3) $p(x, \alpha) \leq p(x, \beta)$ whenever $\alpha \geq \beta$ for any $x \in H$;
4) for any $x \in H$ there exists an $\alpha \in I$ such that $p(x, \alpha)$ is finite.

Consider a linear space H equipped with a polynorm

$$H \times I \to \overline{\mathbb{R}}_+ .$$

Note that property 1) yields $p(0, \alpha) = 0$ for any α; property 3) means that $p(x, \alpha)$ is a nonincreasing function of α for any fixed x. Properties 1) and 2) imply that for each α the set $H_\alpha \subset H$ of elements x with finite $p(x, \alpha)$ is a linear space and that $p(\cdot, \alpha)$ is a seminorm on H_α. Hence, H_α can be considered as a topological linear space with topology determined by the seminorm $p(\cdot, \alpha)$. As follows from property 3), there is a continuous embedding $H_\beta \subset H_\alpha$ for any $\alpha > \beta$.

Let $x \in H$. It follows from property 4) that the subset $I(x) \subset I$ of indices α such that $p(x, \alpha)$ is finite is nonempty. By property 3), if $\alpha_0 \in I(x)$, then $\alpha \in I(x)$ for any $\alpha \geq \alpha_0$, so that $I(x)$ is a section of I in the sense of Definition IV.4. $I(x)$ will be referred to as the finiteness set for x.

Let $\{x_1, \ldots, x_n\}$ be a finite subset of H. The intersection $I(x_1) \cap \cdots \cap I(x_n)$ is nonempty; indeed, if $\alpha_1 \in I(x_1)$ and $\alpha_2 \in I(x_2)$, then there exists a $\gamma \in I$ such that $\gamma \geq \alpha_1$ and $\gamma \geq \alpha_2$; all such γ belong to $I(x_1) \cap I(x_2)$, since $I(x_1)$ and $I(x_2)$ are sections of I; induction over n accomplishes the proof. Moreover, we see that $I(x_1) \cap \cdots \cap I(x_n)$ is a section of I. The finite intersections of the sets $I(x)$, $x \in H$, thus clearly form a filter base for some filter Λ_0 in I; namely, Λ_0 contains a subset A if and only if $A \supset I(x_1) \cap \cdots \cap I(x_n)$ for some $x_1, \ldots, x_n \in H$.

The filter Λ_N is said to be a *filter of sections* of I since it has a base consisting of sections.

Let Λ be an arbitrary filter of sections of I such that $\Lambda \supset \Lambda_0$ (in this case one says that Λ *majorizes* Λ_0; we do not exclude the case $\Lambda = \Lambda_0$). Then the following representation is valid:

$$H = \bigcup_{A \in \Lambda} \Big(\bigcap_{\alpha \in A} H_\alpha \Big). \tag{IV.2}$$

Indeed, to verify this equation we need only to prove that for any $x \in H$ one can find a set $A \in \Lambda$ such that $x \in H_\alpha$ for all $\alpha \in A$, or, equivalently, $p(x, \alpha)$ is finite for any $\alpha \in A$. Since $\Lambda \supset \Lambda_0$, it suffices to find such a set in Λ_0. This, however, is trivial, because Λ_0 contains $I(x)$ and we can set $A = I(x)$.

Denote

$$H^A = \bigcap_{\alpha \in A} H_\alpha .$$

Each H^A is a locally convex space with respect to the topology determined by the system of seminorms $\{p(\cdot, \alpha)\}_{\alpha \in A}$. We introduce a convergence in H as follows. A generalized sequence $\{x_\mu\}_{\mu \in \Gamma}$ in H is said to be *convergent to zero* in H if there exists an $A \in \Lambda$ and $\mu_0 \in \Gamma$ such that $x_\mu \in H^A$ for all $\mu \geq \mu_0$ and x_μ converges to zero in H^A (in other words, $p(x_\mu, \alpha) \to 0$ for each $\alpha \in A$).

Accordingly, we say that a generalized sequence $\{x_\mu\}_{\mu \in \Gamma}$ in H converges to $x \in H$ if $x_\mu - x \to 0$, that is, for some $A \in \Lambda$ and $\mu_0 \in \Gamma$ one has $x \in H^A$, $x_\mu \in H^A$ for $\mu \geq \mu_0$, and $x_\mu \to x$ in H^A.

Definition IV.6 The space H equipped with the convergence described above is called a *polynormed space over the filter* Λ.

Clearly, linear operators on H are continuous with respect to the convergence introduced.

The usual method of constructing polynormed spaces in applications is as follows. Suppose that I is a bidirected set, and let a family $\{V_\alpha\}_{\alpha \in I}$ of linear spaces be given such that $V_\alpha \subset V_\beta$ whenever $\alpha \leq \beta$. Moreover, suppose that each V_α is equipped with a seminorm $p_\alpha(\cdot)$ in such a way that all these embeddings are continuous with norm ≤ 1,

$$p_\beta(x) \leq p_\alpha(x) \quad \text{for} \quad \alpha \leq \beta \quad \text{and} \quad x \in V_\alpha.$$

Now let Λ be an arbitrary filter of sections of I. Set

$$H = \bigcup_{A \in \Lambda} \Big(\bigcap_{\alpha \in A} V_\alpha \Big)$$

(note that this formula, though it is quite like (IV.2), is, in fact, essentially different, since $V_\alpha \not\subset H$ in general). For any $x \in H$ we define the finiteness set $I(x)$ by the formula

$$I(x) = \Big\{ \alpha \in I \mid \alpha \in A \text{ and } x \in \bigcap_{\gamma \in A} V_\gamma \text{ for some } A \in \Lambda \Big\}$$

and the polynorm

$$p(x, \alpha) = \begin{cases} p_\alpha(x) & \text{if } \alpha \in I(x), \\ +\infty & \text{otherwise} \end{cases}$$

(one should resist the temptation to define $p(x, \alpha) = p_\alpha(x)$ for *all* $x \in V_\alpha \cap H$ since then it could not be guaranteed that Λ majorizes the filter Λ_0 generated by the finiteness sets). Such a definition makes H into a polynormed space over Λ, and representation (IV.2) is valid with

$$H_\alpha = \big\{ x \in V_\alpha \cap H \mid \alpha \in I(x) \big\}.$$

Indeed, H_α is clearly a linear space; properties 1) – 4) in Definition IV.5 are obviously satisfied; H_α is the space of elements $x \in H$ for which $p(x, \alpha)$ is finite. It remains to show that Λ majorizes Λ_0. By the definition of a filter, it suffices to prove that $I(x) \in \Lambda$ for each $x \in H$. But this is evident since the definition of $I(x)$ can be rewritten as follows:

$$I(x) = \bigcup_{A \in \Lambda : x \in \bigcap_{\alpha \in A} V_\alpha} A.$$

Thus, $I(x)$ contains a subset belonging to Λ and therefore belongs to Λ (recall that Λ is a filter).

Let us now consider two examples of polynormed spaces. These examples play a crucial role in noncommutative analysis and its applications to differential equations.

Example IV.1 Let $S_k^m(\mathbb{R}^n)$ denote the space of C^k-functions

$$f(y) = f(y_1, \dots, y_n)$$

on $\mathbb{R}^n$ satisfying the estimates

$$\|f\|_{m,k} = \sup_{y \in \mathbb{R}^n} (1 + |y|)^{-m} \sum_{|\alpha|=0}^{k} |f^{(\alpha)}(y)| < \infty$$

(here $\alpha = (\alpha_1, \dots, \alpha_n)$ is a multi-index, $f^{(\alpha)}(y) = \partial^{|\alpha|} f(y)/\partial y^\alpha$). Consider the following partial order on the set $\mathbb{Z} \times \mathbb{Z}^+ \ni (m, k)$:

$$(m, k) \le (m', k') \text{ if and only if } m \le m' \text{ and } k \ge k'.$$

Clearly, if $(m, k) \le (m', k')$, then

$$S_k^m(\mathbb{R}^n) \subset S_{k'}^{m'}(\mathbb{R}^n) \text{ and } \|f\|_{m,k} \ge \|f\|_{m',k'}, \quad f \in S_k^m(\mathbb{R}^n),$$

so that the system of spaces $S_k^m(\mathbb{R}^n)$ equipped with the seminorms $\|\cdot\|_{m,k}$ satisfies the assumptions used in the construction above. Next, let Λ be the filter with filter base formed by the subsets

$$I_{m_0} = \{m \ge m_0, \ k \in \mathbb{Z}^+\} \subset \mathbb{Z} \times \mathbb{Z}^+,$$

where m_0 ranges over $\mathbb{Z}$. Clearly, each I_{m_0} is a section of $\mathbb{Z} \times \mathbb{Z}^+$, so that Λ is a filter of sections. The corresponding polynormed space is denoted by $S^\infty(\mathbb{R}^n)$ and defined by

$$\begin{aligned} S^\infty(\mathbb{R}^n) &= \bigcup_{A \in \Lambda} \Big(\bigcap_{(m,k) \in A} S_k^m(\mathbb{R}^n) \Big) \\ &= \bigcup_{m_0 \in \mathbb{Z}} \Big(\bigcap_{(m,k) \in I_{m_0}} S_k^m(\mathbb{R}^n) \Big) \\ &= \bigcup_{m_0 \in \mathbb{Z}} \Big(\bigcap_{k \in \mathbb{Z}^+} S_k^{m_0}(\mathbb{R}^n) \Big) \end{aligned}$$

(the union over all elements of the filter can clearly be replaced by the union over the filter base, and the intersection over all (m, k) with $m \geq m_0$ can be replaced by the intersection over all k with $m = m_0$ owing to the cited embeddings of the spaces $S_k^{m_0}(\mathbb{R}^n)$).

Furthermore, the spaces $S^m(\mathbb{R}^n) = \bigcap_{k \in \mathbb{Z}^+} S_k^m(\mathbb{R}^n)$ form an increasing sequence, and it is easy to see that for any $f \in S^m(\mathbb{R}^n)$ the finiteness set has the form

$$I(f) = \{(l, k) \mid l \geq m,\ k \in \mathbb{Z}\}.$$

Hence, the convergence on $S^\infty(\mathbb{R}^n)$ can be described as follows. A generalized sequence $\{f_\mu\}_{\mu \in \Gamma}$ converges to $f \in S^\infty(\mathbb{R}^n)$ if there exists a $\mu_0 \in \Gamma$ and $m \in \mathbb{Z}$ such that $f_\mu \in S^m(\mathbb{R}^n)$ for $\mu \geq \mu_0$ and $f - f_\mu \to 0$ in $S^m(\mathbb{R}^n)$, that is,

$$\|f - f_\mu\|_{m,k} = \sup_{y \in \mathbb{R}^n} (1 + |y|)^{-m} \sum_{|\alpha|=0}^{k} |f^{(\alpha)}(x) - f_\mu^{(\alpha)}(x)| \to 0, \ \ k = 0, 1, 2, \ldots$$

In other words , starting from some index μ_0, the functions f_μ and all their derivatives grow at infinity not faster than $(1 + |y|)^m$ and $f_\mu/(1 + |y|)^m \to f/(1 + |y|)^m$ with all derivatives in the uniform metric.

Example IV.2 Let $K \subset \mathbb{C}^n$ be a compact set. For any neighborhood U of K in $\mathbb{C}^n$ denote by $\mathcal{O}_U$ the locally convex space of holomorphic functions on U with the topology of uniform convergence on compact subsets of U. Next, denote by $\mathcal{O}_K$ the space

$$\mathcal{O}_K = \bigcup \mathcal{O}_U,$$

where the union is taken over all open sets U containing K. The space $\mathcal{O}_K$ can be interpreted as a polynormed space in the following manner. Denote by I the set of pairs (U, L), where U is an open neighborhood of K and $L \Subset U$ a compact subset of U. We write $(U, L) \leq (U', L')$ if $U' \subset U$ and $L' \subset L$. Clearly, I is a poset. We denote by $\mathcal{O}_{(U,L)}$ the space $\mathcal{O}_U$ equipped with the seminorm

$$\|f\|_{(U,L)} = \sup_{z \in L} |f(z)|;$$

then, for $(U, L) \leq (U', L)$ we have $\mathcal{O}_{U,L} \subset \mathcal{O}_{U',L'}$ (the embedding is given by the restriction from U to U') and

$$\|f\|_{(U',L')} = \sup_{z \in L'} |f(z)| \leq \sup_{z \in L} |f(z)| = \|f\|_{(U,L)}.$$

Next, we take the filter Λ with base formed by the sets $\{(U, L)\}$ with U fixed and L arbitrary. We may write

$$\mathcal{O}_K = \bigcup_U \Big(\bigcap_{L \Subset U} \mathcal{O}_{(U,L)} \Big)$$

and define the convergence in $\mathcal{O}_K$ by the following condition: a generalized sequence f_μ converges to zero in $\mathcal{O}_K$ is there exists a μ_0 and a neighborhood U of K such that for all $\mu \geq \mu_0$ we have $f_\mu \in \mathcal{O}_U$ and f_μ converges to zero uniformly on compact subsets of U.

A polynormed space H is called a Hausdorff space if each generalized sequence in H has at most one limit.

Let $\{x_\mu\}_{\mu\in\Gamma}$ be a generalized sequence in a polynormed space H. Consider the set $\Gamma \times \Gamma$ with the partial order given by

$$(\mu, \nu) \leq (\mu', \nu') \Leftrightarrow \mu \leq \mu' \quad \text{and} \quad \nu \leq \nu'.$$

Clearly, $\Gamma \times \Gamma$ is a directed set. Consider the generalized sequence

$$\{x_\mu - x_\nu\}_{(\mu,\nu)\in\Gamma\times\Gamma}.$$

One says that x_μ is a Cauchy sequence in H if the sequence $x_\mu - x_\nu$ is convergent to zero.

Definition IV.7 A polynormed space H is called a *complete polynormed space* (or a *poly-Banach space*) if H is a Hausdorff space and each Cauchy sequence in H is convergent.

In particular, if all H^A, $A \in \Lambda$, are complete Hausdorff locally convex spaces, as is the case in Examples IV.1 and IV.2 then H is a poly-Banach space.

Suppose that a polynormed space H is not complete. Under certain conditions, it is possible to define the completion of H, which is a poly-Banach space. These conditions are summarized in the following statement.

Lemma IV.1 *Let H be a Hausdorff polynormed space satisfying the following property: if $A, B \in \Lambda$, $A \subset B$, $\{x_\mu\}_{\mu\in\Gamma}$ is a Cauchy sequence in H^B, and $x_\mu \to 0$ in H^A, then $x_\mu \to 0$ in H^B. Then the completion of H exists and is a poly-Banach space.*

We omit the proof but mention the important particular case in which the polynormed space

$$H = \bigcup_{A\in\Lambda} \Big(\bigcap_{\alpha\in A} H_\alpha \Big)$$

satisfies the following property. For each $\alpha \in I$ the seminorm $p(\cdot, \alpha)$ is a norm on H_α and for any $\alpha \leq \beta \in I$ the embedding $H_\alpha \subset H_\beta$ is consistent, i.e., if a generalized Cauchy sequence $x_\mu \in H_\alpha$ is convergent to zero in H_β, then it is convergent to zero in H_α.

In this case the completion $\bar{H}_\alpha$ of each H_α is a Banach space, there are continuous embeddings $\bar{H}_\alpha \subset \bar{H}_\beta$ for any $\alpha \leq \beta$, and the completion of H is

$$\bar{H} = \bigcup_{A\in\Lambda} \Big(\bigcap_{\alpha\in A} \bar{H}_\alpha \Big),$$

with the natural convergence: $x_\mu \to x$ in $\bar{H}$ if, for some $A \in \Lambda$ and some μ_0, we have $x_\mu \in \bigcap_{\alpha \in A} \bar{H}_\alpha$ for $\mu \geq \mu_0$ and $x_\mu \to x$ in each $\bar{H}_\alpha, \alpha \in A$.

We now pass to the consideration of continuous linear mappings of polynormed spaces. Let

$$H = \bigcup_{A \in \Lambda} \Big(\bigcap_{\alpha \in A} H_\alpha\Big), \quad G = \bigcup_{B \in \Xi} \Big(\bigcap_{\beta \in B} G_\beta\Big)$$

be polynormed spaces indexed by I and J and equipped with polynorms $p(\cdot, \alpha)$ and $q(\cdot, \beta)$, respectively.

Definition IV.8 A linear mapping

$$\psi : H \to G$$

is said to be *continuous* if it takes convergent generalized sequences into convergent generalized sequences.

Clearly, ψ is continuous if for any $A \in \Lambda$ it maps the locally convex space H^A continuously into some G^B. Furthermore,

$$\psi : H^A \to G^B$$

is continuous if and only if for any seminorm $q(\cdot, \beta)$, $\beta \in B$, on G^B there exists a seminorm $p(\cdot, \alpha), \alpha \in A$, on H^A such that

$$q(\psi(x), \beta) \leq C p(x, \alpha), \quad x \in H_\alpha.$$

Hence, the set $\mathcal{L}(H, G)$ of all continuous mappings from H to G can be described as

$$\mathcal{L}(H, G) = \bigcap_{A \in \Lambda} \bigcup_{B \in \Xi} \bigcap_{\beta \in B} \bigcup_{\alpha \in A} \mathcal{L}(H_\alpha, G_\beta), \tag{IV.3}$$

where $\mathcal{L}(H_\alpha, G_\beta)$ is the space of continuous mappings from H_α to G_β. We introduce a seminorm in $\mathcal{L}(H_\alpha, G_\beta)$ by setting

$$p_{\alpha\beta}(\psi) = \sup_{x \in H_\alpha,\, p(x, \alpha) \leq 1} q(\psi(x), \beta).$$

It turns out that $\mathcal{L}(H, G)$ possesses a natural structure of polynormed space. Consider the indexing set $I' \times J$ (where as above I' is the set I equipped with the reverse order). By definition, $(\alpha, \beta) \leq (\alpha_1, \beta_1)$ if and only if $\alpha \geq \alpha_1$ and $\beta \leq \beta_1$. Clearly,

$$\|p_{\alpha\beta}(\psi)\| \geq \|p_{\alpha_1\beta_1}(\psi)\|$$

whenever $(\alpha, \beta) \leq (\alpha_1, \beta_1)$, so that $p_{\alpha\beta}(\psi)$ is a polynorm on $\mathcal{L}(H, G)$.

It remains to reduce the number of intersection and union signs in (IV.3) so as to give the "canonical" representation of $\mathcal{L}(H, G)$ as a polynormed space. Let Ω be the filter of sections of $I' \times J$ defined as follows. For any $A \in \Lambda$ and $B \in \Xi$ denote by $K(A, B)$ the set of all sections C of $I' \times J$ such that for any $\beta \in B$ there exists an $\alpha \in A$ such that $(\alpha, \beta) \in C$. The filter Ω is generated by the family $\bigcap_{A \in \Lambda} \bigcup_{B \in \Xi} K(A, B)$.

Then the filter Ω majorizes the filter in $I' \times J$ generated by the finiteness sets for the polynorm $p_{\alpha\beta}$, we have

$$\mathcal{L}(H, G) = \bigcup_{K \in \Omega} \bigcap_{(\alpha,\beta) \in K} \mathcal{L}(H_\alpha, G_\beta),$$

and $\mathcal{L}(H, G)$ is a polynormed space over the filter Ω.

It can be proved that $\mathcal{L}(H, G)$ is a Hausdorff space whenever G is so, and that if G is a poly-Banach space, then so is $\mathcal{L}(H, G)$.

In what follows we always consider $\mathcal{L}(H, G)$ endowed with the structure of a poly-Banach space defined above.

Let us now consider the particular case in which $H = G$. We will denote $\mathcal{L}(H, H) = \mathcal{L}(H)$. The polynormed space $\mathcal{L}(H)$ is equipped with the natural multiplication defined by the composition of mappings. We have

$$p_{\alpha\beta}(\psi \varphi) \le p_{\alpha\gamma}(\varphi) p_{\gamma\beta}(\psi)$$

whenever the right-hand side is finite. This suggests that the multiplication is continuous on $\mathcal{L}(H) \times \mathcal{L}(H)$, which is indeed true. In fact, for any $K_1, K_2 \in \Omega$ we can find a $K \in \Omega$ such the multiplication is continuous from $\mathcal{L}(H)^{K_1} \times \mathcal{L}(H)^{K_2}$ to $\mathcal{L}(H)^K$, that is, for any $\delta = (\alpha, \beta) \in K$ there exists a $\delta_1 = (\alpha_1, \beta_1) \in K_1$ and $\delta_2 = (\alpha_2, \beta_2) \in K_2$ such that

$$p_\delta(\psi \varphi) \le \text{const} \cdot p_{\delta_1}(\psi) p_{\delta_2}(\varphi)$$

(the constant, of course, can be taken equal to 1).

Furthermore, $\mathcal{L}(H)$ contains the identity element represented by the identity mapping 1 of H, and 1 is separated from 0 in the sense that $p_\gamma(1) \neq 0$ for some $\gamma \in I' \times J$ (it suffices to choose γ of the form $\gamma = (\alpha, \alpha)$). We obtain the definition of a polynormed algebra by taking the abstract version of these properties.

Definition IV.9 An (associative) *polynormed algebra* (with 1) is a polynormed space $\mathcal{A}$ over a filter Λ, equipped with a bilinear associative operation (multiplication) such that

(a) multiplication is continuous, i.e. for any $A_1, A_2 \in \Lambda$ there exists an $A \in \Lambda$ such that multiplication is continuous in the space

$$\mathcal{A}^{A_1} \times \mathcal{A}^{A_2} \to \mathcal{A}^A;$$

in other words, for any $\alpha \in A$ there exist $\alpha_i \in A_i$, $i = 1, 2$, such that

$$p(a_1 a_2, \alpha) \le p(a_1, \alpha) p(a_2, \alpha)$$

for any a_1 and a_2;

(b) $\mathcal{A}$ is an algebra with 1;

(c) 1 is separated from 0 by the polynorm, i.e., $p(1, \alpha) > 0$ for some α.

If a polynormed algebra is Hausdorff and complete, then it is called a poly-Banach algebra.

Thus, if H is a poly-Banach space, then $\mathcal{L}(H)$ is a poly-Banach algebra.

The following lemma is obvious.

Lemma IV.2 *Let $\mathcal{A}$ be a polynormed algebra. The mappings*

$$\begin{aligned} L : \mathcal{A} &\to \mathcal{L}(\mathcal{A}) \\ A &\mapsto L_A \end{aligned}$$

and

$$\begin{aligned} R : \mathcal{A} &\to \mathcal{L}(\mathcal{A}) \\ A &\mapsto R_A, \end{aligned}$$

where L_A and R_A are the operators of left and right multiplication by A in $\mathcal{A}$, are a continuous homomorphism and a continuous antihomomorphism of polynormed algebras, respectively.

In particular, each polynormed algebra can be realized as an algebra of continuous endomorphisms of a polynormed space; such a realization is given by the left regular representation $\mathcal{A} \mapsto \mathcal{L}(\mathcal{A})$, which is faithful since $L_A(1) = A \cdot 1 = A \neq 0$ if $A \neq 0$.

Example IV.3 The polynormed spaces $S^\infty(\mathbb{R}^n)$ and $\mathcal{O}^K$ described in Examples IV.1 and IV.2 are poly-Banach algebras with respect to pointwise multiplication.

Remark IV.1 The convergence of a polynormed space H is defined by the triple (I, p, Λ), where I is the indexing set, p the polynorm, and Λ the filter involved in Definition IV.6. However, different triples may define the same convergence on H (in this case the identity mapping of H is two-sided continuous between these convergences).

More generally, we can consider two polynormed spaces H_1 and H_2 with a continuous continuously invertible linear mapping $i : H_1 \to H_2$. In this case we say that H_1 and H_2 are equivalent and do not distinguish between them.

1.3 Tensor Products

Let H_1 and H_2 be poly-Banach spaces over filters Λ_1 and Λ_2 with polynorms p_1 and p_2 and indexing sets I_1 and I_2, respectively. We intend to define the tensor product of H_1 and H_2 in the category of poly-Banach spaces.

Following the usual practice, we define the tensor product by the universal mapping property.

Definition IV.10 The tensor product of the poly-Banach spaces H_1 and H_2 is the poly-Banach space $H_1 \hat{\otimes} H_2$ such that for any poly-Banach space H and any continuous bilinear mapping $\tau : H_1 \times H_2 \to H$ there exists a unique mapping $\tau : H_1 \hat{\otimes} H_2 \to H$ such that the diagram

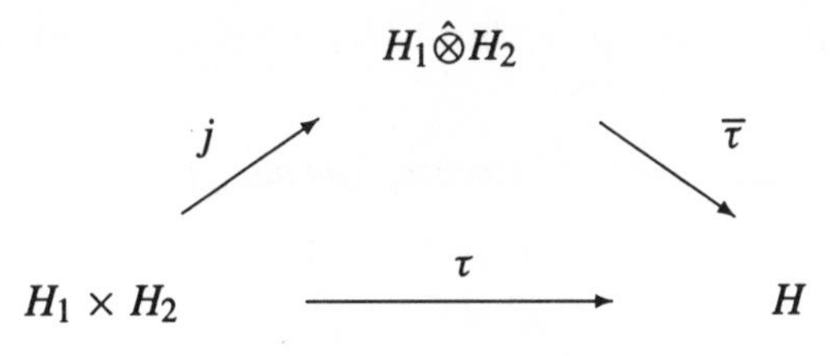

commutes. Here j is a predefined continuous bilinear mapping.[2]

Theorem IV.1 *The tensor product of poly-Banach spaces is well-defined, that is, it always exists and is unique up to a uniquely determined equivalence (see Remark* IV.1). *(The explicit structure of $H_1 \hat{\otimes} H_2$ will be given below.)*

Remark IV.2 The tensor product thus defined is usually referred to as the *projective* tensor product of H_1 and H_2.

Proof. Uniqueness. Suppose that $H_1 \hat{\otimes} H_2$ and $H_1 \tilde{\hat{\otimes}} H_2$ are two different tensor products of H_1 and H_2. According to Definition IV.10, there are unique morphisms $\tilde{\tau}$ and $\tilde{\tilde{\tau}}$ such that the diagrams

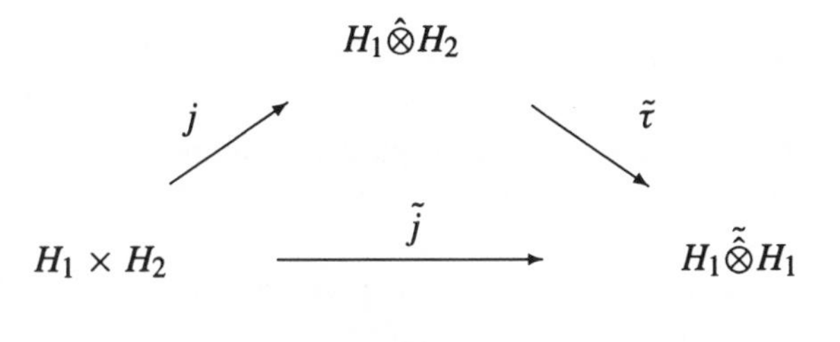

and

$$\begin{array}{ccc} & H_1 \tilde{\hat{\otimes}} H_2 & \\ \tilde{j} \nearrow & & \searrow \tilde{\tilde{\tau}} \\ H_1 \times H_2 & \xrightarrow{\ j\ } & H_1 \hat{\otimes} H_1 \end{array}$$

[2]Purists would probably object to our definition by saying that j should be included in the notion of the tensor product, i.e., the tensor product of H_1 and H_2 is the *pair* $(H_1 \hat{\otimes} H_2, j : H_1 \times H_2 \to H_1 \hat{\otimes} H_2)$.

commute. Then so does the diagram

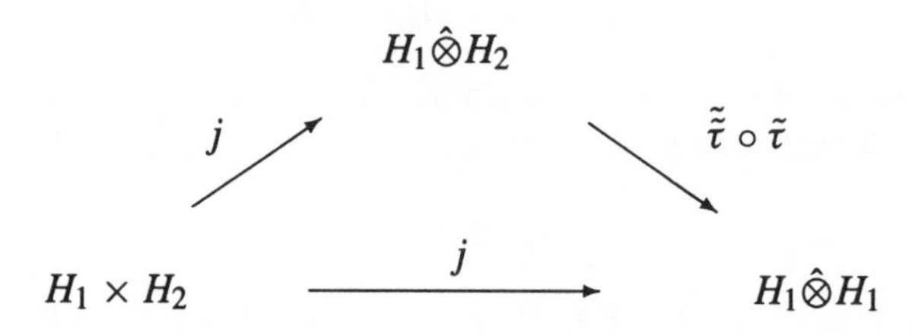

and since $\tilde{\tilde{\tau}} \circ \tilde{\tau}$ is unique, we must have $\tilde{\tilde{\tau}} \circ \tilde{\tau} = \mathrm{id}$. Similarly, $\tilde{\tau} \circ \tilde{\tilde{\tau}} = \mathrm{id}$, that is, $\tilde{\tau}$ is an isomorphism.

Existence. Consider the algebraic tensor product $H_1 \otimes H_2$. Clearly,

$$\begin{aligned} H_1 \otimes H_2 &= \bigcup_{A \in \Lambda_1, B \in \Lambda_2} \{(\bigcap_{\alpha \in A} H_{1\alpha}) \otimes (\bigcap_{\beta \in B} H_{2\beta})\} \\ &= \bigcup_{(A,B) \in \Lambda_1 \times \Lambda_2} \left\{ \bigcap_{(\alpha,\beta) \in A \times B} H_{1\alpha} \otimes H_{2\beta} \right\}. \end{aligned}$$

We equip $I_1 \times I_2$ with the product partial order and introduce a polynorm $p_{\alpha\beta}(\cdot)$, $(\alpha, \beta) \in I_1 \times I_2$, on $H_1 \otimes H_2$ by setting

$$p_{\alpha\beta} = \inf \sum \|f_i\|_\alpha \|g_i\|_\beta,$$

where $\varphi \in H_{1\alpha} \otimes H_{2\beta}$ and the infimum is taken over all finite presentations

$$\varphi = \sum_{j=1}^{N} f_j \otimes g_j$$

with $f_j \in H_{1\alpha}$ and $g_j \in H_{2\beta}$ (here N is not fixed). The space $H_1 \otimes H_2$ is polynormed over the filter $\Lambda_1 \times \Lambda_2$; its completion is a poly-Banach space, which will be denoted by $H_1\hat{\otimes}H_2$ Let us prove that this space is the tensor product of H_1 and H_2.

Suppose $H = \bigcup_{A \in \Lambda} \bigcap_{\alpha \in A} H_\alpha$ is a poly-Banach space and

$$\mu : H_1 \times H_2 \to H$$

a continuous bilinear mapping. By the properties of the algebraic tensor product, there is a unique mapping

$$\tilde{\mu} : H_1 \otimes H_2 \to H$$

such that $\mu = \tilde{\mu} \circ j$, where $j : H_1 \times H_2 \to H_1 \otimes H_2$ is the natural embedding, $j(x, y) = x \otimes y$.

Since $H_1\hat{\otimes}H_2$ is the completion of $H_1 \otimes H_2$, the mapping $\tilde{\mu}$ can be extended to $H_1\hat{\otimes}H_2$ uniquely provided that $\tilde{\mu}$ is continuous. Thus, it remains to prove the continuity of $\tilde{\mu}$.

Since μ is continuous, it is continuous in the spaces

$$\mu : H_{1\alpha} \times H_{2\beta} \to H_\gamma$$

for certain triples (α, β, γ); the structure of the set of such triples agrees with the filters Λ_1, Λ_2, and Λ. Hence it suffices to prove that $\tilde{\mu}$ is continuous in the normed spaces

$$\tilde{\mu} : H_{1\alpha} \otimes H_{2\beta} \to H_\gamma$$

for the same triples (α, β, γ). But the proof of this fact is essentially that used for Banach spaces (see the discussion at the beginning of Subsection 1.1). Thus, the theorem is proved. □

In a similar way one can define tensor products $H_1 \hat{\otimes} H_2 \hat{\otimes} \cdots \hat{\otimes} H_n$ of $n \geq 2$ poly-Banach spaces. This definition involves no new ideas and is therefore omitted.

Remark IV.3 If H_1 and H_2 are poly-Banach algebras, then $H_1 \hat{\otimes} H_2$ is also a poly-Banach algebra.

2 Symbol Spaces and Generators

Symbol spaces and their properties are a topic of interest in noncommutative analysis. They were widely used throughout the exposition, but at an intuitive level: we appealed to the notion of convergence in symbol spaces without knowing exactly what is meant. Now that we are aware of the notion of poly-Banach algebras, we are in a position to give precise definitions.

2.1 Definitions

The main properties of symbol spaces are those used in Definitions I.1 and I.2 and in Theorems I.1, I.4, and I.5. Here we collect these properties and express them in the language of poly-Banach spaces, which yields the desired definition.

Let $Z \subset \mathbb{C}$ be a subset without isolated points, and let $\mathcal{F}$ be a linear subspace (over $\mathbb{C}$) of the space of $\mathbb{C}$-valued continuous functions on Z. We impose some conditions on $\mathcal{F}$.

Condition A. The space $\mathcal{F}$ is a poly-Banach algebra with 1 with respect to the pointwise multiplication.

That is, $\mathcal{F}$ is equipped with a polynorm and is Hausdorff and complete with respect to this polynorm; moreover, the product of any two functions in $\mathcal{F}$ belongs to $\mathcal{F}$, and the multiplication is continuous; the function identically equal to one on Z belongs to $\mathcal{F}$.

Condition B. The function $f(x) = x$, $x \in Z$, belongs to $\mathcal{F}$.

Condition C. The poly-Banach convergence in $\mathcal{F}$ implies (i.e., is stronger than) locally uniform convergence on Z.

Consider the space F_Z of all continuous functions $f : Z \to \mathbb{C}$ (we assume that Z is equipped with the topology induced from $\mathbb{C}$). The space F_Z can be equipped the structure of a poly-Banach space as follows. Let I be the set of all subsets of Z (the powerset of Z). Let I be partially ordered by inclusion,

$$\alpha \geq \beta \Leftrightarrow \alpha \subset \beta, \quad \alpha, \beta \in I.$$

We set

$$p(f, \alpha) = \sup_{x \in \alpha} |f(x)|, \quad f \in F_Z.$$

Clearly, p is a polynorm on F_Z and $p(f, \alpha)$ is finite whenever α is a compact subset of Z. Let Λ be the filter in I determined by the condition that each $A \in \Lambda$ contains all compact subsets of Z. Then, since each finiteness set satisfies the same property, it follows that Λ majorizes the filter Λ_0 generated by finiteness sets, and we can equip F_Z with the structure of a polynormed space over the filter Λ. It is easy to see that F_Z with this convergence is a poly-Banach space and the convergence in F is just the locally uniform convergence.

Condition C means that the embedding $\mathcal{F} \subset F_Z$ is a continuous mapping.

Let $F_{Z \times Z}$ be the poly-Banach space of all continuous functions on $Z \times Z$ constructed by analogy with F_Z. There is a sequence of continuous mappings

$$\begin{array}{ccc} \mathcal{F} \times \mathcal{F} & \to & F_Z \times F_Z \to F_{Z \times Z} \\ (f, g) & \mapsto & (f, g) \mapsto f(x)g(y). \end{array}$$

The composite mapping is bilinear and, passing to the projective tensor product, we obtain the natural mapping $\mathcal{F} \hat{\otimes} \mathcal{F} \to F_{Z \times Z}$.

Condition D. The natural mapping $\mathcal{F} \hat{\otimes} \mathcal{F} \to F_{Z \times Z}$ is an embedding.

The meaning of Condition D is that the elements of $\mathcal{F} \hat{\otimes} \mathcal{F}$ can be unambiguously identified with some continuous functions on $Z \times Z$. Technically, no extension of this condition is really necessary to construct functions of operators. However, in all practical examples the following condition is satisfied:

Condition D′. The natural mapping

$$\underbrace{\mathcal{F} \hat{\otimes} \mathcal{F} \hat{\otimes} \cdots \hat{\otimes} \mathcal{F}}_{n \text{ copies}} \to F_{\underbrace{Z \times Z \times \cdots \times Z}_{n \text{ factors}}}$$

is an embedding.

Conditions A–D are routine ones. Finally, we introduce the most important condition.

Condition E. For any $f \in \mathcal{F}$ the difference derivative

$$\frac{\delta f}{\delta x}(x, y) = \begin{cases} \dfrac{f(x) - f(y)}{x - y}, & x \in Z,\ y \in Z,\ x \neq y, \\ \lim\limits_{v \to x,\, v \in Z} \dfrac{f(v) - f(y)}{v - y}, & x = y \in Z \end{cases}$$

exists and is continuous. Moreover, $\delta f / \delta x \in \mathcal{F} \hat{\otimes} \mathcal{F}$ (this assertion makes sense in view of Condition D) and the mapping

$$\frac{\delta}{\delta x} : \mathcal{F} \to \mathcal{F} \hat{\otimes} \mathcal{F}$$

is a continuous mapping of poly-Banach algebras.

Definition IV.11 The space $\mathcal{F}$ is called a *(proper unary) symbol space* if it satisfies Conditions A–D.[3]

Given a symbol space $\mathcal{F}$, we define the spaces of n-ary symbols as the projective tensor products

$$\mathcal{F}_n = \underbrace{\mathcal{F} \hat{\otimes} \mathcal{F} \hat{\otimes} \cdots \hat{\otimes} \mathcal{F}}_{n \text{ copies}} = \mathcal{F}^{\hat{\otimes} n}.$$

Each $\mathcal{F}_n$ is a poly-Banach algebra. For $n = 2$ elements of $\mathcal{F}$ are naturally interpreted as functions of two arguments $y_1, y_2 \in Z$, and the same is true for arbitrary n provided that Condition D$'$ is satisfied.

Let $\mathcal{A}$ be a poly-Banach algebra and $\mathcal{F}$ a symbol space.

Definition IV.12 An element $A \in \mathcal{A}$ is called an $\mathcal{F}$-generator if there exists a continuous homomorphism

$$\mu_A : \mathcal{F} \to \mathcal{A}$$

of poly-Banach algebras such that

$$\mu_A(x) = A$$

(here x is the function taking each point $x \in Z$ into $x \in \mathbb{C}$). We denote $\mu_A(f) = f(A)$.

Let $A_1, \dots, A_n \in \mathcal{A}$ be $\mathcal{F}$-generators. The mapping

$$\begin{aligned} \mathcal{F} \times \cdots \times \mathcal{F} &\to \mathcal{A} \\ (f_1, \dots, f_n) &\mapsto f_n(A_n) \dots f_1(A_1) \end{aligned}$$

factors through $\mathcal{F}_n = \mathcal{F}^{\hat{\otimes} n}$, and we denote the associated mapping by

$$\mu_A : \mathcal{F}_n \to \mathcal{A}$$

where $A = (\overset{1}{A_1}, \dots, \overset{n}{A_n})$.

[3] The words "proper" and "unary" will usually be omitted.

Definition IV.13 A tuple $A = (\overset{1}{A_1}, \dots, \overset{n}{A_n})$ of $\mathcal{F}$-generators equipped with Feynman indices is called a *Feynman tuple*. For any $f \in \mathcal{F}_n$ we define the function of A with symbol f by setting

$$f(A) \equiv f(\overset{1}{A_1}, \dots, \overset{n}{A_n}) = \mu_A(f).$$

We have reproduced all definitions given in Subsection 1.2 in the rigorous context of polynormed spaces. As to propositions, lemmas, theorems, corollaries, etc. concerning functions of Feynman tuples, we point out that no one can object to the rigor of the argument given in their proofs, with the understanding that convergence, continuity, etc. is considered in poly-Banach spaces. For this reason, we do not dwell on these assertions here and leave the subject.

Remark IV.4 Our definitions are flexible enough to cover most of the applications. Nevertheless, there is an important case that does not match the definitions literally. We mean functions of matrices. Indeed, given a matrix A, the natural symbol space for functions of A is comprised by functions on the spectrum $\sigma(A)$ together with their derivatives up to a certain order (depending on the size of Jordan blocks of A). The set $\sigma(A)$ is discrete, so that the derivatives of a symbol cannot be reconstructed from its values but must be determined separately. Hence we can take $Z = \sigma(A)$, but in this case jets should be considered instead of functions (the order of the jet may vary from point to point).

Having finished with the general theory, we proceed to the consideration of concrete symbol classes. The symbol classes most frequently used are S^∞ and $\mathcal{O}_K$. The former is used to define functions of tempered generators and the latter to define functions of bounded operators in a Banach space. We consider S^∞ in some detail and leave the consideration of $\mathcal{O}_K$ as an exercise.

2.2 S^∞ Is a Proper Symbol Space

Theorem IV.2 *The poly-Banach space* $S^\infty(\mathbb{R}^1)$ *satisfies Conditions* A–D *and* D′ *of the preceding subsection. Moreover,*

$$\left[S^\infty(\mathbb{R}^1)\right]^{\hat{\otimes} n} \simeq S^\infty(\mathbb{R}^n),$$

where the $\simeq$ *stands for the isomorphism of poly-Banach spaces and is given by the natural interpretation of elements of* $[S^\infty(\mathbb{R}^1)]^{\hat{\otimes} n}$ *as functions of* n *variables.*

Proof. The validity of Conditions A,B,C, and E follows directly from the definition of $S^\infty(\mathbb{R}^1)$. The main difficulty is, of course, to prove D and D′. We carry out the proof only for $n = 2$ (Condition D), since the argument for $n > 2$ (Condition D′) involves no additional ideas. Following the argument used in Subsection 1.3 to prove the equality

$C^\infty(\mathbb{R}^1)\hat{\otimes}C^\infty(\mathbb{R}^1) = C^\infty(T^2)$, we subdivide the proof into three stages expressed by the following lemmas.

Lemma IV.3 *There is a natural embedding*

$$j : S^\infty(\mathbb{R}^1) \otimes S^\infty(\mathbb{R}^1) \subset S^\infty(\mathbb{R}^2),$$

and this embedding is continuous with respect to the projective inf-seminorm on $S^\infty(\mathbb{R}^1) \otimes S^\infty(\mathbb{R}^1)$.

Lemma IV.4 *The space* $S^\infty(\mathbb{R}^1) \otimes S^\infty(\mathbb{R}^1)$ *is dense in* $S^\infty(\mathbb{R}^2)$.

Lemma IV.5 *The embedding* j *has a continuous inverse*

$$j^{-1} : j(S^\infty(\mathbb{R}^1) \otimes S^\infty(\mathbb{R}^1)) \to S^\infty(\mathbb{R}^1) \otimes S^\infty(\mathbb{R}^1)$$

on its range (the range $j(S^\infty(\mathbb{R}^1) \otimes S^\infty(\mathbb{R}^1)$ *is equipped with the polynorm inherited from* $S^\infty(\mathbb{R}^2)$*).*

Combining these lemmas, we readily obtain the assertion of the theorem. Indeed, Lemmas IV.4 and IV.5 imply that j^{-1} can be extended by continuity to the entire space $S^\infty(\mathbb{R}^2)$ (of course, the range of the extension $\overline{j^{-1}}$ will be in the completion $S^\infty(\mathbb{R}^1)\hat{\otimes}S^\infty(\mathbb{R}^1)$ of $S^\infty(\mathbb{R}^1) \otimes S^\infty(\mathbb{R}^1)$. Furthermore j extends by continuity to a mapping $\overline{j} : S^\infty(\mathbb{R}^1)\hat{\otimes}S^\infty(\mathbb{R}^1) \to S^\infty(\mathbb{R}^2)$. Since

$$jj^{-1} = j^{-1}j = \mathrm{id},$$

the same is true of their extensions by continuity:

$$\overline{j}\,\overline{j^{-1}} = \overline{j^{-1}}\,\overline{j} = \mathrm{id},$$

so that

$$\overline{j^{-1}} = (\overline{j})^{-1}.$$

Thus,

$$\overline{j} = S^\infty(\mathbb{R}^1)\hat{\otimes}S^\infty(\mathbb{R}^1) \to S^\infty(\mathbb{R}^2)$$

is an isomorphism of polynormed spaces, and the theorem is proved. □

It remains to prove Lemmas IV.3 – IV.5.

Proof of Lemma IV.3. Obvious, since if the functions $f(x)$ and $g(y)$ grow, together with all their derivatives, not faster than some polynomial, then the same is true of $f(x)g(y)$. □

Proof of Lemma IV.4. As with $C^\infty(S^1)$, trigonometric polynomials are dense in $S^\infty(\mathbb{R}^k)$, $k = 1, 2$.[4] However the periods of the exponentials are not exhausted by

[4] And for $k > 2$ as well.

multiples of 2π. Let us prove this assertion. Recall that $S^\infty(\mathbb{R}^k)$ is the union

$$S^\infty(\mathbb{R}^k) = \bigcup_{m\in\mathbb{Z}} S^m(\mathbb{R}^k)$$

of the Fréchet spaces $S^m(\mathbb{R}^k)$ with locally convex topology determined by the system of seminorms

$$\|f\|_{m,l} = \sup_{y\in\mathbb{R}^k}(1+|y|)^{-m}\sum_{|\alpha|=0}^{l}|f^{(\alpha)}(x)|.$$

We need to prove that any $f \in S^m(\mathbb{R}^k)$ can be approximated by trigonometric polynomials *in the convergence of* $S^\infty(\mathbb{R}^k)$ (not in the topology of $S^m(\mathbb{R}^k)$; this is exactly the point where the apparatus of polynormed spaces works). In fact, such an approximation already exists in $S^r(\mathbb{R}^k)$, where $r = \max\{m, 1\}$. Indeed, let $\varphi(y)$ be a smooth compactly supported function of $y \in \mathbb{R}^k$ equal to 1 in the neighborhood of the origin. Then the sequence

$$f_n(y) = \varphi(y/n)f(y)$$

of smooth compactly supported functions converges to $f(y)$ in $S^r(\mathbb{R}^k)$ as $n \to \infty$, since the norm of the "tails" will be killed by the extra factor $(1+|y|)^{-1} \sim n^{-1}$ in the definition of the norm. Hence, it suffices to obtain trigonometric approximations for smooth compactly supported functions. However, this is trivial: any $f \in C_0^\infty(\mathbb{R}^k)$ can be expanded into the Fourier integral

$$f(y) = \left(\frac{i}{2\pi}\right)^{k/2}\int_{\mathbb{R}^k} e^{ipy}\tilde{f}(p)\,dp$$

with smooth rapidly decaying Fourier transform $\tilde{f}(p)$, this integral converges absolutely in any $S^r(\mathbb{R}^k)$ with $r > 0$, and the desired approximations can be taken in the form of finite Riemann sums

$$f(y) \simeq \sum c_\alpha e^{ip_\alpha y}.$$

Take $k = 2$. Then

$$e^{ip_\alpha y} = e^{ip_{\alpha 1}y_1}e^{ip_{\alpha 2}y_2} \in S^\infty(\mathbb{R}^1)\otimes S^\infty(\mathbb{R}^1).$$

The lemma is proved. □

Proof of Lemma IV.5. As was shown just a few lines above, smooth compactly supported functions are dense in $S^\infty(\mathbb{R}^k)$ for any k and hence in $S^\infty(\mathbb{R}^2)$ and in $S^\infty(\mathbb{R}^1)\otimes S^\infty(\mathbb{R}^1)$. Consequently, it suffices to check the boundedness of j^{-1} on the subset consisting of finite sums

$$\varphi(x, y) = \sum_s f_s(x)g_s(y), \tag{IV.4}$$

where $f_s, g_s \in C_0^\infty(\mathbb{R}^1)$. Let φ be such a function. Let us prove that for any $m' > m > 0$,

$$\|\varphi\| \equiv \|\varphi\|_{S_0^{m'}(\mathbb{R}^1)\otimes S_0^{m'}(\mathbb{R}^1)} \leq \text{const}\, \|\varphi\|_{S_2^{m-2}(\mathbb{R}^2)} \equiv \text{const}\, |||\varphi|||. \tag{IV.5}$$

Assuming that (IV.5) is valid, it is easy to prove the inequalities

$$\|\varphi\|_{S_r^{m'}(\mathbb{R}^1)\otimes S_t^{2m'}(\mathbb{R}^1)} \leq \text{const}\, \|\varphi\|_{S_{r+t+2}^{m-2}(\mathbb{R}^2)}$$

for all $r, t \geq 0$ by induction on $r + t$. The collection of these inequalities implies that for any $m' > m > 0$ the mapping j^{-1} is continuous between the spaces

$$j^{-1} : S^{m-2}(\mathbb{R}^2) \to S^{m'}(\mathbb{R}^1) \otimes S^{m'}(\mathbb{R}^1),$$

on the range of j, as desired.

We will prove inequality (IV.5) and omit the trivial induction step.

We have

$$|||\varphi||| = \sup_{(x,y)\in\mathbb{R}^2} \sum_{l+k\leq 2} \left| \frac{\partial^{l+k}\varphi(x,y)}{\partial x^l \partial y^k} \right| (1 + |x| + |y|)^{-m},$$

$$\|\varphi\| = \inf \sum_l \|\varphi_l\|_{S_0^m(\mathbb{R}^1)} \|\psi_l\|_{S_0^m(\mathbb{R}^1)},$$

where

$$\|\psi\|_{S_0^m(\mathbb{R}^1)} = \sup_{x\in\mathbb{R}^1} (1 + |x|)^{-m} |\psi(x)|$$

and the infimum is taken over all finite representations $\varphi = \sum_l \varphi_l \otimes \psi_l$ (without the assumption that φ_l and ψ_l are compactly supported).

Let us represent $\varphi(x, y)$ defined in (IV.4) by the Fourier integral

$$\varphi(x, y) = \frac{i}{2\pi} \iint_{\mathbb{R}^2} e^{i(px+qy)} \sum_s \tilde{f}_s(p)\tilde{g}_s(q)\, dp\, dq,$$

where $\tilde{f}_s$ and $\tilde{g}_s$ are the Fourier transforms of f and g, respectively.

Denote

$$f_{sm}(x) = (1 + x^2)^{-m/2} f_s(x), \quad g_{sm}(y) = (1 + y^2)^{-m/2} g_s(y).$$

Then we can write

$$\varphi(x, y) = \frac{i}{2\pi} \iint_{\mathbb{R}^2} (1 + x^2)^{m/2} (1 + y^2)^{m/2} \sum_s \tilde{f}_{sm}(p)\tilde{g}_{sm}(q) e^{i(px+qy)}\, dp\, dq.$$

The last integral can be approximated by Riemann sums in $S_0^{m'}(\mathbb{R}^1) \otimes S_0^{m'}(\mathbb{R}^1)$ for any $m' > m$:

$$\frac{2\pi}{i}\varphi(x, y) = \sum_s \Big[\int (1+x^2)^{m/2} f_{sm}(p) e^{ipx}\, dp\Big]\Big[\int (1+y^2)^{m/2} g_{sm}(q) e^{iqx}\, dq\Big]$$

$$= \sum_s \lim \lambda^2 \Big[\sum_{|k|\le R} (1+x^2)^{m/2} e^{i\lambda kx} f_{sm}(\lambda k)\Big] \lim \Big[\sum_{|l|\le R} (1+y^2)^{m/2} e^{i\lambda ly} g_{sm}(\lambda l)\Big]$$

$$= \lim \lambda^2 \sum_{|k|,|l|\le R} \tilde{\varphi}_m(\lambda k, \lambda l)(1+x^2)^{m/2} e^{i\lambda kx}(1+y^2)^{m/2} e^{i\lambda ly},$$

where the limits as $\lambda \to 0$, $R \to \infty$, and $\lambda R \to \infty$ are in $S_0^{m'}(\mathbb{R}^1)$ on the second line and in $S_0^{m'}(\mathbb{R}^1) \otimes S_0^{m'}(\mathbb{R}^1)$ on the third and fourth lines (we have used the fact that the mapping $S_0^{m'}(\mathbb{R}^1) \otimes S_0^{m'}(\mathbb{R}^2) \to S_0^{m'}(\mathbb{R}^2) \otimes S_0^{m'}(\mathbb{R}^1)$ is continuous. Here we used the notation

$$\tilde{\varphi}_m(p, q) = \sum_s \tilde{f}_{sm}(p) \tilde{g}_{sm}(q).$$

It follows that

$$\|\varphi\| = \frac{1}{2\pi} \lim \lambda^2 \Big\| \sum_{|k|,|l|\le R} (1+x^2)^{m/2}(1+y^2)^{m/2} e^{i\lambda(kx+ly)} \tilde{\varphi}_m(\lambda k, \lambda l) \Big\|.$$

However, we can write out the following estimate:

$$\Big\| \sum_{|k|,|l|\le R} (1+x^2)^{m/2}(1+y^2)^{m/2} e^{i\lambda(kx+ly)} \tilde{\varphi}_m(\lambda k, \lambda l) \Big\| \le \sum_{|k|,|l|\le R} |\tilde{\varphi}_m(\lambda k, \lambda l)|,$$

since

$$\|(1+x^2)^{m/2} e^{i\lambda kx}\|_{S_0^{m'}(\mathbb{R}^1)} = 1.$$

Furthermore,

$$\lambda^2 \sum_{|k|,|l|\le R} |\tilde{\varphi}_m(\lambda k, \lambda l)|$$

$$\le \Big\{ \sum_{|k|,|l|\le R} \lambda^2 |\tilde{\varphi}_m(\lambda k, \lambda l)|^2 (1+\lambda^2 k^2)(1+\lambda^2 l^2)$$

$$\times \sum_{|k|,|l|\le R} \frac{\lambda^2}{(1+\lambda^2 k^2)(1+\lambda^2 l^2)} \Big\}^{1/2}$$

by the Cauchy–Schwarz–Bunyakovskii inequality. The sum

$$\sum_{|k|,|l|\le R} \frac{\lambda^2}{(1+\lambda^2 k^2)(1+\lambda^2 l^2)} = \left(\sum_{|k|\le R} \frac{\lambda}{1+\lambda^2 k^2} \right)^2$$

is uniformly bounded, and

$$\begin{aligned}\lim \sum_{|k|,|l|} \lambda^2 |\tilde{\varphi}_m(\lambda k, \lambda l)|^2 (1+\lambda^2 k^2)(1+\lambda^2 l^2) &= \int (1+p^2)(1+q^2)|\tilde{\varphi}_m(p,q)|^2\, dp dq \\ &= \int \Big(1 - \frac{\partial^2}{\partial x^2}\overline{\big[(1+x^2)^{-m/2}(1+y^2)^{-m/2}\varphi(x,y)\big]} \\ &\quad \times \int \Big(1 - \frac{\partial^2}{\partial y^2}\overline{\big[(1+x^2)^{-m/2}(1+y^2)^{-m/2}\varphi(x,y)\big]}\, dx dy\end{aligned}$$

by the Parseval identity. Since

$$|\varphi^{(\alpha)}(x,y)| \le \text{const} \cdot \|\varphi\| (1+x^2+y^2)^{(m-2)/2}, \quad |\alpha| = 0, 1, 2,$$

the last integral converges and is bounded by const $|||\varphi|||^2$. Thus we obtain

$$\|\varphi\| \le \text{const} \cdot |||\varphi|||.$$

The lemma is proved. □

2.3 S^∞-Generators

Now let us describe the structure of S^∞-generators in poly-Banach algebras.

Theorem IV.3 *Let A be an S^∞-generator in a poly-Banach algebra*

$$\mathcal{A} = \bigcup_{K \in \Lambda} \bigcap_{\alpha \in K} \mathcal{A}_\alpha.$$

Then A is a tempered generator *in $\mathcal{A}$, that is, A generates a one-parameter group $\{\exp(iAt)\}$, $t \in \mathbb{R}$, satisfying the polynomial growth estimates: there exists a $K \in \Lambda$ such that for any $\alpha \in K$ the estimate*

$$\|\exp(iAt)\| \le C_\alpha (1+|t|)^{m_\alpha}$$

is valid with nonnegative constants C_α and m_α independent of t. Similar estimates are valid for all t-derivatives of $\exp(iAt)$.

Proof. The function e^{ity} belongs to $S^0(\mathbb{R}^1)$ for any $t \in \mathbb{R}^1$ and is continuously differentiable with respect to t as a function with values in $S^1(\mathbb{R}^1)$. Hence the mapping $U(\cdot,t) : t \mapsto e^{ity}$ is a continuously differentiable mapping of $\mathbb{R}^1$ into $S^\infty(\mathbb{R}^1)$ and satisfies the Cauchy problem

$$\begin{cases} \dfrac{dU(y,t)}{dt} = yU(y,t), \\ U\big|_{t=0} = 1. \end{cases}$$

Applying the continuous mapping

$$\mu_A : S^\infty(\mathbb{R}^1) \to \mathcal{A}$$

to both sides of the equation and the initial condition, we obtain

$$\begin{cases} \dfrac{dU(A,t)}{dt} = AU(A,t), \\ U(A,0) = 1, \end{cases}$$

that is, $U(A,t) = \exp(iAt)$ is the one-parameter subgroup generated by A in $\mathcal{A}$.

Let us now prove the polynomial estimates. The mapping $\mu_A : S^\infty(\mathbb{R}^1) \to \mathcal{A}$ is continuous, in particular, there exists a $K \subset \Lambda$ such that μ_A is continuous in the spaces

$$\mu_A : S^0(\mathbb{R}^1) \to \mathcal{A}^K.$$

This implies that for any $\alpha \in K$ there exists an l such that μ_A is continuous in the spaces

$$S_l^0(\mathbb{R}^1) \to \mathcal{A}_\alpha.$$

But

$$\|e^{iyt}\|_{S_l^0(\mathbb{R}^1)} = \sup_{y\in\mathbb{R}^n} \sum_{j=0}^{l} \left| \left(\frac{\partial}{\partial y}\right)^j e^{iyt} \right| = \sup_{y\in\mathbb{R}^n} \sum_{j=0}^{l} |t|^j \le C_l(1+|t|)^l.$$

The estimates for $(d/dt)^r \exp(iAt)$ can be proved in a similar way. The theorem is proved. □

Corollary IV.1 *Functions of S^∞-generators can be defined via the Fourier transform: if $A_1, \ldots, A_n$ are S^∞-generators in a poly-Banach algebra $\mathcal{A}$ and $f \in S^\infty(\mathbb{R}^n)$, then*

$$f(\overset{1}{A_1}, \ldots, \overset{n}{A_n}) = \left(\frac{i}{2\pi}\right)^n \int_{\mathbb{R}^n} \tilde{f}(p_1, \ldots, p_n) \exp(ip_nA_n) \ldots \exp(ip_1A_1) dp_1 \ldots dp_n,$$

where

$$\tilde{f}(p_1, \ldots, p_n) = \left(\frac{-i}{2\pi}\right)^n \int_{\mathbb{R}^n} f(y_1, \ldots, y_n) e^{-i(y_1p_1+\cdots+y_np_n)} dy_1 \ldots dy_n$$

is the Fourier transform of f.

The proof is obvious.

Theorem IV.3 and Corollary IV.1 give a convincing example of how the structure of the symbol space determines the properties of generators and the method of defining the mapping symbol $\mapsto$ operator.

3 Functions of Operators in Scales of Spaces

Functions of noncommuting operators were defined in Section 1 and Section 2 in the situation of abstract poly-Banach algebras. However, in applications (particularly, to differential equations) one usually deals with algebras of operators acting on a scale of spaces (say, on the Sobolev scale W_s^2). Algebras of continuous operators on Banach scales are special case of general poly-Banach algebras, and in this section we consider such algebras in some detail.

3.1 Banach Scales

Let I be a poset and $\{B_i\}_{i\in I}$ a family of Banach spaces with continuous embeddings

$$B_i \subset B_j$$

defined whenever $i \le j$. Such a family $\{B_i\}_{i\in I}$ is called a *Banach scale*. If I is a directed set, then the scale is said to be *inductive*; and if I' is a directed set, then the scale is said to be *projective*. Thus, in an inductive scale for each two spaces B_i and B_j there always exists a space B_k such that $B_i \subset B_k$ and $B_j \subset B_k$, whereas in a projective scale for each two spaces B_i and B_j there always exists a space B_k such that $B_k \subset B_i \cap B_j$.

In applications, most frequently one has $I = \mathbb{Z}$ or $I = \mathbb{R}$, and we restrict ourselves to one of these two situations.

Let $\{B_i\}_{i\in I}$ be a Banach scale. According to definition IV.6, to any filter of sections of I there corresponds a polynormed space, namely, $\bigcup_{A\in\Lambda}\bigcap_{i\in A} B_i$. Among these spaces, there is a maximal space

$$B_\infty = \bigcup_{i\in I} B_i$$

and a minimal space

$$B_{-\infty} = \bigcap_{i\in I} B_i.$$

Lemma IV.6 *One has*

$$\mathcal{L}(B_\infty) = \bigcap_i \bigcup_j \mathcal{L}(B_i, B_j),$$

$$\mathcal{L}(B_{-\infty}) = \bigcap_j \bigcup_i \mathcal{L}(B_i, B_j).$$

The elements of the former algebra are called operators continuous to the left, and the elements of the latter algebra are called operators continuous to the right in the scale $\{B_j\}$.

Proof. Trivial. □

One of the usual methods of constructing a scale is to take some Banach space $\mathcal{B}$ and a tuple of unbounded operators in $\mathcal{B}$; the scale is constructed with the help of graph or similar norms. The case in which $\mathcal{B}$ is a Hilbert space and $\mathcal{B}_i = D(A^{i/2})$, $\|u\|_i = \|A^{i/2}u\|_{\mathcal{B}}$, where A is a positive definite unbounded self-adjoint operator in $\mathcal{B}$, is considered in Chapter II in the context of representations of Lie algebras and groups (see also the following subsection). Here we deal with the case in which $\mathcal{B}$ is a Banach space and the norms in $\mathcal{B}_i$ are defined according to a tuple of unbounded operators in $\mathcal{B}$.

Let B be a Banach space and $D \subset B$ a dense linear subset. Let also $A_1, \dots, A_m$ be unbounded closed operators on B such that for each $j = 1, \dots, m$ the set D is invariant under A_i and moreover, D is a core of A_i (that is, A_i is the closure of its restriction to D). There are numerous possible ways to define a scale using these data. Here we consider the simplest one. Let $\|\cdot\|$ denote the norm of B. For any positive integer k set

$$\|x\|_k \equiv \|x\|_{A,k} = \sum_{0\le s\le k} \|A_{j_1}A_{j_2}\dots A_{j_s}x\|,$$

where the sum is taken over all sequences $j_1, \dots, j_s$ of integers satisfying the condition $1 \le j_l \le m, l = 1, \dots, s$. Denote by B_k the completion of D with respect to the norm $\|\cdot\|_k$.

Proposition IV.1 *The collection $\{B_k\}_{k\ge0}$ is a Banach scale. Each of the operators A_j, $j = 1, \dots, m$, is continuous in the spaces*

$$A_j : B_k \to B_{k-1}, \quad k = 1, 2, \dots$$

Proof. Clearly,

$$\|x\|_k \sim \|x\|_{k-1} + \sum_{j=1}^{n} \|A_j x\|_{k-1}, \tag{IV.6}$$

where the tilde stands for equivalence of norms. In particular, the identity operator on D is continuous in the pair of norms $(\|\cdot\|_k, \|\cdot\|_{k-1})$, and so it extends by continuity to a bounded operator $\varepsilon_{k,k-1} : \mathcal{B}_k \to \mathcal{B}_{k-1}$. Let us prove by induction on k that $\varepsilon_{k,k-1}$ has trivial kernel and each A_j is closable from D in $\mathcal{B}_k$. The induction hypothesis ($k = 0$) is obviously valid since there is no $\varepsilon_{0,-1}$ at all (and so nothing to check), whereas the A_j are closed in $\mathcal{B} = \mathcal{B}_0$ by our assumption. Let us carry out the inductive step. Assume that the statement is proved for $k < k_0$ and set $k = k_0$. First we prove that $\varepsilon_{k,k-1}$ has trivial kernel. Let $x \in \operatorname{Ker} \varepsilon_{k,k-1}$. Then there exists a sequence $x_l \in D, l = 1, 2, \dots$ such that $x_l \to x$ in $\mathcal{B}_k$ (we write $x_l \overset{k}{\to} x$ for short) and $x_l \overset{k-1}{\to} 0$. It follows from (IV.6) that for each j the sequence $A_j x_l, l = 1, 2, \dots,$ is convergent in B_{k-1}. Since, by the induction assumption, A_j is closable in B_{k-1}, it follows that $A_j x_l \overset{k-1}{\to} 0$. But now, again by (IV.6), we see that $\|x_l\|_k \to 0$, that is, $x = 0$. Let us now prove that each A_j is closable from D in B_k. Let $x_l \in B_k$, $l = 1, 2, \ldots$, be a sequence such that $x_l \overset{k}{\to} 0$ and $A_j x_l \overset{k}{\to} y$. Then $A_j x_l \overset{k-1}{\to} \varepsilon_{k,k-1}(y)$ and $x_l \overset{k-1}{\to} 0$; since A_j is closable in B_{k-1}, we necessarily have $\varepsilon_{k,k-1}(y) = 0$ and hence $y = 0$ due to the triviality of Ker $\varepsilon_{k,k-1}(y)$. The proposition is proved. □

Now let

$$A : D \to D$$

be a linear operator. We are interested in conditions that would guarantee that A extends to a bounded linear operator in the scale $\{B_k\}$. To state the following theorem conveniently, we introduce some terminology. We will assign *length* s to any product $A_{j_1} A_{j_2} \ldots A_{j_s}$ as well as to the commutator

$$K_j(A) = [A_{j_1}[A_{j_2}[\ldots[A_{j_s}, A]\ldots]]].$$

Theorem IV.4 *Suppose that the operator A satisfies the following condition: there exists a function $\varphi : \mathbb{Z}_+ \cup \{0\} \to \mathbb{Z}_+ \cup \{0\}$ such that*

$$\|K_j(A)\| \le C_r \|x\|_{\varphi(r)}, \quad x \in D$$

for any commutator $K_j(A)$ of length r, where C_r depends only on r. Then A extends to a bounded operator in the scale B_k. Namely,

$$\|Ax\|_k \le C_k \|x\|_{\psi(k)}, \quad x \in E, \quad k = 0, 1, 2, \ldots$$

(with some other constant C_k), where

$$\psi(k) = \max_{0 \le k \le r} (k + \varphi(r) - r).$$

Proof. One has

$$\|Ax\|_k = \sum_{0 \le s \le k} \|A_{j_1} A_{j_2} \ldots A_{j_s} Ax\|.$$

However, $A_{j_1} A_{j_2} \ldots A_{j_s} Ax$ is a linear combination of terms of the form $K_l(A) A_{\pi_1} \ldots A_{\pi_r} x$ with $l + r = s$ (this can readily be proved by induction). Each such term can be estimated by

$$\|K_l(A) A_{\pi_1} \ldots A_{\pi_r} x\| \le \|A_{\pi_1} \ldots A_{\pi_r} x\|_{\Phi(l)} \le C \|x\|_{\Phi(l)+r}.$$

But $\Phi(l) + r = \Phi(l) + s - l$ and the desired estimate follows immediately. The theorem is proved. □

3.2 S^∞-Generators in Banach Scales

In Theorem IV.3 we used S^∞-generators in a poly-Banach space to produce one-parameter groups of tempered growth in this space. In applications (at least to differential equations) one mostly deals with scales obtained from a Banach (or even Hilbert) space; the commonest way to construct groups of tempered growth in scales is to consider such a semigroup in the "parent" Banach space and then to use some additional information in order to prove that the group extends appropriately to the entire scale.

Thus, we start from one-parameter groups in Banach spaces and their generators.

Definition IV.14 Let X be a Banach space. A family $U(t)$, $t \in \mathbb{R}$, of bounded linear operators on X is called a strongly continuous one-parameter group if

(i) $U(t)$ is strongly continuous with respect to t;

(ii) $U(0) = 1$ and $U(t)U(\tau) = U(t+\tau)$ for any $t, \tau \in \mathbb{R}$. One says that $U(t)$ is a one-parameter group of tempered growth if

$$\|U(t)\| \le K(1+|t|)^m$$

for some K and m (note that an exponential bound for $\|U(t)\|$ is always guaranteed).

The strong limit

$$C = \lim_{\varepsilon \to 0} \frac{T(\varepsilon) - T(0)}{i\varepsilon}$$

exists on a dense subset D_C and is a closed operator. It is easy to show that for $x \in D_C$ the family $u(t) = U(t)x$ can be defined as the unique solution to the Cauchy problem

$$\frac{du}{dt} = iCu, \quad u|_{t=0} = x.$$

Let $\mathcal{B}$ be a Banach space, $\mathcal{D} \subset \mathcal{B}$ a dense linear subset, and $\{\mathcal{B}_s\}$ the scale generated by the tuple $(A_1, \dots, A_n)$ of closed operators with invariant core $\mathcal{D}$ on $\mathcal{B}$ (recall that $\mathcal{D}$ is called an invariant core of a closed operator A if $A\mathcal{D} \subset \mathcal{D}$ and A is the closure of the restriction $A|_{\mathcal{D}}$). Thus, the spaces $\mathcal{B}_s$ are defined for all integer $s \ge 0$, $\mathcal{B}_0 = \mathcal{B}$, and $\mathcal{B}_s$ is the completion of $\mathcal{D}$ with respect to the norm

$$\|u\|_1 = \|u\|_{s-1} + \sum_{j=1}^{n} \|A_j u\|_{s-1}, \quad s = 1, 2, \dots$$

We intend to consider generators of tempered one-parameter groups in $\{\mathcal{B}_s\}$. The main idea is as follows. Suppose that C is the generator of a semigroup e^{iCt} in $\mathcal{B}$. In view of the definition of the norm in $\mathcal{B}_s$, the estimation of e^{iCt} in $\mathcal{B}_1, \mathcal{B}_2, \dots, \mathcal{B}_s, \dots$ requires permuting A_j and e^{iCt}. In doing so the following lemma can prove useful.

Lemma IV.7 *Let X_1 and X_2 be Banach spaces, and let C_1 and C_2 be generators of strongly continuous one-parameter groups in X_1 and X_2, respectively. Also, let*

$A : X \to Y$ be a closed operator with dense domain $\mathcal{D}_A$. Suppose that there exists a core $\mathcal{D} \subset \mathcal{D}_A$ of the operator A such that, whenever

$$x \in \mathcal{D}_0, \text{ that is, } x \in \mathcal{D} \cap \mathcal{D}_{C_1} \text{ and } C_1 x \in \mathcal{D},$$

we have $Ax \in \mathcal{D}_{C_2}$ and

$$C_2 A x = A C_1 x.$$

Finally, suppose that one of the following conditions is satisfied:

(i) *$\mathcal{D}_0$ is a core of A, $\mathcal{D}$ is $e^{iC_1 t}$-invariant, and $Ae^{iC_1 t}x$ is continuous in X_2 for each $x \in \mathcal{D}$.*

(ii) *$\mathcal{D}$ is invariant under the resolvent $R_\lambda(C_1)$ for $|\operatorname{Im}\lambda|$ large enough.*

Then for any $x \in \mathcal{D}_A$ we have $e^{iC_1 t}x \in \mathcal{D}_A$ and

$$Ae^{iC_1 t}x = e^{iC_2 t}Ax, \quad t \in \mathbb{R}.$$

Remark IV.5 We cannot derive Lemma IV.7 from Theorem I.4 because C_1 and C_2 are not known to be included in an operator algebra as $\mathcal{F}$-generators with some symbol class $\mathcal{F}$ containing the exponentials e^{ity}.

Proof. We omit the consideration of case (i) (see [134]). Suppose that (ii) is satisfied. Suppose $\operatorname{Im}\lambda < 0$ and $|\operatorname{Im}\lambda|$ is large enough. Since $\mathcal{D}$ is $R_\lambda(C_1)$-invariant, it follows that $R_\lambda(C_1)x \in \mathcal{D}_0$ for $x \in \mathcal{D}$, and

$$(C_2 - \lambda)AR_\lambda(C_1)x = A(C_1 - \lambda)R_\lambda(C_1)x = Ax, \quad x \in \mathcal{D},$$

whence it follows that

$$AR_\lambda(C_1)x = R_\lambda(C_2)Ax, \quad x \in \mathcal{D}.$$

Fix $x \in \mathcal{D}$ and $t \geq 0$ and set

$$x_n = (-inR_{-in}(C_1))^{[nt]}x,$$

where $[nt]$ is the integral part of nt. Obviously, $x_n \in \mathcal{D}$ and

$$Ax_n = (-inR_{-in}(C_2))^{[nt]}Ax.$$

Passing to the limit as $n \to \infty$ we obtain

$$x_n \to e^{iC_1 t}x, \quad Ax_n \to e^{iC_2 t}Ax$$

(see [197]). Since A is closed, it follows that

$$Ae^{iC_1 t}x = e^{iC_2 t}Ax.$$

This identity extends to $\mathcal{D}_A$ by closure. The case $t < 0$ is treated similarly. The lemma is proved. □

Now let C be the generator of a strongly continuous one-parameter group in $\mathcal{B}$. We will prove that C is a generator in each of the $\mathcal{B}_s$ provided that the commutator C with each of the A_i can be expressed linearly via the operators A_i.

Proposition IV.2 *Let C be the generator of a strongly continuous one-parameter group in $\mathcal{B}$. Suppose that $\mathcal{D} \subset \mathcal{D}_C$, $\mathcal{D}$ is invariant under C, and there exist numbers λ_{jk}, $j, k = 1, \dots, n$, such that*

$$[C, A_j]x = \sum_{k=1}^{n} \lambda_{jk} A_k x, \quad x \in \mathcal{D}.$$

Suppose also that either $\mathcal{D}$ is $e^{iC_1 t}$-invariant and $A_j e^{iC_1 t} x$ is a strongly continuous function of t for $x \in \mathcal{D}$, or $\mathcal{D}$ is invariant under the resolvent $R_\lambda(C)$ for $|\operatorname{Im}\lambda|$ sufficiently large. Then C is the generator of a strongly continuous one-parameter group in $\mathcal{B}_s$ for each s.

Proof. In order to apply Lemma IV.7, let us represent the cited commutation relations in the form

$$ACx = (C \otimes I - I \otimes \Lambda)Ax, \quad x \in \mathcal{D}$$

where

$$A = {}^t(A_1, \dots, A_n) : \mathcal{B} \to \underbrace{\mathcal{B} \oplus \cdots \oplus \mathcal{B}}_{n \text{ copies}} = \mathcal{B} \otimes \mathbb{C}^n,$$

I is the identity operator in either factor, and Λ is the matrix with entries λ_{jk}. Let us check that Lemma IV.7 applies with $C_1 = C$ and $C_2 = C \otimes I - I \otimes \Lambda$. The operator A is closable from $\mathcal{D}$ since if $x_n \in \mathcal{D}$, $x \to 0$, and $Ax \to y = (y_1, \dots, y_n)$, i.e., $A_j x \to y_j$, $j = 1, \dots, n$, then $y_j = 0$ because A_j is closed. Furthermore, C_2 generates a strongly continuous one-parameter group in $\mathcal{B} \otimes \mathbb{R}^n$, namely,

$$e^{iC_2 t} \equiv e^{i(C \otimes I - I \otimes \Lambda)t} = e^{iCt} \otimes e^{-i\Lambda t}$$

(the reader is advised to check this identity by direct computation).

Thus we may conclude that

$$Ae^{iC_1 t}x = e^{iC_2 t}Ax, \quad x \in \mathcal{D},$$

or, in more detail,

$$A_j e^{iCt} x = e^{iCt} \sum_{k=1}^{n} \exp[-\Lambda t]_{jk} A_k x.$$

Let us now estimate e^{iCt} in $\mathcal{B}_1$. For $x \in \mathcal{D}$, we have

$$\begin{aligned} \|e^{iCt}x\|_1 &= \|e^{iCt}x\|_0 + \sum_{j=1}^{n} \|A_j e^{iCt} x\|_0 \\ &\le \|e^{iCt}x\|_1 + \sum_{k,j=1}^{n} \|\exp[-\Lambda t]_{jk}\| \, \|e^{iCt} A_k x\|_0. \end{aligned}$$

Let $R(t)$ be a bound for $\|e^{iCt}x\|_{\mathcal{B}\to\mathcal{B}}$; then

$$\begin{aligned}\|e^{iCt}x\|_1 &\leq R(t)\big(\|x\|_0 + \sum_{j,k=1}^{n} \|\exp[-\Lambda t]_{jk}\|\,\|x\|_1\big)\\ &\leq MR(t)\|e^{-\Lambda t}\|\,\|x\|_1\end{aligned}$$

for some constant M, where $\|e^{-\Lambda t}\|$ is any matrix norm of $\|e^{-\Lambda t}\|$ (clearly, only M depends on the particular choice of norm).

Proceeding by induction on s, we easily prove that

$$\|e^{iCt}x\|_s \leq M_s R(t)\|e^{-\Lambda t}\|^s \|x\|_1,$$

that is, e^{iCt} extends to a continuous operator in each $\mathcal{B}_s$. Next, $Ae^{iCt}x = e^{iCt} \otimes e^{i\Lambda t}(Ax)$ is strongly continuous in $\mathcal{B}$ for $x \in \mathcal{D}$, whence it easily follows that e^{iCt} is strongly continuous $\mathcal{B}_1$ and, by induction, on each $\mathcal{B}_s$. The proposition is proved. □

Corollary IV.2 *If, under the conditions of Proposition* IV.2, *C is a tempered generator in $\mathcal{B}$ (that is, $|R(t)| \leq C(1+|t|)^m$ for some $m \geq 0$), the matrix Λ has purely imaginary eigenvalues, and the operator C satisfies the conditions of Theorem* IV.4, *then C is a tempered generator in $\mathcal{L}(\mathcal{B}_\infty) = \mathcal{L}(\bigcap_s \mathcal{B}_s)$.*

Proof. If the spectrum of Λ is pure imaginary, then

$$\|e^{-\Lambda t}\| \leq \text{const}\cdot(1+|t|)^{l-1},$$

where l is the largest size of Jordan blocks in the Jordan normal form of Λ. Hence,

$$\|e^{iCt}\|_{\mathcal{B}_s\to\mathcal{B}_s} \leq \text{const}\cdot(1+|t|)^{m+(l-1)s}.$$

By Theorem IV.4, for any s there exists an s_1 such that $\|C\|_{\mathcal{B}_{s_1}\to\mathcal{B}_s} < \infty$.

Since

$$e^{iCt}x - e^{iC\tau}x = i\int_\tau^t e^{iC\theta}Cx\,d\theta,$$

we see that e^{itC} is a continuous function of t with values in $\mathcal{L}(\mathcal{B}_{s_1}, \mathcal{B}_s)$, and its norm grows polynomially as $|t| \to \infty$. □

Remark IV.6 We cannot make C be a generator in $\mathcal{L}(\mathcal{B}_{-\infty}) = \mathcal{L}(\bigcup_s \mathcal{B}_s)$ since the scale is bounded from below, the spaces $\mathcal{B}_s$ with $s < 0$ are not defined. In particular, the operator C cannot be realized as a bounded operator acting from $\mathcal{B}$ to any function space. In the situation of Hilbert spaces one usually deals with the case in which $A_1, \dots, A_n$ are self-adjoint operators in $\mathcal{B}$, and then the "negative spaces" $\mathcal{B}_{-s}$ can be defined by duality, $\mathcal{B}_{-s} = \mathcal{B}_s^*$ (Chapter II above). Then the operator C (under certain additional assumptions) can be proved to be a generator in $\mathcal{L}(\mathcal{B}_{-\infty})$.

Example IV.4 Let $\mathcal{B} = L_2(\mathbb{R})^n$, $n = 1$, $A_1 = -i\partial/\partial x$. Then the scale $\mathcal{B}_s$ is (up to norm equivalence) the usual Sobolev scale $\mathcal{B}_s = W_2^s(\mathbb{R})$, and A_1 is a tempered generator in $\mathcal{L}(W_2^\infty(\mathbb{R}))$ (as well as in $\mathcal{L}(W_2^{-\infty}(\mathbb{R}))$); in fact, A_1 is self-adjoint in each of the spaces $W_2^s(\mathbb{R})$). Let C be the operator of multiplication by x in $L_2(\mathbb{R})$. It generates the strongly continuous group of multiplication by e^{ixt} in $L_2(\mathbb{R})$. We have

$$[C, A_1] = i$$

and the hypotheses of Proposition IV.2 can be satisfied by the following trick: set $n = 2$ and add the operator $A_2 = i$. The norm $\|\cdot\|$ is thus replaced by an equivalent one (since A_2 is bounded in $L_2(\mathbb{R}^n)$) and the scale remains unchanged. We have

$$[C, A_1] = iA_2,$$
$$[C, A_1] = 0,$$

that is,

$$\Lambda = \begin{pmatrix} 0 & i \\ 0 & 0 \end{pmatrix}$$

is a Jordan block of size 2 with eigenvalue zero. Thus,

$$\|e^{-\Lambda t}\| \le C(1 + |t|)$$

and the group e^{iCt} grows as $(1 + |t|)^s$ in $W_2^s(\mathbb{R})$. However, C is *not* a generator in $\mathcal{L}(W_2^\infty)$ (in particular, it does not satisfy the conditions of Theorem IV.4). Therefore, it is not very surprising that S^∞-functions of C are not well-defined in $\mathcal{L}(\mathcal{B}^\infty)$ (in particular, the operator C itself does not belong to $\mathcal{L}(\mathcal{B}^\infty)$ — multiplication by x is an unbounded operator in any pair of spaces $W_2^s(\mathbb{R})$, $W_2^s(\mathbb{R}^1)$). However, we can devise another scale of spaces in which x is a generator.

Let

$$A_1 = -\frac{\partial}{\partial x}, \quad A_2 = C = x, \quad A_3 = 1,$$

so that

$$\|u\|_s = 2\|u\|_{s-1} + \|xu\|_{s-1} + \|\partial u/\partial x\|_{s-1}.$$

The matrix Λ has the form

$$\Lambda = \begin{pmatrix} 0 & 0 & i \\ 0 & 0 & 0 \\ 0 & 0 & 0 \end{pmatrix},$$

the size of the Jordan block is still 2, so that the growth estimates remain the same. However, C is now a bounded operator in the scale $\mathcal{B}_s$ and hence is proved to be a generator in $\mathcal{L}(\mathcal{B}_\infty)$ (and, in fact, in $\mathcal{L}(\mathcal{B}_{-\infty})$).

In some applications (for example, to representations of stratified Lie algebras (see [159], [61], [62], etc.), a slightly more complicated situation occurs. Namely, operators $A_1, \dots, A_n, A_{n+1}, \dots, A_{m+n}$ are given and each of the A_j, $j = n+1, \dots, m+n$, can be represented as a linear combination of commutators of length $\le r$ of $A_1, \dots, A_n$. An operator C is given satisfying the conditions of Proposition IV.2 except for the fact that the commutation relations have the form

$$[C, A_j]x = \sum_{k=1}^{m+n} \lambda_{jk} A_k x, \quad x \in \mathcal{D}, \quad j = 1, \dots, m+n.$$

Proposition IV.3 *Under the above conditions, the following estimate is valid:*

$$\|e^{iCt} x\|_k \le \text{const}\, R(t) \|e^{-\Lambda t}\|^k \|x\|_{rk}.$$

The proof, which can be obtained from that of Proposition IV.2 by purely technical manipulations, is omitted.

Using this proposition, one can prove various results concerning generators in Banach scales.

3.3 Functions of Feynman-Ordered Selfadjoint Operators

In this subsection we prove some estimates for functions of a Feynman tuple of self-adjoint operators in the scale generated by (part of) these operators. These estimates use a technique specific to Hilbert spaces and are more precise than the general estimates that can be derived from the results of the preceding subsection. Our main goal is to consider the case in which the Feynman tuple considered has the left ordered representation in S^∞. However, we begin by proving some auxiliary results.

Let $\mathcal{H}$ be a Hilbert space, and let $A_1, \dots, A_n$ be unbounded selfadjoint operators on $\mathcal{H}$. We assume that the intersection $D_{A_1} \cap \dots \cap D_{A_n}$ of their domains contains a linear manifold D dense in $\mathcal{H}$ and invariant with respect to A_j and the corresponding groups $\exp(itA_j)$, $j = 1, \dots, n$, of unitary operators. Also, we assume that for an arbitrary sequence $j_1, \dots, j_m \in \{1, \dots, n\}$ the product

$$\exp(it_m A_{j_m}) \dots \exp(it_1 A_{j_1})$$

is strongly infinitely differentiable with respect to $t \in \mathbb{R}^m$ on D and all its derivatives are of tempered growth,

$$\left\| \frac{\partial^{|\alpha|}}{\partial t^\alpha} (\exp(it_m A_{j_m}) \dots \exp(it_1 A_{j_1}) u \right\| \le C(1 + |t|)^N, \quad u \in D,$$

where N may depend only on $|\alpha|$, whereas C depends on $|\alpha|$ and on $u \in D$. Under these conditions, for any $f \in S^\infty(\mathbb{R}^n)$ the formula

$$f(A) \equiv f(\overset{1}{A_1}, \dots, \overset{n}{A_n}) = \langle \tilde{f}, Q_A \rangle,$$

defines a linear operator on a dense linear manifold $D_\infty \supset D$ invariant with respect to all operators $f(A)$, $f \in S^\infty(\mathbb{R}^n)$. Here

$$\tilde{f}(t) = (2\pi)^{-n} \int f(\xi) e^{-it\xi}\, d\xi, \quad t, \xi \in \mathbb{R}^n$$

is the Fourier transform of f,

$$Q_A \equiv A_A(t) = \exp(it_n A_n) \ldots \exp(it_1 A_1),$$

and the brackets $\langle \cdot, \cdot \rangle$ denote the pairing of distributions and test functions of t in the weak topology on D.

The operator $f(A)$ is necessarily bounded if the Fourier transform of f is a limit of compactly supported continuous functionals over the space $C(\mathbb{R}^n)$ of bounded continuous functions on $\mathbb{R}^n$ with the norm

$$\|f\|_C = \sup_{y \in \mathbb{R}^n} f(y).$$

Under our assumptions $f(A)$ can also be represented by the iterated Stieltjes integral

$$f(\overset{1}{A_1}, \ldots, \overset{n}{A_n}) = \int f(\lambda_1, \ldots, \lambda_n)\, dE_{\lambda_n}(A_n) \ldots dE_{\lambda_1}(A_1),$$

where $dE_\lambda(A_i)$ is the spectral measure of A_i and the integral is understood in the sense of the weak topology in $\mathcal{H}$. This formula can also be used for symbols $f \notin S^\infty(\mathbb{R}^n)$, but the resultant operator need not be densely defined if $f \notin S^\infty(\mathbb{R}^n)$.

Theorem IV.5 *Assume the operators $A_{k+1}, \ldots, A_n$ are pairwise commuting (i.e., their spectral families commute), and suppose that the function $f(y_1, \ldots, y_n)$ satisfies the estimates*

$$\left| \frac{\partial^{\alpha_1 + \cdots + \alpha_k}}{\partial y_1^{\alpha_1} \ldots \partial y_k^{\alpha_k}} f(y_1, \ldots, y_n) \right| \leq C(1 + |y_1| + \cdots + |y_k|)^{-k-\varepsilon}$$

for some $\varepsilon > 0$ and every $\alpha = (\alpha_1, \ldots, \alpha_k)$ with $|\alpha| = \alpha_1 + \cdots + \alpha_k \leq k + 1$. Then $f(\overset{1}{A_1}, \ldots, \overset{n}{A_n})$ is bounded in $\mathcal{H}$ (and, consequently, extends by continuity to the entire space $\mathcal{H}$).

The proof is divided into several lemmas.

Lemma IV.8 *Let $B_1, \ldots, B_m$ be pairwise commuting selfadjoint operators on a Hilbert space $\mathcal{H}$ and let the sequence $g_n(x)$ of bounded continuous functions of $(x_1, \ldots, x_m) \in \mathbb{R}^m$ be uniformly bounded (that is, $|g_n(x)| \leq M$ for all n and x) and convergent as $n \to \infty$ to some function $g(x)$ locally uniformly with respect to $x \in \mathbb{R}^m$. Then the sequence of operators $g_n(B_1, \ldots, B_m)$ strongly converges as $n \to \infty$ to $g(B_1, \ldots, B_m)$.*

Proof. Since the spectral measures

$$dE_{\lambda_1}(B_1), \dots, dE_{\lambda_m}(B_m)$$

commute, we have the bound

$$\|g_n(B_1, \dots, B_m)\| \le \sup_{x\in\mathbb{R}^m} |g_n(x)| \le M,$$

and it suffices to verify the convergence on the dense subset in $\mathcal{H}$ consisting of the vectors

$$u = E_{\Delta_1}(B_1)E_{\Delta_2}(B_2)\dots E_{\Delta_1}(B_m)v, \quad v \in \mathcal{H},$$

where $\Delta_1, \dots, \Delta_m$ are compact Borel subsets of the real axis. For any such u we have

$$\|g_n(B_1, \dots, B_m)u - g(B_1, \dots, B_m)u\| \le \sup_{x\in\Delta_1\times\dots\times\Delta_m} |g_n(x) - g(x)|,$$

and, consequently, $g_n(B_1, \dots, B_m)u \to g(B_1, \dots, B_m)u$. The lemma is proved. □

Lemma IV.9 *Suppose that the hypotheses of Theorem* IV.5 *are satisfied. Then the Fourier transform $\tilde{\varphi}_v(t_1, \dots, t_k)$ of the $\mathcal{H}$-valued function*

$$\varphi_v(y_1, \dots, y_k) = f(y_1, \dots, y_k, A_{k+1}, \dots, A_n)v, \quad v \in \mathcal{H}$$

is continuous and

$$\int_{\mathbb{R}^k} \|\tilde{\varphi}_v(t_1, \dots, t_k)\| \, dt_1 \dots dt_k \le C\|v\|,$$

where C is independent of U.

Proof. Since $f(y_1, \dots, y_n)$ is jointly continuous and bounded, it follows from Lemma IV.8 and from the estimate in Theorem IV.5 that $\varphi_v(y_1, \dots, y_k)$ is a continuous integrable function. Hence the Fourier transform

$$\tilde{\varphi}_v(t_1, \dots, t_k) = (2\pi)^{-k} \int \varphi_v(y_1, \dots, y_k) e^{-ity} \, dy_1 \dots dy_k$$

exists and is a bounded continuous $\mathcal{H}$-valued function. Furthermore, we have

$$t^\alpha \tilde{\varphi}_v(t_1, \dots, t_k) = (2\pi)^{-k} \int \left[\left(-i\frac{\partial}{\partial y}\right)^\alpha \varphi_v(y_1, \dots, y_k) \right] e^{-ity} \, dy_1 \dots dy_k$$

for any multi-index α with $|\alpha| \le k+1$ (the differentiability of $\varphi_v(y_1, \dots, y_k)$ also follows from Lemma IV.8 and, consequently, the norm of $t^\alpha \tilde{\varphi}_v$ is bounded for $|\alpha| \le k+1$ by const $\cdot \|v\|$. We now have

$$\left\| (1+t^2)^{(k+1)/2} \tilde{\varphi}_v(t) \right\| \le \text{const} \cdot \sum_{|\alpha| \le k+1} \|t^\alpha \tilde{\varphi}_v(t)\| \le C\|v\|,$$

whence the desired estimate follows immediately. □

Lemma IV.10 *Under the conditions of the theorem we have*

$$(f(\overset{1}{A_1}, \dots, \overset{n}{A_n})u, v) = \int_{\mathbb{R}^k} (\exp(it_k A_k) \dots \exp(it_1 A_1)u, \tilde{\varphi}_v(t_1, \dots, t_k))dt_1 \dots dt_k$$

(here $(\cdot, \cdot)$ is the inner product on $\mathcal{H}$).

Proof. For any function $f \in S^\infty(\mathbb{R}^n)$ the required formula can be proved by direct computation.

Later on, if $f \notin S^\infty(\mathbb{R}^n)$ still satisfies the estimates in Theorem IV.5, then we can approximate f by a sequence $f_s \in S^\infty(\mathbb{R}^n)$, $s = 1, 2, \dots$, in such a way that $f - f_s$ satisfies the same estimates with $C_s \to 0$. Passing to the limit as $s \to \infty$, we complete the proof. □

Theorem IV.5 follows from Lemmas IV.9 and IV.10.

Now let $A_1, \dots, A_n$ be a tuple of selfadjoint operators on a Hilbert space $\mathcal{H}$. We assume that these operators satisfy the conditions stated at the beginning of this subsection and also the following conditions:

(a) The operators $A_{k+1}, \dots, A_n$ are pairwise commuting ($k \le n$ is fixed).

(b) The left ordered representation $l_1, \dots, l_n$ of the Feynman tuple $(\overset{1}{A_1}, \dots, \overset{n}{A_n})$ in $S^\infty(\mathbb{R}^n)$ exists, and the operators $l_1, \dots, l_n$ are S^∞-generators in S^∞. In other words, for any two symbols $f, g \in S^\infty(\mathbb{R}^n)$ we have

$$[\![f(\overset{1}{A_1}, \dots, \overset{n}{A_n})]\!][\![g(\overset{1}{A_1}, \dots, \overset{n}{A_n})]\!] = h(\overset{1}{A_1}, \dots, \overset{n}{A_n})$$

on $\mathcal{D}$, where

$$h(y) = f(\overset{1}{l_1}, \dots, \overset{n}{l_n})(g(y)),$$

$l_1, \dots, l_n$ are linear operators on $S^\infty(\mathbb{R}^n)$, and the operator $f(\overset{1}{l_1}, \dots, \overset{n}{l_n})$ is defined on S^∞ by the formula

$$f(\overset{1}{l_1}, \dots, \overset{n}{l_n}) = \int \tilde{f}(t_1, \dots, t_n) \exp(il_n t_n) \dots \exp(il_1 t_1).$$

Condition (b) is satisfied e.g. if $A_1, \dots, A_n$ form a representation of nilpotent Lie algebra and satisfy the conditions of the Krein–Shikhvatov theorem.

Denote

$$y = (y', y''), \quad y' = (y_1, \dots, y_k), \quad y'' = (y_{k+1}, \dots, y_n),$$

and introduce similar notation for multi-indices $\alpha = (\alpha_1, \dots, \alpha_n)$. Let

$$S^m_\rho(\mathbb{R}^k, \mathbb{R}^{n-k}) \equiv S^m_\rho, \quad m \in \mathbb{R}, \quad \rho > 0,$$

be the space of functions $f(y) \in C^\infty(\mathbb{R}^n)$ such that

$$\left|\frac{\partial^{|\alpha|} f}{\partial y^\alpha}(y)\right| \leq C_\alpha(1+|y'|)^{m-\rho|\alpha'|}, \quad |\alpha| = 0, 1, 2, \ldots,$$

and let G_ρ^m be the set of operators $f(\overset{1}{A}_1, \ldots, \overset{n}{A}_n)$ with symbols $f \in S_\rho^m$. We introduce the following condition on the representation operators.

Condition IV.1 For any $m \in \mathbb{R}$, $f \in S_\rho^m$, and a permutation $\pi = (\pi_1, \ldots, \pi_n)$ of $(1, \ldots, n)$ the operator $f(\overset{\pi_1}{l_1}, \ldots, \overset{\pi_n}{l_n})$ is a pseudodifferential operator on $S^\infty(\mathbb{R}^n)$,

$$f(\overset{\pi_1}{l_1}, \ldots, \overset{\pi_n}{l_n}) = \mathcal{H}(\overset{2}{y}, \overset{1}{-i\partial/\partial y}),$$

whose symbol $\mathcal{H}(y, \xi)$ satisfies the following estimates:

$$(\mathcal{H}(y, 0) - f(y)) \in S_\rho^{m-\rho}$$

$$\left|\frac{\partial^{|\alpha+\beta|}\mathcal{H}}{\partial y^\alpha \partial \xi^\beta}(y, \xi)\right| \leq C_{\alpha\beta}(\Phi(y, \xi))^{m-\rho(|\alpha'|+|\beta''|)},$$

where $\Phi(y, \xi)$ is a function on $\mathbb{R}^{2n} \ni (y, \xi)$ such that for some $N_0 \geq 0$ the estimates

$$1 \leq \Phi(y, \xi) \leq C(1+|y'|)(1+|\xi|)^{N_0},$$

$$(\Phi(y, \xi))^{-1} \leq C(1+|y'|)^{-1}(1+|\xi|)^{N_0}$$

are valid with some constant C independent of (y, ξ).

Theorem IV.6 *The following holds under the above conditions.*

(a) *The union* $\bigcup_{m\in\mathbb{R}} G_\rho^m$ *is a filtered algebra. More precisely, if* $f_i \in S_\rho^{m_i}$, $i = 1, 2$, *then*

$$[\![f_1(\overset{1}{A}_1, \ldots, \overset{n}{A}_n)]\!][\![f_2(\overset{1}{A}_1, \ldots, \overset{n}{A}_n)]\!] = g(\overset{1}{A}_1, \ldots, \overset{n}{A}_n),$$

where $g \in S_\rho^{m_1+m_2}$. *Furthermore,*

$$(f_1(y)f_2(y) - g(y)) \in S_\rho^{m_1+m_2-\rho}.$$

(b) *For positive integer* m *let* $\mathcal{H}^m$ *be the completion of* $\mathcal{D}$ *with respect to the norm* $\| \ \|_m$ *defined by*

$$\|u\|_0 = \|u\|,$$

where $\|\cdot\|$ *is the Hilbert norm on* $\mathcal{H}$;

$$\|u\|_m = \|u\|_{m-1} + \sum_{j=1}^k \|A_j u\|_{m-1}, \quad m = 1, 2, \ldots$$

Then each operator $f(\overset{1}{A}_1, \dots, \overset{n}{A}_n) \in G^m_\rho$ *extends by continuity to a bounded operator from* $\mathcal{H}^s$ *to* $\mathcal{H}^{s+[-m]}$ *(here* $[-m]$ *is the integral part of* $-m$*) for*

$$s \geq \max(-[-m], 0).$$

In particular, the operators with symbols in S^0_ρ *are bounded in* $\mathcal{H}$.

Remark IV.7 One can define $\mathcal{H}^{-s}$ for $s \leq 0$ as the dual of $\mathcal{H}^s$ with respect to the inner product on $\mathcal{H}$. Then any operator $\hat{f} \in G^m_\rho$ has order $[-m]$ in the entire scale $\{\mathcal{H}^s\}$.

Proof of Theorem IV.6. To begin with, we formulate and prove the following auxiliary assertion.

Proposition IV.4 *Let* $f_i \in S^{m_i}_\rho$, $i = 1, 2$, *and let* $(\pi_1, \dots, \pi_n)$ *be a permutation of* $(1, \dots, n)$. *Then*

$$g \overset{\text{def}}{=} f_1(\overset{\pi_1}{l_1}, \dots, \overset{\pi_n}{l_n})(f_2(y_1, \dots, y_n)) \in S^{m_1+m_2}_\rho$$

and

$$g(y) - f_1(y) f_2(y) \in S^{m_1+m_2-\rho}_\rho.$$

Proof. It suffices to prove the latter inclusion. By Condition IV.1, we have

$$\begin{aligned} g(y) &= \mathcal{H}\left(\overset{2}{y}, -i\frac{\overset{1}{\partial}}{\partial y}\right) f_2(y) = \mathcal{H}(y, 0) f_2(y) \\ &\quad - i\int_0^1 d\tau \left\{\left\langle \mathcal{H}_\xi\left(\overset{2}{y}, -i\tau\frac{\overset{1}{\partial}}{\partial y}\right), f_{2y}(y)\right\rangle\right\} \\ &= f_1(y) f_2(y) - i\int_0^1 d\tau \left\{\left\langle \mathcal{H}_\xi\left(\overset{2}{y}, -i\tau\frac{\overset{1}{\partial}}{\partial y}\right), f_{2y}(y)\right\rangle\right\} + \psi(y) \\ &\equiv f_1(y) f_2(y) - iG(y) + \psi(y), \end{aligned}$$

where $\psi(y) \in S^{m_1+m_2-\rho}_\rho$ and the angle brackets $\langle\ ,\ \rangle$ denote summation over the coordinate indices from 1 to n. Let us estimate the derivatives of $G(y)$. Set $\varepsilon_j = 0$, $j \in \{1, \dots, k\}$, $\varepsilon_j = 1$, $j \in \{k+1, \dots, n\}$. We have

$$\left|\frac{\partial^{|\alpha+\beta|}}{\partial y^\alpha \partial \xi^\beta} \mathcal{H}_{\xi_j}(y, \tau\xi)\right| \leq C(\Phi(y, \xi))^{m_1 - \rho(|\alpha'| + |\beta''| + \varepsilon_j)}$$

uniformly with respect to $\tau \in [0, 1]$. (Here and below C denotes different constants). Using the inequalities for Φ and the consequent inequality $(\Phi(y, \xi))^{-1} \leq 1$, we obtain

$$\left|\frac{\partial^{|\alpha+\beta|}}{\partial y^\alpha \partial \xi^\beta} \mathcal{H}_{\xi_j}(y, \tau\xi)\right| \leq C(1 + |y'|)^{m_1 - \rho\varepsilon_j - \rho|\alpha'|}(1 + |\xi|)^{N_0|m_1 - \rho\varepsilon_j - \rho|\alpha'||}$$

with some constant C independent of $\tau \in [0, 1]$. Set

$$F_{\alpha,N,j}(y,\xi) = \left[\int_0^1 \frac{\partial^{|\alpha|}}{\partial y^\alpha}\mathcal{H}_{\xi_j}(y,\tau\xi)\,d\tau\right](1+\xi^2)^{-N}.$$

If N is sufficiently large, then for all $\alpha \le \alpha_0$, where α_0 is an arbitrary fixed multi-index, and for any $|\beta| = 1, 2, \dots$, we have

$$\left|\frac{\partial^{|\beta|}}{\partial \xi^\beta}F_{\alpha,N,j}(y,\xi)\right| \le C(1+|y'|)^{m_1-\rho\varepsilon_j-\rho|\alpha'|}(1+|\xi|)^{-N-1}.$$

These estimates imply that the Fourier transform $\hat{F}_{\alpha,N,j}(y,\tau)$ with respect to ξ is integrable and

$$\int |\hat{F}_{\alpha,N,j}(y,\tau)|\,d\eta \le C(1+|y'|)^{m_1-\rho\varepsilon_j-\rho|\alpha'|}.$$

We have

$$\begin{aligned} G^{(\alpha_0)} &= \sum_{j=1}^N \sum_{\beta+\gamma=\alpha_0} \frac{\alpha_0!}{\beta!\gamma!} F_{\beta,N,j}\left(\overset{2}{y}, -i\overset{1}{\frac{\partial}{\partial y}}\right)(1-\Delta)^N f_{2y_j}^{(\gamma)}(y) \\ &= \frac{1}{(2\pi)^n} \sum_{\beta+\gamma=\alpha_0} \sum_{j=1}^N \frac{\alpha_0!}{\beta!\gamma!} \hat{F}_{\beta,N,j}(y, y-\eta)(1-\Delta)^N f_{2y_j}^{(\gamma)}(\eta)\,d\eta. \end{aligned}$$

By the preceding inequality and the fact that $f_2 \in S_\rho^{m_2}$ each term in this sum can be estimated by

$$C(1+|y'|)^{m_1+m_2-\rho-\rho|\beta'|-\rho|\gamma'|} = C(1+|y'|)^{m_1+m_2-\rho-\rho|\alpha'|}.$$

Hence we have shown that $G(y) \in S_\rho^{m_1+m_2-\rho}$. Proposition IV.1 is proved. □

Now we remark that statement (a) of Theorem IV.6 is a particular case of Proposition IV.4.

Let us now prove assertion (b) of the theorem.

First, we establish the boundedness of operators with symbols in S_ρ^0.

Let $f \in S_\rho^0$. We will find a symbol $g \in S_\rho^0$ such that

$$(f(A))^* f(A) + (g(A))^* g(A) = M^2 + \psi(A), \tag{IV.7}$$

where $M = 1 + \sup_{y\in\mathbb{R}^n} |f(y)|$, $\psi(y) \in S_\rho^{-k-1}$, and the asterisk denotes the adjoint operator on $\mathcal{H}$. Equation (IV.7) implies the boundedness of $f(A)$ in $\mathcal{H}$ immediately since $\psi(A)$ is bounded by Theorem IV.5. Since $A_1, \ldots, A_n$ are selfadjoint operators, the adjoint of $f(A)$ is given on $\mathcal{D}$ by

$$(f(A))^* = \overline{f}(\overset{n}{A_1}, \ldots, \overset{1}{A_n}),$$

where $\overline{f}$ is the complex conjugate of f and the Feynman indices occur in the inverse order. If $g \in S^0_\rho$, then, by Proposition IV.4, we obtain

$$(f(A))^*(f(A)) + (g(A))^*(g(A)) = (|f|^2 + |g|^2)(A) + \psi(A), \quad \psi \in S^{-\rho}_\rho.$$

Let us solve equation (IV.7) by the method of successive approximations. Set

$$g_0(y) = \sqrt{M^2 - |f(y)|^2}.$$

Clearly, $g_0 \in S^0_\rho$ and we have (the uppercase letters stand for operators and the lower case letters for the corresponding symbols):

$$F^*F + G_0^*G_0 = M^2 + \psi_0, \quad \psi_0 \in S^{-\rho}_\rho.$$

We now set

$$g_j(y) = -\frac{1}{2}\frac{\operatorname{Re}\psi_{j-1}(y)}{g_{j-1}(y)} + g_{j-1}(y) \equiv r_j(y) + g_{j-1}(y)$$

for $j = 1, 2, \ldots$, where ψ_{j-1} is determined from the conditions

$$F^*F + G_{j-1}^*G_{j-1} = M^2 + \psi_{j-1},$$

$$\psi_j \in S^{-\rho(j+1)}_\rho, \quad \operatorname{Im}\psi_j \in S^{-\rho(j+2)}_\rho. \tag{IV.8}$$

It turns out that such a choice of ψ is always possible. Indeed, it follows from (IV.8) that $\psi_j = \psi_j(A)$ is symmetric on $\mathcal{D}$. We put $f_1 = \overline{\psi}_j(y_1, \ldots, y_n)$, $\pi_l = n + 1 - l$, $l = 1, \ldots, n$, $f_2 = 1$, and derive from Proposition IV.4 the equation

$$\psi_j(A) = \overline{\psi}_j(\overset{n}{A_1}, \ldots, \overset{1}{A_n}) = \overline{\psi}_j(\overset{1}{A_1}, \ldots, \overset{n}{A_n}) + \chi(\overset{1}{A_1}, \ldots, \overset{n}{A_n}),$$

where $\chi \in S^{-\rho(j+2)}_\rho$ provided that $\psi_j \in S^{-\rho(j+1)}_\rho$. Set

$$\psi_j'(y) = (\psi_j(y) + \overline{\psi}_j(y) + \chi(y))/2;$$

then $\psi_j'(A) = \psi_j(A)$, and $\psi_j'(A)$ satisfies both conditions in (IV.8). Thus, it suffices to find symbols ψ_j satisfying the first condition in (IV.8), and then the second condition can also be guaranteed.

The argument goes by induction over j. Suppose that the symbols g_{j-1}, ψ_{j-1} satisfying the induction hypothesis are given.

Note that g_{j-1} and r_j are real-valued. Due to the definition of $g_j(y)$, we have

$$\begin{aligned} F^*F + G_j^*G_j &= F^*F + G_{j-1}^*G_{j-1} + G_{j-1}^*R_j + R_j^*G_{j-1} + R_j^*R_j \\ &= M^2 + \psi_{j-1} + G_{j-1}^*R_j + R_j^*G_{j-1} + R_j^*R_j, \end{aligned}$$

where $r_j \in S_\rho^{-\rho_j}$, $g_{j-1} \in S_\rho^0$ and $\overline{g}_{j-1}r_j + \overline{r}_j g_{j-1} = 2g_{j-1}r_j = -\psi_{j-1}$. By virtue of Proposition IV.4,

$$F^*F + G_j^*G_j = M^2 + \psi_j, \quad \psi_j \in S_\rho^{-\rho(j+1)},$$

and the inductive step is complete.

Thus we have shown that operators with symbols in S_ρ^0 are bounded. Let us now prove that if $F \in G_\rho^m$ (m is an integer), then $F : \mathcal{H}^s \to \mathcal{H}^{s-m}$ is continuous. The following facts are evident:

(a) $A_j \in G_\rho^1$, $j = 1, \dots, k$;

(b) $A_j : \mathcal{H}^s \to \mathcal{H}^{s-1}$ are continuous for $j = 1, \dots, k$;

(c) there is an implication

$$F \in G_\rho^m \Rightarrow [A_j, F] \overset{\text{def}}{\equiv} A_jF - FA_j \in G_\rho^{m-\rho+1}, \quad j = 1, \dots, k;$$

(d) if $F \in G_\rho^0$, then F is bounded on $\mathcal{H}$;

(e) $G_\rho^m G_\rho^{m'} \subset G_\rho^{m+m'}$.

The boundedness of $F \in G_\rho^k$ from $\mathcal{H}^s$ to $\mathcal{H}^{s-m}$ follows from (a) – (e) by a standard argument, which we do not reproduce here. The theorem is proved. □

Remark IV.8 It may be difficult to check the validity of Condition IV.1. However, it is valid for the usual pseudodifferential operators (here $n = 2k$, $A_{j+k} = x_j$, $A_j = -i\partial/\partial x_j$ in $L_2(\mathbb{R}_k)$; we have $l_j = y_j - i\partial/\partial y_{k+j}$ and $l_{k+j} = y_{k+j}$, $j = 1, \dots, k$, so that

$$\mathcal{H}(y, \xi) = f(y_1 + \xi_1, \dots, y_k + \xi_k, y_{k+1}, \dots, y_n)$$

and one can set $\Phi(y, \xi) = 1 + |y' + \xi'|$, $N_0 = 1$.)

Corollary IV.3 *Let a function* $f(y^{(1)}, y^{(2)}, y^{(3)})$, $y^{(i)} \in \mathbb{R}^n$, $i = 1, 2, 3$, *satisfy the estimates*

$$\left| \frac{\partial^{|\beta+\gamma+\delta|} f}{\partial y^{(1)\beta} \partial y^{(2)\gamma} \partial y^{(3)\delta}} \right| \leq C_{\beta\gamma\delta} \left(1 + \sqrt{(y^{(1)})^2 + (y^{(2)})^2} \right)^{-|\beta|-|\gamma|},$$

$$|\beta| + |\gamma| + |\delta| = 0, 1, 2, \dots$$

Then the operator $f(\overset{1}{-i\partial/\partial\xi}, \overset{2}{\xi}, \overset{2}{\xi})$ *is bounded on* $L_2(\mathbb{R}^n)$ *and the upper bound of its norm is completely determined by the constants* $C_{\beta\gamma\delta}$.

Appendix A
Representation of Lie Algebras and Lie Groups

Here we present some well-known elementary facts concerning Lie algebras, Lie groups and their representations. For further information on these topics, see e.g. [162], [184], [86].

1 Lie Algebras and Their Representations

1.1 Lie Algebras, Bases, Structure Constants, Subalgebras

A *Lie algebra over* $\mathbb{R}$ (or $\mathbb{C}$) is a linear space L over the corresponding field endowed with a bilinear operation (the *Lie bracket* or *commutator*)

$$\begin{aligned} [\,,\,] \;:\; L \times L &\longrightarrow L, \\ (x, y) &\longmapsto [x, y], \end{aligned}$$

such that

$$[x, y] = -[y, x] \qquad x, y \in L \tag{A.1}$$

(*Antisymmetry*) and

$$[[x, y], z] + [[y, z], x] + [[z, x], y] = 0, \quad x, y, z \in L \tag{A.2}$$

(*Jacobi identity*).

Let L be a Lie algebra of finite dimension n. Let $\{a_1, \dots, a_n\}$ be a basis in L. The commutator of any two elements of the basis may be expanded in this basis,[1]

$$[a_j, a_k] = \lambda^l_{jk} a_l.$$

The numbers λ^l_{jk} are called the *structure constants* of the algebra L with respect to the basis $\{a_1, \dots, a_n\}$. Clearly, for any elements $x = x^j a_j$, $y = y^j a_j \in L$, we have

$$[x, y] = \lambda^l_{jk} x^j y^k a_l. \tag{A.3}$$

---

[1] Here and below we use the *summation convention.* If a superscript coincides with a subscript in a product, then summation is performed over this index. In particular, $\lambda^l_{jk} a_l$ means $\sum_l \lambda^l_{jk} a_l$.

A *Lie subalgebra* in L is a linear subspace $W \subseteq L$ such that $[W, W] \subset W$. In particular, W itself is a Lie algebra.

1.2 Examples of Lie Algebras

The simplest example of a Lie algebra is an *abelian algebra.* One takes any linear space L and defines the Lie bracket on L by the equality

$$[x, y] = 0, \quad x, y \in L.$$

However, Lie algebras induced by associative algebras are of particular interest for us. Let $\mathcal{A}$ be an associative algebra. We define the Lie bracket on $\mathcal{A}$ by setting

$$[x, y] = xy - yx, \quad x, y \in \mathcal{A} \tag{A.4}$$

that is, $[x, y]$ is the conventional commutator of elements of $\mathcal{A}$. The antisymmetry and the Jacobi identity can be easily verified for this definition of Lie bracket. Thus the commutator (A.3) defines the structure of a Lie algebra on $\mathcal{A}$. By taking Lie subalgebras in $\mathcal{A}$, one can provide a large number of examples.

Let $\mathcal{A} = \mathrm{Mat}_n(\mathbb{R})$ be the algebra of real $(n \times n)$ matrices. The corresponding Lie algebra is denoted by $\mathrm{gl}(n, \mathbb{R})$.

There is a Lie subalgebra $\mathrm{so}(n, \mathbb{R})$ in $\mathrm{gl}(n, \mathbb{R})$ consisting of the skew-symmetric matrices, that is, matrices $\mathcal{A}$ such that

$${}^tA = -A$$

(here, as above, ${}^tA$ denotes the transpose of the matrix A). Indeed, we have

$$\begin{aligned} {}^t[A, B] &= {}^t(AB - BA) = {}^tB^tA - {}^tA^tB \\ &= [{}^tB, {}^tA] = [B, A] = -[A, B] \end{aligned} \tag{A.5}$$

so that $\mathrm{so}(n, \mathbb{R})$ is closed under commutation.

Let $\mathcal{A}$ denote the algebra of all differential operators in $\mathbb{R}^n$ with smooth real coefficients. Its elements have the form

$$D = \sum_{|\alpha|=0}^{m(D)} a_\alpha(x) \left(\frac{\partial}{\partial x}\right)^\alpha,$$

where $m(D)$ is the *order* of the operator D. It is clear that

$$m([D_1, D_2]) \leq m(D_1) + m(D_2) - 1 \tag{A.6}$$

so that the set $\mathcal{A}_1 \subset \mathcal{A}$ of differential operators of order $m(D) \leq 1$ is a Lie subalgebra of $\mathcal{A}$.

In turn, one may consider the subalgebra of $\mathcal{A}_1$ consisting of operators of the form

$$D = a_i x^i + b^j \frac{\partial}{\partial x^j} + \lambda. \tag{A.7}$$

This is a Lie algebra of dimension $2n + 1$. It is called the *Heisenberg algebra* and denoted by $\mathrm{hb}(n)$. The elements

$$x^1, \dots, x^n, \quad \frac{\partial}{\partial x^1}, \dots, \frac{\partial}{\partial x^n}, 1$$

form a basis in $\mathrm{hb}(n)$, with the commutation relations given by

$$\begin{aligned} &[x^j, x^k] = [\frac{\partial}{\partial x^j}, \frac{\partial}{\partial x^k}] = 0, \\ &[x^j, \frac{\partial}{\partial x^k}] = -\delta^j_k 1, \\ &[x^j, 1] = [\frac{\partial}{\partial x^k}, 1] = 0, \qquad j, k = 1, \dots, n. \end{aligned} \tag{A.8}$$

Here δ^j_k is the Kronecker symbol.

Let now $\mathcal{A}$ be some algebra. Consider the space $\mathrm{Der}(\mathcal{A})$ of linear mappings $D : \mathcal{A} \to \mathcal{A}$ such that

$$d(u, v) = uD(v) + D(u)v\,, \quad u, v \in \mathcal{A}$$

(the "Leibniz rule"). A mapping D satisfying (A.4) is called a *derivation* of the algebra $\mathcal{A}$. Let D_1 and D_2 are derivations of $\mathcal{A}$. Then, as one can easily verify, their commutator is also a derivation. Thus, $\mathrm{Der}(\mathcal{A})$ is a Lie algebra.

Take now $\mathcal{A} = C^\infty(M)$, the algebra of smooth functions on a manifold M, with the usual pointwise multiplication. Its derivations are nothing other than vector fields on M, so that

$$\mathrm{Der}(\mathcal{A}) = \mathrm{Vect}(M)$$

is the algebra of vector fields of the manifold M, with the Lie bracket given by the commutator of vector fields.

1.3 Homomorphisms, Ideals, Quotient Algebras

Let L_1 and L_2 be Lie algebras. A linear mapping

$$\varphi : L_1 \to L_2$$

is called a *homomorphism* (of Lie algebras) if

$$\varphi([x, y]) = [\varphi(x), \varphi(y)] \tag{A.9}$$

for any $x, y \in L_1$.

The kernel

$$\mathcal{G} = \operatorname{Ker} \varphi = \{x \in L_1 | \varphi(x) = 0\} \tag{A.10}$$

of a homomorphism φ is an *ideal* of L_1, that is, a linear subspace with the property

$$[x, y] \in \mathcal{G}, \quad \text{whenever} \quad x \in L_1 \quad \text{and} \quad y \in \mathcal{G}. \tag{A.11}$$

This is a direct consequence of formula (A.2).

If $J \subset L$ is an ideal of a Lie algebra L, the quotient space L/J is naturally endowed with the structure of a Lie algebra. Indeed, it follows from (A.11) that the equivalence class of $[x, y]$ depends only on those of x and y. The Lie algebra L/J is called the *quotient algebra of the Lie algebra* L with respect to the ideal J.

1.4 Representations

A *representation* is a homomorphism of a special type. Namely, if L is a Lie algebra and V is a linear space, any homomorphism

$$\varphi : L \longrightarrow \operatorname{End}(V),$$

(where $\operatorname{End}(V)$ is the algebra of linear operators on V), that is, a linear mapping such that

$$\varphi([x, y]) = \varphi(x)\varphi(y) - \varphi(y)\varphi(x),$$

is called a *representation* of the Lie algebra L in the linear space V. The representation is said to be *faithful* if $\operatorname{Ker} \varphi = \{0\}$.

Examples of Representations. The algebras $\mathrm{gl}(n, \mathbb{R})$, $\mathrm{so}(n, \mathbb{R}) \subset \mathrm{gl}(n, \mathbb{R})$, obviously act on the space $\mathbb{R}^n$ by multiplication:

$$\varphi(a)x = ax,$$
$$a \in \mathrm{gl}(n, \mathbb{R}), \quad x \in \mathbb{R}^n,$$

which gives us natural representations of these algebras in the linear space $\mathbb{R}^n$.

The algebra $\mathrm{Vect}(M)$ of vector fields on a manifold M possesses a natural representation φ in the space $C^\infty(M)$ of smooth function on M; this representation is given by

$$\varphi(X)\,(f(x)) = Xf\,(x).$$

The Heisenberg algebra $\mathrm{hb}(n, \mathbb{R})$ possesses a natural representation in the space $C^\infty(\mathbb{R})$. Elements of $\mathrm{hb}(n, \mathbb{R})$ act as the corresponding differential operators:

$$\varphi\left(\frac{\partial}{\partial x^j}\right) f(x) \quad = \quad \frac{\partial f}{\partial x^j},$$

$$\varphi(x^j)f(x) = x^j f(x), \quad j = 1, \dots, n,$$
$$\varphi(1)f(x) = f(x).$$

1.5 The Associated Representation ad. The Center of a Lie Algebra

Let L be a Lie algebra. Its *associated representation* is the representation in the linear space L given by the formula

$$\begin{aligned} \mathrm{ad} : L &\longrightarrow \mathrm{End}(L), \\ x &\longmapsto \mathrm{ad}_x, \end{aligned} \tag{A.12}$$

where

$$\mathrm{ad}_x(y) = [x, y].$$

The mapping (A.12) is necessarily a representation since by the Jacobi identity we have

$$\begin{aligned} \mathrm{ad}_{[x,y]}(z) &= [[x, y], z] = -[[y, z], x] - [[z, x], y] \\ &= [x, [y, z]] - [y, [x, z]] = [\mathrm{ad}_x, \mathrm{ad}_y](z). \end{aligned} \tag{A.13}$$

The kernel $Z \subset L$ of the representation ad consists of all $x \in L$ satisfying the property $[x, y] = 0$ for all $y \in L$. Z is called the *center* of the Lie algebra L.

Let L be a Lie algebra of finite dimension n. Let us fix some basis $\{a_1, \dots, a_n\}$ in L. Let $x = x^i a_i$. The operator ad_x in this basis is given by the matrix

$$\mathrm{ad}_x = x^{i\Lambda_s},$$

where

$$(\Lambda_s)_j^k = \lambda_{sj}^k,$$

and λ_{sj}^k are the structure constants of L with respect to the basis $\{a_1, \dots, a_n\}$.

1.6 The Ado Theorem

The representation ad of a finite-dimensional Lie algebra L is a representation of L in a finite-dimensional space.

Generally, this representation is not faithful (unless the center of L is trivial). However, the following theorem is valid.

Theorem A.1 *Any finite-dimensional Lie algebra possesses faithful finite-dimensional representations.*

Thus we may consider any finite-dimensional Lie algebra as a matrix Lie algebra (though the size of the matrices may be very large).

The proof of the Ado theorem is very complicated, and we do not consider it here.

1.7 Nilpotent Lie Algebras

A Lie algebra L is said to be *nilpotent* if for any $x \in L$ the operator ad_x is nilpotent (that is,

$$(\mathrm{ad}_x)^N = 0$$

for some N).

The *Engel theorem* says that a Lie algebra is nilpotent if and only if there exists a number N such that for $x_1, x_2, \dots, x_N \in L$ one has

$$[\dots[[x_1, x_2], x_3], \dots, x_N] = 0.$$

The minimal N with this property is called the *nilpotency rank* of L.

Let L be a nilpotent Lie algebra. Let us define a decreasing sequence of ideals in L by setting

$$L^{(0)} = L, L^{(1)} = [L, L^{(0)}], L^{(2)} = [L, L^{(1)}], \dots, L^{(k)} = [L, L^{(k-1)}].$$

We have

$$L = L^{(0)} \supset L^{(1)} \supset L^{(2)} \supset \dots \supset L^{(k-1)} \supset L^{(k)} \supset \dots.$$

It is easy to see that $L^{(N)} = \{0\}$, where N is the nilpotency rank of L. Let $\dim L^{(k)} = r_k,\ r_0 = n$. Let us choose a basis in L, by using the following procedure: choose some basis $\{a_1, \dots, a_{r_{N-1}}\}$ in $L^{(N-1)}$, extend it to a basis $\{a_1, \dots, a_{r_{N-2}}\}$ in $L^{(N-2)}$ and so on, until a basis $\{a_1, \dots, a_n\}$ in L is obtained. Now set

$$b_j = a_{n+1-j}, \quad j = 1, \dots, n.$$

It is clear that

$$[b_j, b_k] = \sum_{l > \max(j,k)} \lambda^l_{jk} b_l;$$

thus, $\lambda^l_{jk} = 0$ for $k \geq l$. In other words, all matrices Λ_j, $j = 1, \dots, n$, of the associated representation are strictly upper-triangular in the basis $\{b_1, \dots, b_n\}$.

2 Lie Groups and Their Representations

2.1 Lie Groups, Subgroups, the Gleason–Montgomery–Zippin Theorem

A *Lie group* is a smooth manifold G together with a group structure such that the group operations

$$G \times G \longrightarrow G, \quad (x, y) \mapsto xy;$$
$$G \longrightarrow G, \quad x \longmapsto x^{-1}$$

are smooth mappings of manifolds.

The *Gleason–Montgomery–Zippin theorem* states that it suffices to require that the manifold and the group operations be of class C^0 in the above definition. Under this condition, one may define the structure of a real-analytic manifold on G such that the group operations are real-analytic mappings.

Let G be a Lie group. A *Lie subgroup* in G is an embedded submanifold $N \subset G$ which is at the same time a subgroup of G.

2.2 Examples of Lie Groups

The simplest example of a Lie group is the real line $\mathbb{R}$ with the multiplication $(x, y) \mapsto x + y$ and the inverse element $x^{-1} = -x$. It is easy to check all the conditions of the definition.

Another example is the group T of unimodular complex numbers with the usual multiplication. The mapping

$$\varphi \longmapsto e^{i\varphi} \tag{A.14}$$

introduces the (local) coordinate φ on T, and the group operations may be written in the form

$$\begin{aligned} \varphi_1, \varphi_2 \quad &\longmapsto \quad \varphi_1 + \varphi_2 \pmod{2\pi} \\ \varphi \quad &\longmapsto \quad -\varphi. \end{aligned}$$

Let us now consider the space $\mathbb{R}^{2n+1}$ with the coordinates $y^1, \dots, y^n, z_1, \dots, z_n$, c and define the multiplication law by setting

$$\begin{aligned} &(y^{(1)}, z^{(1)}, c^{(1)})\,(y^{(2)}, z^{(2)}, c^{(2)}) \\ &= (y^{(1)} + y^{(2)}, z^{(1)} + z^{(2)}, c^{(1)} + c^{(2)} + z^{(1)} y^{(2)}), \end{aligned} \tag{A.15}$$

where $z \cdot y = z_i y^i$. We leave to the reader the easy verification of the fact that the multiplication law (A.15) is associative. Further, this law possesses the neutral element $(0, 0, 0)$, the inverse element always exists and is given by $(y, z, c)^{-1} = (-y, -z, -c - zy)$. Thus, $\mathbb{R}^{2n+1}$ with this operation is a Lie group. It will be denoted by $\mathrm{HB}(n, \mathbb{R})$ and called the *Heisenberg group*.

Consider now the set of nondegenerate $n \times n$ matrices with real elements. (Of course, they can be identified with nondegenerate linear operators in $\mathbb{R}^n$.) It is clear that this set is in fact a Lie group (the product is the conventional matrix product, whereas the smooth structure is inherited from the space $\mathbb{R}^{n^2}$, in which nondegenerate matrices form an open subset). This Lie group is denoted by $\mathrm{gl}(n, \mathbb{R})$ and called the *full linear group*. The group $\mathrm{gl}(n, \mathbb{R})$ and its subgroups are called *matrix Lie groups*.

As an example, consider the group $\mathrm{SO}(n, \mathbb{R}) \subset \mathrm{gl}(n, \mathbb{R})$ of matrices $\mathcal{A}$ such that

$$\begin{aligned} &\det A = 1, \\ &{}^t A\, A = E \end{aligned}$$

(E is the identity matrix). Thus, $\mathrm{SO}(n, \mathbb{R})$ consists of orthogonal matrices with unit determinant. The group $\mathrm{SO}(n, \mathbb{R})$ is called the *special orthogonal group*.

2.3 Local Lie Groups

Let $(U, \varphi : U \to \mathbb{R}^n)$ be a coordinate chart on a Lie group G in a neighborhood of the unit element $e \in G$. We assume that $\varphi(e) = 0$. Then the group operations are represented by the mappings

$$\begin{aligned} \psi :&\quad V \times V \longrightarrow \mathbb{R}^n, \\ \varepsilon :&\quad V \longrightarrow \mathbb{R}^n \end{aligned} \tag{A.16}$$

in some neighborhood $V \subset \varphi(U)$ of the point $0 \in \mathbb{R}^n$, and

$$\begin{aligned} &\psi(0, x) = \psi(x, 0) = x, \ \varepsilon(0) = 0, \\ &\psi(x, \psi(y, z)) = \psi(\psi(x, y), z), \\ &\psi(x, \varepsilon(x)) = \psi(\varepsilon(x), x) = 0 \end{aligned} \tag{A.17}$$

(these equalities are satisfied whenever all the expressions involved are defined).

A neighborhood of zero $V \subset \mathbb{R}^n$ together with a pair of mappings (A.16) satisfying (A.17) is called a *local Lie group*. Thus, any Lie group gives a local Lie group via the above construction. It turns out that, vice versa, any local Lie group gives rise to a Lie group. We do not prove the latter statement here.

2.4 Homomorphisms of Lie Groups, Normal Subgroups, Quotient Groups

Let G_1 and G_2 be Lie groups. The mapping

$$\varphi : G_1 \longrightarrow G_2$$

is called a *homomorphism of Lie groups*, if φ is a smooth mapping of manifolds an a group homomorphism at the same time.

If $\varphi : G_1 \to G_2$ is a Lie group homomorphism, its kernel

$$N = \operatorname{Ker} \varphi = \{x \in G_1 | \varphi(x) = e\}$$

is a Lie subgroup in G_1. It is a *normal* subgroup, that is, $x \in N$ implies $yxy^{-1} \in N$ for any $y \in G$. Indeed, if $x \in N$ then

$$\begin{aligned} &\varphi(yxy^{-1}) = \varphi(y)\varphi(x)\varphi(y^{-1}) = \varphi(y)\, e\, \varphi(y^{-1}) \\ &= \varphi(y)\varphi(y^{-1}) = \varphi(yy^{-1}) = \varphi(e) = e. \end{aligned}$$

The quotient group G/N clearly possesses the structure of a smooth manifold concordant with the group structure, so that G/N is a Lie group.

3 Left and Right Translations. The Haar Measure

Let G be a Lie group. Then for any $g \in G$ the following mappings are defined:

$$\begin{aligned} L_g : \quad & G \longrightarrow G, \\ & h \longmapsto gh \end{aligned}$$

(left translation), and

$$\begin{aligned} R_g : \quad & G \longrightarrow G, \\ & h \longmapsto hg \end{aligned}$$

(right translation). Both mappings are diffeomorphisms.

The mappings (A.10), (A.11) induce the corresponding mappings of tangent and cotangent spaces,

$$\begin{aligned} L_{g*} &: T_h G \longrightarrow T_{gh} G, \\ R_{g*} &: T_h G \longrightarrow T_{hg} G \end{aligned}$$

and

$$\begin{aligned} L_g^* &: T_{gh}^* G \longrightarrow T_h^* G, \\ R_g^* &: T_{hg}^* G \longrightarrow T_h^* G \end{aligned}$$

for any $h \in G$.

A nondegenerate volume form dw on G is called a *right (left) Haar measure* on G, if it is right- (left-)invariant, that is,

$$R_g^*(dw) = dw, \quad \text{or} \quad L_g^*(dw) = dw,$$

respectively, for any $g \in G$.

Right and left Haar measures on a Lie group G always exist and are defined up to a nonzero factor. Indeed, we have

$$(d_r g)_h = \left(R_h^*\right)^{-1} (d_r g)_e \quad \text{or} \quad (d_l g)_h = \left(L_h^*\right)^{-1} (d_l g)_e \,.$$

Here d_r is a right Haar measure, $d_l g$ is a left Haar measure, e is the unit element of the group.

3.1 Left and Right Regular Representations

Let G be a Lie group, C^∞ the space of smooth functions of G. For any $g \in G$ let us define the operators $\mathcal{L}_g$ and $\mathcal{R}_g$ in $C^\infty(G)$ by setting

$$\begin{aligned} \left(\mathcal{L}_g f\right)(h) &= f(g^{-1}h), \\ \left(\mathcal{R}_g f\right)(h) &= f(hg) \end{aligned}$$

for any $h \in G$. The introduced operators possess the following properties:

$$\begin{aligned} \mathcal{L}_e &= \mathcal{R}_e = \mathrm{id}; \\ \left(\mathcal{L}_{g_1}\mathcal{L}_{g_2} f\right)(h) &= \left(\mathcal{L}_{g_1} f\right)(g_2^{-1}h) = f\left(g_2^{-1}g_1^{-1}h\right) \\ &= f\left((g_1g_2)^{-1}h\right) = \left(\mathcal{L}_{g_1g_2} f\right)(h), \\ \left(\mathcal{R}_{g_1}\mathcal{R}_{g_2} f\right)(h) &= \left(\mathcal{R}_{g_1} f\right)(hg_2) = f(hg_1g_2) \\ &= \left(\mathcal{R}_{g_1g_2} f\right)(h). \end{aligned}$$

Thus, the mappings

$$\begin{aligned} \mathcal{L} &: g \longmapsto \mathcal{L}_g, \\ \mathcal{R} &: g \longmapsto \mathcal{R}_g \end{aligned}$$

are homomorphisms of G into the group $\mathrm{Aut}(C^\infty(G))$ of automorphisms of the linear space $C^\infty(G)$.

These homomorphisms are called the *left* and the *right regular representation*, respectively.

3.2 Representations of Lie Groups

The homomorphisms considered in the preceding subsection are nothing other than special examples of representations of Lie groups.

A *representation* of a Lie group G in a linear space V is a homomorphism

$$T : G \longrightarrow \mathrm{Aut}(V)$$

from the group G to the group of automorphisms of V.

We assume that V is a topological linear space and that $\mathrm{Aut}(V)$ is the set of linear continuous operators on V with continuous inverses. Further, we introduce a certain topology in the set $\mathrm{Aut}(V)$, and the mapping T is assumed to be continuous in this topology.

The following situation is of particular interest for us. Let V be a Banach space, $\mathrm{Aut}(V)$ be a set of bounded operators in V with bounded inverses, with the topology of strong convergence. The corresponding representations are called *strongly continuous representations of the group G in the Banach space V*. In particular,

(1) the operators $T(g)$ are bounded operators in V for all $g \in G$;

(2) $T(e) = \mathrm{id}_V$; $T(g_1g_2) = T(g_1)T(g_2)$ for any $g_1, g_2 \in G$;

(3) for any $v \in V$ the V-valued vector function $g \mapsto T(g)\,v$ on G is continuous on G with respect to the norm in V.

A representation T is said to be *faithful* if $T(g) \neq \mathrm{id}_V$ for any $g \neq e$.

Examples of Representations. We consider only two examples here.

A. Any matrix group acts in $\mathbb{R}^n$ by left multiplications.

B. Consider the Heisenberg group $\mathrm{HB}(n, \mathbb{R})$. We define the mapping

$$T_\lambda : \mathrm{HB}(n, \mathbb{R}) \longrightarrow \mathrm{Aut}(L^2(\mathbb{R}^n)) \tag{A.18}$$

by the formulas

$$(T_\lambda(y, z, c)f)\ (x) = e^{i\lambda(yx+c)} f(x + z), \tag{A.19}$$

where $\lambda \in \mathbb{R}$ is a fixed number.

The mapping T_λ is a strongly continuous representation of the group $\mathrm{HB}(n, \mathbb{R})$ in the space $L^2(\mathbb{R}^n)$. Indeed, we have

$$||T_\lambda(y, z, c)||_{L_2(\mathbb{R}^n)\to L_2(\mathbb{R}^n)} = 1 \tag{A.20}$$

for any $(y, z, c) \in \mathrm{HB}(n, \mathbb{R})$.

Further, if $f(x)$ is a smooth finite function, it is obvious that formula (A.19) defines an element of the space $L^2(\mathbb{R})$ which depends on $(y, z, c) \in \mathrm{HB}(n, \mathbb{R})$ continuously.

Since $C_0^\infty(\mathbb{R}^n)$ is dense in $L^2(\mathbb{R}^n)$, and the norms of the operators (A.18) are uniformly bounded (A.20), one may conclude that the element (A.19) of $L^2(\mathbb{R}^n)$ depends on $(y, z, c) \in \mathrm{HB}(n, \mathbb{R})$ continuously for any $f \in L^2(\mathbb{R}^n)$.

Finally, $T_\lambda(0, 0, 0)$ is the identity operator in $f \in L^2(\mathbb{R}^n)$, and

$$T_\lambda(g)\, T_\lambda(h) = T_\lambda(gh),$$

as one can easily verify. Thus our statement is proved.

4 The Relationship between Lie Groups and Lie Algebras

4.1 The Lie Algebra of a Lie Group

Let G be a Lie group. A vector field $X \in \mathrm{Vect}(G)$ is said to be *right-invariant* if

$$R_{g*}X_h = X_{gh} \tag{A.21}$$

for any $g, h \in G$. It is easy to see that right-invariant vector fields form a finite-dimensional linear space naturally isomorphic to the tangent space T_eG of the group G at the point e. Indeed, (A.21) implies

$$X_g = R_{g*}X_e$$

for any $g \in G$, thus proving the claimed isomorphism.

The commutator of right-invariant vector fields is itself a right-invariant vector field. Indeed, let us consider any field X as an operator $X : C^\infty(G) \to C^\infty(G)$. The fact that X is right-invariant means exactly that

$$X\mathcal{R}_g = \mathcal{R}_g X$$

for any $g \in G$. Now let X_1, X_2 be right-invariant vector fields. We have

$$\begin{aligned}[X_1, X_2]\mathcal{R}_g &= X_1X_2\mathcal{R}_g - X_2X_1\mathcal{R}_g \\ &= \mathcal{R}_g X_1X_2 - \mathcal{R}_g X_2X_1 \\ &= \mathcal{R}_g[X_1, X_2],\end{aligned}$$

so that $[X_1, X_2]$ is also right-invariant.

Thus, right-invariant vector fields on G form an n-dimensional Lie algebra $\mathcal{G}$. It will be called the *Lie algebra of the Lie group* G. We use the above isomorphism to identify $\mathcal{G}$ with the tangent space T_eG.

4.2 Examples

A. Let us construct the Lie algebra of the group $\mathbb{R}$. Any vector field on $\mathbb{R}$ has the form

$$X = a(x)\frac{d}{dx}.$$

Right translations act in $\mathbb{R}$ according to the formula

$$R_y x = x + y.$$

Hence

$$\left(R_{g*}X\right)_{x+y} = a(x)\frac{d}{dx}$$

and the right-invariance condition means that

$$a(x + y) = a(x),$$

that is, $a(x)$ is a constant. We see that any element of the considered Lie algebra has the form

$$X = a\frac{d}{dx}, \quad a = \text{const},$$

and the Lie algebra is one-dimensional, with the trivial Lie bracket.

B. Let us now consider the Lie algebra of the group $\mathrm{HB}(1, \mathbb{R})$. Recall that the standard coordinates on $\mathrm{HB}(1, \mathbb{R})$ were denoted by (y, z, c). Any vector field on $\mathrm{HB}(1, \mathbb{R})$ has the form

$$X = a_1(y, z, c)\frac{\partial}{\partial y} + a_2(y, z, c)\frac{\partial}{\partial z} + a_3(y, z, c)\frac{\partial}{\partial c}.$$

Right translations on $\mathrm{HB}(1,\mathbb{R})$ are given by

$$R_{(\tilde{y},\tilde{z},\tilde{c})}(y,z,c) = (y+\tilde{y}, z+\tilde{z}, c+\tilde{c}+z\tilde{y}),$$

so that the matrix of the operator $R_{(\tilde{y},\tilde{z},\tilde{c})*}$ has the form

$$R_{(\tilde{y},\tilde{z},\tilde{c})*} = \begin{pmatrix} 1 & 0 & 0 \\ 0 & 1 & 0 \\ 0 & \tilde{y} & 1 \end{pmatrix}.$$

Thus,

$$\left(R_{(\tilde{y},\tilde{z},\tilde{c})*}X\right)_{(y+\tilde{y},z+\tilde{z},c+\tilde{c}+z\tilde{y})} =$$
$$= a_1(y,z,c)\frac{\partial}{\partial y} + a_2(y,z,c)\frac{\partial}{\partial z} + (a_3(y,z,c)+\tilde{y}a_2(y,z,c))\frac{\partial}{\partial c}.$$

The right-invariance conditions imply

$$\begin{aligned} a_1(y+\tilde{y}, z+\tilde{z}, c+\tilde{c}+z\tilde{y}) &= a_1(y,z,c), \\ a_2(y+\tilde{y}, z+\tilde{z}, c+\tilde{c}+z\tilde{y}) &= a_2(y,z,c), \\ a_3(y+\tilde{y}, z+\tilde{z}, c+\tilde{c}+z\tilde{y}) &= a_3(y,z,c)+\tilde{y}a_2(y,z,c), \end{aligned}$$

and so

$$\begin{aligned} a_1 &= \mathrm{const}, \\ a_2 &= \mathrm{const}, \\ a_3 &= ya_2 + b, \quad b = \mathrm{const}\,. \end{aligned}$$

Thus, any right-invariant vector field X on $\mathrm{HB}(1,\mathbb{R})$ has the form

$$X = a_1\frac{\partial}{\partial y} + a_2\left(\frac{\partial}{\partial z} + y\frac{\partial}{\partial c}\right) + b\frac{\partial}{\partial c}$$

with some constants a_1, a_2 and b.

Example A.1 The Lie algebra of $\mathrm{HB}(1,\mathbb{R})$ is isomorphic to $\mathrm{hb}(1,\mathbb{R})$.

4.3 The Exponential Mapping, One-Parameter Subgroups, Coordinates of I and II Genera

Let G be a Lie group, $\mathcal{G}$ its Lie algebra, $X \in \mathcal{G}$ an arbitrary element, considered as a right-invariant vector field on G. Consider an ordinary differential equation on G,

$$\dot{g} = X_g \tag{A.22}$$

with the initial data

$$g(0) = e. \tag{A.23}$$

Lemma A.1 *The Cauchy problem* (A.22) – (A.23) *possesses a unique solution* $g(\tau)$ *defined for all* $\tau \in \mathbb{R}$ *and satisfying*

$$g(\tau)\, g(t) = g(t+\tau) \tag{A.24}$$

for all $t, \tau \in \mathbb{R}$.

Proof. The existence theorem for ordinary differential equations guarantees that the solution of (A.22) – (A.23) exists on the interval $\tau \in (-\varepsilon, \varepsilon)$. The curve $g(t+\tau) = \tilde{g}(\tau)$ satisfies the equation (A.22) with the initial condition $\tilde{g}(0) = g(t)$. Since X is right-invariant, the curve

$$\tilde{\tilde{g}}(\tau) = g(\tau)\, g(t) = \mathcal{R}_{g(t)} g(\tau)$$

satisfies the same equation, and the same initial condition. By the uniqueness theorem, $\tilde{\tilde{g}}(\tau) = \tilde{g}(\tau)$ on their common domain. Thus, (A.24) is valid for $|t|, |\tau|, |t+\tau| < \varepsilon$. The identity (A.24) allows us to extend the definition of $g(\tau)$ for all $\tau \in \mathbb{R}$, by setting

$$g(\tau) = \left(g\left(\frac{\tau}{N}\right)\right)^N$$

for sufficiently large integer N. It is easy to verify that (A.22) remains valid, the lemma being thereby proved.

A curve $g : \mathbb{R} \to G$ satisfying the conditions (A.23) – (A.24) is called a *one-parameter subgroup* of G; the one-parameter subgroup constructed in the above lemma is said to *correspond to the element* $X \in \mathcal{G}$. It will be denoted by $\mathcal{U}_X(t)$.

Let us now define the *exponential mapping*

$$\exp : \mathcal{G} \to G,$$

by setting

$$\exp(X) = \mathcal{U}_X(1).$$

It is clear that exp is a smooth mapping. Let us show that this mapping is nondegenerate in a neighborhood of zero. To do this, let us compute the derivative

$$\exp_*(0) : T_0\mathcal{G} \longrightarrow T_eG.$$

The tangent space $T_0\mathcal{G}$ may be identified with $\mathcal{G} = T_eG$. Then, for any

$$X \in T_eG,$$

we have

$$\begin{aligned} \exp_*(0)X &= \frac{d}{dt}\exp(tX)\Big|_{t=0} = \frac{d}{dt}\mathcal{U}_{tX}(1)\Big|_{t=0} \\ &= \frac{d}{dt}\mathcal{U}_X(t)\Big|_{t=0} = \dot{\mathcal{U}}_X(0) = X, \end{aligned}$$

so that $\exp_*(0)$ is the identity operator. Thus, exp is a diffeomorphism from a neighborhood of zero in $\mathcal{G}$ to a neighborhood of e in G.

Special coordinate systems in G are related to the mapping exp. Let $\{a_1, \ldots, a_n\}$ be a basis in $\mathcal{G}$. The coordinates of the I (first) genus are defined in the following way: the element $\exp(x^i a_i) \in G$ is considered as having the coordinates $x = (x^1, \ldots, x^n)$. Since exp is a nondegenerate mapping, we obtain a coordinate system in a neighborhood of $e \in G$.

Next we define the coordinates of the II (second) genus. Consider the mapping

$$\begin{aligned} \exp_2 : \mathbb{R}^n &\longrightarrow G \\ x &\longmapsto \exp(x^n a_n) \ldots \exp(x^1 a_1). \end{aligned} \tag{A.25}$$

The derivative $\exp_2(0)_*$ is nondegenerate since it takes the vectors

$$\frac{\partial}{\partial x^i}, \ldots, \frac{\partial}{\partial x^n}$$

to the vectors $a_1, \ldots, a_n$ which are linearly independent. Consequently, the mapping (A.25) defines a coordinate system in a neighborhood of the point $e = \exp_2(0)$.

We point out that the coordinates of both I and II genera on the group G depend on the choice of the basis $\{a_1, \ldots, a_n\}$ in its Lie algebra $\mathcal{G}$.

4.4 Evaluating the Commutator with the Help of the Mapping exp

Let G be a Lie group. Let $X, Y \in \mathcal{G}$ be elements of the corresponding Lie algebra. Then the following formula for the commutator is valid:

$$[X, Y] = \left(\frac{d}{dt} \exp(\sqrt{t}X) \exp(\sqrt{t}Y) \exp(-\sqrt{t}X) \exp(-\sqrt{t}Y)\right)\bigg|_{t=0}. \tag{A.26}$$

Indeed, let us consider X, Y as vector fields on G, and consider the following linear operator in $C^\infty(G)$

$$B(\tau) = e^{\tau X} e^{\tau Y} e^{-\tau X} e^{-\tau Y}.$$

Here $e^{\tau X}$ is the operator of left translation in $C^\infty(G)$,

$$\left(e^{\tau X} f\right)(g) = f(\exp(\tau X) g), \quad g \in G$$

and consequently, we have

$$\frac{d}{d\tau} e^{\tau X} = X e^{\tau X}.$$

Furthermore, we have

$$\begin{aligned} \frac{d}{d\tau} B(\tau) &= X e^{\tau X} e^{\tau Y} e^{-\tau X} e^{-\tau Y} + e^{\tau X} Y e^{\tau Y} e^{-\tau X} e^{-\tau Y}, \\ &- e^{\tau X} e^{\tau Y} X e^{-\tau X} e^{-\tau Y} - e^{\tau X} e^{\tau Y} e^{-\tau X} Y e^{-\tau Y}; \end{aligned}$$

and hence

$$\left.\frac{d}{d\tau}B(\tau)\right|_{\tau=0} = 0.$$

Differentiating once more with respect to τ we obtain in a similar way

$$\left.\frac{d^2}{d\tau^2}B(\tau)\right|_{\tau=0} = 2[X, Y].$$

Finally, we have proved that

$$\left.\frac{d}{dt}B(\sqrt{t})\right|_{t=0} = [X, Y],$$

which implies (A.26).

4.5 Derived Homomorphisms

Let

$$\varphi : G_1 \longrightarrow G_2$$

be a homomorphism of Lie groups. Consider the corresponding mapping

$$\varphi_* \stackrel{\text{def}}{=} \varphi_*(0) : \mathcal{G}_1 \longrightarrow \mathcal{G}_2 \tag{A.27}$$

of Lie algebras. We claim that the mapping (A.27) is a homomorphism. To prove this, consider the following diagram.

$$\begin{array}{ccc} G_1 & \xrightarrow{\ \varphi\ } & G_2 \\ \uparrow{\scriptstyle \exp} & & \uparrow{\scriptstyle \exp} \\ \mathcal{G}_1 & \xrightarrow{\ \varphi_*\ } & \mathcal{G}_2 \end{array}$$

It is commutative. Indeed, if $X \in \mathcal{G}_1$ and $\mathcal{U}_X(t) = \exp(tX)$ is the corresponding one-parameter subgroup, then $\varphi(\mathcal{U}_X(t))$ is a one-parameter subgroup. Next,

$$\varphi(\exp X) = \varphi(\mathcal{U}_X(1)) = \exp\left(\left.\frac{d}{dt}\varphi(\mathcal{U}_X(t))\right|_{t=0}\right) = \exp(\varphi_* X),$$

as desired. Let us now use formula (A.26). We have, for $X, Y \in \mathcal{G}_1$,

$$\begin{aligned}
&\varphi_*([X, Y]) \\
&= \varphi_* \left(\frac{d}{dt} \exp(\sqrt{t}X) \exp(\sqrt{t}Y) \exp(-\sqrt{t}X) \exp(-\sqrt{t}Y) \Big|_{t=0} \right) \\
&= \frac{d}{dt} \varphi \left(\exp(\sqrt{t}X) \exp(\sqrt{t}Y) \exp(-\sqrt{t}X) \exp(-\sqrt{t}Y) \right) \Big|_{t=0} \\
&= \frac{d}{dt} \exp(\sqrt{t}\varphi_* X) \exp(\sqrt{t}\varphi_* Y) \exp(-\sqrt{t}\varphi_* X) \exp(-\sqrt{t}\varphi_* Y) \Big|_{t=0} \\
&= [\varphi_* X, \varphi_* Y]
\end{aligned}$$

The mapping φ_* is called the *Lie algebra homomorphism corresponding to the homomorphism* φ .

4.6 Derived Representation

Let G be a Lie group,

$$T : G \longrightarrow \mathrm{Aut}(V)$$

be a representation of G in a vector space V. We wish to construct the corresponding representation of the Lie algebra $\mathcal{G}$.

We consider two cases.

A. The space V is finite-dimensional, $\dim V = n$. In this case $\mathrm{Aut}(V) \cong \mathrm{Gl}(n, \mathbb{R})$ is a Lie group, and T is a homomorphism of Lie groups,

$$T : G \longrightarrow \mathrm{Gl}(n, \mathbb{R}).$$

The corresponding representation of the Lie algebra is defined as

$$T_* : \mathcal{G} \longrightarrow \mathrm{gl}(n, \mathbb{R}) \cong \mathrm{End}(V)$$

(see the preceding subsection to verify that T_* is a representation of the Lie algebra).

B. V is a Banach space,

$$T : G \longrightarrow \mathrm{Aut}(V)$$

is a strongly continuous representation of the group G.

Consider the subspace $V^\infty \subset V$ consisting of all vectors $v \in V$ such that the vector-function $g \mapsto T(g)v$ on G is infinitely smooth. The space V^∞ is called the *Gårding space* of the representation T. The subspace V^∞ is dense in V. Indeed, let $\varphi(g)$ be a smooth finite function on G, and $v \in V$ be an arbitrary element. Consider the element

$$\tilde{v} = \int_G \varphi(h)\, T(h)\, v d_l h$$

then

$$T(g)\tilde{v} = \int_G \varphi(h)\, T(g)\, T(h)\, v\, d_l h = \int_G \varphi(h)\, T(gh)\, v\, d_l h = \int_G \varphi(g^{-1}h)\, T(h)\, v\, d_l h$$

(here $d_l h$ is a left Haar measure; see Section 3 above) and, since $T(h)v$ is continuous, $T(g)\tilde{v}$ is a smooth function on G. Let us now take φ converging to the δ-function. Under these conditions, $\tilde{v} \to v$, so we see that V^∞ is dense in V.

Define the mapping

$$T_* : \mathcal{G} \longrightarrow \mathrm{End}(V^\infty)$$

by setting

$$(T_* X)\, x = \frac{d}{dt} T(\exp tX) x \Big|_{t=0}$$

for $v \in V^\infty$, $X \in \mathcal{G}$.

As in Subsection 4.5, one can verify that T_* is a homomorphism of Lie algebras. T_* is called the *derived representation* of $\mathcal{G}$ (associated with T).

Let $X \in \mathcal{G}$, $A = T_* X$. It is evident that A may be considered as an (unbounded) operator in V with domain V^∞. Consider the representation

$$U(t) = T(\exp tX)$$

of the one-parameter semigroup corresponding to the element X. By the properties of T, $U(t)$ is a strongly continuous one-parameter group of bounded linear operators in V (that is, $U(t)$ are bounded operators). $U(t)\, v$ is a continuous (with respect to norm) function of $t \in \mathbb{R}$ for any $v \in V$, and $U(t)\, U(\tau) = U(t + \tau)$, $t, \tau \in \mathbb{R}$).

By the *Gel'fand theorem*, the *generator* $\tilde{A}$ of the group $U(t)$ is an unbounded operator on V of the form

$$\tilde{A}v = \frac{d}{dt}(U(t)v)\Big|_{t=0}$$

with the domain $D_{\tilde{A}}$ consisting of the vectors $v \in V$ for which $U(t)\, v$ is differentiable with respect to t. The operator $\tilde{A}$ is a closed densely defined operator whose resolvent $R_\lambda(\tilde{A})$ satisfies, for some $M > 0$, $\omega > 0$, the estimates

$$\|R_\lambda(\tilde{A})^m\| \le \frac{M}{(\mathrm{Re}\,\lambda - \omega)^m}, \quad m = 1, \ldots, \mathrm{Re}\,\lambda > \omega.$$

(And, vice versa, each closed densely defined operator satisfying such estimates is a generator of a strongly continuous group in V.)

It is clear that $\tilde{A} \supset A$ (that is, $D_{\tilde{A}} \supset D_A$ and $Av = \tilde{A}v$ for $v \in D_A$). In fact, $\tilde{A} = \overline{A}$ (the closure of the operator A). This is a consequence of the following lemma.

Lemma A.2 *Let A be a generator of a strongly continuous one-parameter group of bounded operators $U(t) = e^{At}$ in a Banach space V, and let $D \subset D_A$ be a dense subspace of V such that $e^{At} \subset D$ for all t. Then*

$$A = \overline{(A|_D)}.$$

Proof. Denote $D_\lambda = (A - \lambda E)\, D$. Let us show that D_λ is dense in D for $\operatorname{Re}\lambda > \omega$. (This implies the desired result immediately since we have $R_\lambda(A) = \overline{R_\lambda(A)|_{D_\lambda}}$ and, consequently,

$$A - \lambda E = [R_\lambda(A)]^{-1} = \overline{R_\lambda(A)^{-1}\big|_D} = \overline{(A - \lambda E)|_D}.)$$

Let $v \in D$. Consider the integral

$$w = \int_0^\infty e^{-\lambda t} e^{At} x\, dt.$$

Since $\operatorname{Re}\lambda > \omega$ and $\|e^{At}\| \le M\, e^{\omega t}$, this integral converges; furthermore, we have

$$\begin{aligned}(A - \lambda)\, w &= \int_0^\infty (A - \lambda) e^{-\lambda t} e^{At} v\, dt \\ &= \int_0^\infty \frac{d}{dt}\left(e^{-\lambda t} e^{At} v\right) dt = -v.\end{aligned}$$

Thus, the element v may be approximated by integral sums of the form

$$\sum_j (A - \lambda) e^{-\lambda t_j} e^{At_j} At_j v \in D_\lambda.$$

Since D is dense in V, the same is true of D_λ. The lemma is proved. □

4.7 The Lie Group Corresponding to a Lie Algebra

Let Γ be an n-dimensional Lie algebra with basis $a_1, \ldots, a_n$, so that

$$[a_i, a_j] = \sum_{k=1}^n \lambda_{ij}^k a_k, \quad i, j = 1, \ldots, n,$$

where λ_{ij}^k are the structure constants of Γ with respect to the basis $a_1, \ldots, a_n$.

By the Campbell–Hausdorff theorem, there is a unique, up to isomorphism, local Lie group with Lie algebra Γ, and by the Cartan–Levi–Maltsev theorem, there is a unique connected simply connected Lie group with Lie algebra Γ.

Let G be a Lie group with Lie algebra Γ. We shall use the canonical coordinates of second genus in the neighborhood of the neutral element $e \in G$: if $x = (x_1, \ldots, x_n)$ lies in a neighborhood of zero in $\mathbb{R}^n$, then we set

$$g(x) = g_n(x_n) g_{n-1}(x_{n-1}) \ldots g_1(x_1) \in G,$$

where $g_i(t)$ is the one-parameter subgroup of G corresponding to $a_i \in \Gamma$. The composition law in the coordinate system $(x_1, \dots, x_n)$ has the form

$$g(x)g(y) = g(\psi(x, y)), \tag{A.28}$$

where ψ is a smooth mapping of a neighborhood of the origin in $\mathbb{R}^n \times \mathbb{R}^n$ into $\mathbb{R}^n$, $\psi(y, 0) = \psi(0, y) \equiv y$. The mapping ψ can be expressed explicitly via the Campbell–Hausdorff–Dynkin formula. However, this expression is not needed for our aims, and we omit it. It is easy to calculate the derivative $(\partial\psi/\partial x)(x, y)$. Since our consideration is local we may assume, by the Ado theorem, that Γ is realized as a matrix Lie algebra and the neigbborhood of e in G is that in a matrix Lie group. Then $g(x)$ has the form

$$g(x) = e^{x_n a_n} e^{x_{n-1}a_{n-1}} \cdot \ldots \cdot e^{x_1 a_1},$$

where e^{xa} is the usual matrix exponent. We now calculate the derivative $(\partial/\partial\psi)g(\psi)$. We have

$$\frac{\partial}{\partial\psi_j} g(\psi) = e^{\psi_n a_n} e^{x\psi_{n-1}a_{n-1}} \ldots a_j e^{\psi_j a_j} e^{x\psi_{j-1}a_{j-1}} \ldots e^{\psi_1 a_1}.$$

Since

$$e^{tb} a e^{-tb} = e^{t\,\mathrm{ad}_b}(a),$$

where $\mathrm{ad}_b = [b, \cdot\,]$, we obtain

$$\frac{\partial}{\partial\psi_j} g(\psi) = \left[e^{\psi_n \mathrm{ad}_{a_n}} \cdot \ldots \cdot e^{\psi_j \mathrm{ad}_{a_j}}(a_j)\right] g(\psi).$$

To simplify the expression in the square brackets, we note that in the basis $(a_1, \dots, a_n)$ the operator ad_{a_s} has the matrix Λ_s with entries $(\Lambda_s)_{pq} = \lambda^p_{sq}$ and, consequently, the operator $\exp(\psi_n \mathrm{ad}_{a_n}) \cdot \ldots \cdot \exp(\psi_j \mathrm{ad}_{a_j})$ is represented by the matrix $\exp(\psi_n \Lambda_n) \cdot \ldots \cdot \exp(\psi_j \Lambda_j)$. Thus

$$\begin{aligned} \frac{\partial}{\partial\psi_j} g(\psi) &= \sum_{p=1}^{n} \left[\exp(\psi_n \Lambda_n) \cdot \ldots \cdot \exp(\psi_j \Lambda_j)\right]_{pj} a_p g(\psi) \\ &= \Big(\sum_{p=1}^{n} B_{pj}(\psi) a_p\Big) g(\psi), \end{aligned}$$

the matrix $B(\psi) = B(\psi_1, \dots, \psi_n) = (B_{pq}(\psi))$ being equal to

$$B_{pq}(\psi) = \left[\exp(\psi_n \Lambda_n) \cdot \ldots \cdot \exp(\psi_q \Lambda_q)\right]_{pq}.$$

In particular, $B(0) = I$ (the identity matrix), so that the inverse matrix $C(\psi) = B^{-1}(\psi)$ is defined when ψ is close to zero.

Calculating the derivative with respect to x on both sides of (A.28), we obtain

$$\begin{aligned} \frac{\partial}{\partial x_i} g(\psi(x,y)) &= \sum_{p,j=1}^{n} \frac{\partial \psi_j}{\partial x_i} B_{pj}(\psi) a_p g(\psi(x,y)), \\ &= \frac{\partial}{\partial x_i}(g(x)g(y)) = \sum_{p=1}^{n} B_{pi}(x) a_p g(x) g(y) \\ &= \sum_{p=1}^{n} B_{pi}(x) a_p g(\psi(x,y)). \end{aligned}$$

The matrices a_p, $p = 1, \ldots, n$, are linearly independent and so are $a_p g(\psi(x, y))$, since $g(\psi)$ is invertible. Thus we obtain

$$B_{pi}(x) = \sum_{j=1}^{n} B_{pj}(\psi) \frac{\partial \psi_j}{\partial x_i}, \tag{A.29}$$

whence it follows that

$$\frac{\partial \psi_j}{\partial x_i}(x, y) = \sum_{k=1}^{n} C_{jk}(\psi(x, y)) B_{ki}(x)$$

or simply

$$\frac{\partial \psi}{\partial x} = C(\psi) B(x) = B^{-1}(\psi) B(x).$$

4.8 The Krein–Shikhvatov Theorem

Assume now that a representation of the Lie algebra Γ is given in a Banach space. Under what conditions is it *integrable*, i.e., gives rise to a strongly continuous representation of the Lie group G? The answer is given by the Krein–Shikhvatov theorem for strongly continuous representations in Banach spaces and by the Nelson theorem for unitary representations in Hilbert spaces.

Let us prove a version of the Krein–Shikhvatov theorem suitable for our purposes.

Theorem A.2 *Assume that $A_1, \ldots, A_n$ are the generators of strongly continuous groups of bounded linear operators in a Banach space $\mathcal{B}$. Let $\mathcal{D} \subset \mathcal{B}$ be a dense linear subset such that:*

(a) *$\mathcal{D} \subset \mathcal{D}_{A_1} \cap \cdots \cap \mathcal{D}_{A_n}$ and $\mathcal{D}$ is invariant under the operators A_j, their resolvents $R_\lambda(A_j)$, and the groups e^{itA_j}, $j = 1, \ldots, n$.*

(b) *The operators iA_j form a representation of Γ in $\mathcal{D}$, i.e.,*

$$[A_j, A_k]h = -i \sum_{s=1}^{n} \lambda_{jk}^{s} A_s h, \quad h \in \mathcal{D}, \quad j, k = 1, \ldots, n.$$

Then there exists a representation T of G in $\mathcal{B}$ such that A_j are its generators, $A_j = T_*(a_j)$.

Proof. Since G is connected and simply connected, it suffices to construct $T(g)$ for g in a neighborhood of e, namely in the neighborhood covered by a canonical coordinate system. For such g we set

$$T(g(x)) \stackrel{\text{def}}{=} T(x) = e^{iA_n x_n} \cdot \ldots \cdot e^{iA_1 x_1}.$$

Then $T(x)$ is a bounded strongly continuous function, and moreover,

$$T_{a_i}(t) = \exp(iA_i t)$$

is a strongly continuous group with generator A_i, so to prove that the operators $T(x)$ satisfy the group law in the neighborhood of zero, i.e.,

$$T(x)T(y) = T(\psi(x, y))$$

for x, y small enough, it suffices to check this identity on the dense subset $\mathcal{D} \subset \mathcal{B}$. Note that $\mathcal{D}$ is a core of each A_i.

Lemma A.3 *For $h \in \mathcal{D}$ we have*

$$e^{iA_s t} A_k h = \sum_{p=1}^{n} [\exp(t\Lambda_s)]_{pk} A_p e^{iA_s t} h, \quad s, k = 1, \ldots, n.$$

Proof. Consider the Banach space

$$Y = \mathcal{B} \otimes \mathbb{C}^n = \underbrace{\mathcal{B} \oplus \mathcal{B} \oplus \cdots \oplus \mathcal{B}}_{n \text{ summands}}.$$

The operators in Y may be represented as matrices with operators in $\mathcal{B}$ as their elements. We introduce the operator $B : \mathcal{B} \to Y$ which is the closure[1] from $\mathcal{D}$ of the operator $h \to (A_1 h, \ldots, A_n h) \in Y$, and the operator $C_s : Y \to Y$, which has the form

$$\begin{aligned} C_s &= A_s \otimes I + iI \otimes \Lambda_s' \\ &= \begin{pmatrix} A_s & & 0 \\ & \ddots & \\ 0 & & A_s \end{pmatrix} + i \begin{pmatrix} \lambda_{s1}^1 I & \ldots & \lambda_{s1}^n I \\ \vdots & & \vdots \\ \lambda_{sn}^1 I & \ldots & \lambda_{sn}^n I \end{pmatrix}, \\ D_{C_s} &= D_{A_s} \oplus \cdots \oplus D_{A_s}, \end{aligned}$$

[1] B is closable since $h_n \to 0$, $Bh_n \to \tilde{h} = (\tilde{h}_1 \ldots, \tilde{h}_n)$ implies $A_i h_n \to \tilde{h}_i$, $i = 1, \ldots, n$, so that $\tilde{h}_i = 0$, since A_i are closed operators.

where Λ_s' is the transpose of Λ_s. Then G_s is a generator of the strongly continuous group $\exp(iC_s t)$ in the space Y. Indeed, direct computation shows that

$$\exp(iC_s t) = \exp(iA_s t) \otimes \exp(-\Lambda_s' t).$$

Since λ_{jk}^s are antisymmetric with respect to the subscripts j, k, it follows that

$$BA_s h = (A_1 A_s h, \dots, A_n A_s h) = (A_s A_1 h, \dots, A_s A_n h)$$

$$-i\Big(\sum_{l=1}^{n} \lambda_{1s}^l A_l h, \dots, \sum_{l=1}^{n} \lambda_{ns}^l A_l h\Big) = (A_s \otimes I)Bh + i(I \times \Lambda_s')Bh = C_s Bh, \; h \in \mathcal{D}.$$

Since $\mathcal{D}$ is invariant under the resolvent $R_\lambda(A_s)$, we can apply Lemma IV.7 and obtain

$$B \exp(iA_s t)h = \exp(iC_s t)Bh, \quad h \in \mathcal{D}_B \supset \mathcal{D}.$$

This identity can be written in the form

$$\begin{aligned} A_p \exp(A_s t)h &= \exp(iA_s t) \sum_{k=1}^{n} [\exp(-\Lambda_s' t)]_{pk} A_k h \\ &= \exp(iA_s t) \sum_{k=1}^{n} [\exp(-\Lambda_s t)]_{kp} A_k h. \end{aligned}$$

Since $\exp(-\Lambda_s t)$ is the inverse of $\exp(\Lambda_s t)$, the lemma is proved. □

Lemma A.4 *For any $h \in \mathcal{D}$ the function $T(\lambda)$ is differentiable with respect to $\lambda \in \mathbb{R}^n$ and*

$$-i \frac{\partial}{\partial \lambda_j} T(\lambda)h = \sum_{p=1}^{n} B_{pj}(\lambda) A_p T(\lambda)h.$$

Proof. Set $h(\lambda) = T(\lambda)h$. For any $\varepsilon = (\varepsilon_1, \dots, \varepsilon_n) \in \mathbb{R}^n$, we have

$$h(\lambda + \varepsilon) - h(\lambda) = \sum_{j=1}^{n} [h(\lambda_1, \dots, \lambda_{j-1}, \lambda_j + \varepsilon_j, \dots, \lambda_n + \varepsilon_n)$$

$$-h(\lambda_1, \dots, \lambda_j, \lambda_{j+1} + \varepsilon_{j+1}, \dots, \lambda_n + \varepsilon_n)] = i \sum_{j=1}^{n} \varepsilon_j e^{iA_n(\lambda_n + \varepsilon_n)} \dots$$

$$\times e^{iA_{j+1}(\lambda_{j+1} + \varepsilon_{j+1})} \int_0^1 d\tau e^{iA_j(\lambda_j + \tau\varepsilon_j)} A_j e^{iA_{j-1}\lambda_{j-1}} \cdot \dots \cdot e^{iA_1\lambda_1} h$$

$$= \; i \sum_{j=1}^{n} \varepsilon_j e^{iA_n\lambda_n} \cdot \dots \cdot e^{iA_j\lambda_j} A_j e^{iA_{j-1}\lambda_{j-1}} \cdot \dots \cdot e^{iA_1\lambda_1} h + O(\|\varepsilon\|)$$

as $\|\varepsilon\| = (\varepsilon_1^2 + \cdots + \varepsilon_n^2)^{1/2} \to 0$. Here we used the strong continuity of $\exp(iA_n\Lambda_n)\ldots\exp(iA_j\Lambda_j)$ and the invariance of $\mathcal{D}$ under $\exp(iA_it)$. Thus $h(\lambda)$ is differentiable and

$$-i\frac{\partial}{\partial\lambda_j}h(\lambda) = e^{iA_n\lambda_n}\cdot\ldots\cdot e^{iA_j\lambda_j}A_j e^{iA_{j-1}\lambda_{j-1}}\cdot\ldots\cdot e^{iA_1\lambda_1}h.$$

Successive applications of Lemma A.3 yield

$$-i\frac{\partial}{\partial\lambda_j}h(\lambda) = \sum_{p=1}^{n}[\exp(\lambda_n\Lambda_n)\cdot\ldots\cdot\exp(\lambda_j\Lambda_j)]_{pj}A_ph(\lambda).$$

The lemma is proved. □

For $h \in \mathcal{D}$, set

$$h_1(x, y) = T(x)T(y)h, \quad h_2(x, y) = T(\psi(x, y))h.$$

We have

$$h_1(0, y) = h_2(0, y) = T(y)h.$$

Be Lemma A.4

$$\begin{aligned} -\frac{\partial h_1}{\partial x_j}(x, y) &= \sum_{p=1}^{n} B_{pj}(x)A_ph_1(x, y), \\ -\frac{\partial h_2}{\partial x_j}(x, y) &= \sum_{s=1}^{n}\frac{\partial\psi_s(x, y)}{\partial x_j}\sum_{p=1}^{n} B_{ps}(\psi(x, y))A_ph_2(x, y) \\ &= \sum_{p=1}^{n} B_{pj}(x)A_ph_2(x, y) \end{aligned}$$

by (A.29). We note that

$$B_{pj}(x_1, \ldots, x_j, 0, \ldots, 0) = [\exp(x_j\Lambda_j)]_{pj} = \delta_{pj},$$

since the j-th column of Λ_j consists of zeros ($\lambda_{jj}^p \equiv 0$), and so we have

$$-i\frac{\partial h_i}{\partial x_j}(x_1, \ldots, x_j, 0, \ldots, 0) = A_jh_i(x_1, \ldots, x_j, 0, \ldots, 0, y),$$

$$i = 1, 2, \quad j = 1, \ldots, n.$$

We can prove that

$$h_1(x_1, \ldots, x_j, 0, \ldots, 0, y) = h_2(x_1, \ldots, x_j, 0, \ldots, 0, y),$$

by induction on j. This is valid for $j = j_0$; we see that for $j = j_0 + 1$ both h_1 and h_2 satisfy the same Cauchy data at $x_j = 0$. Since A_j is the generator of a strongly continuous semigroup, the solution of the Cauchy problem is unique, and we obtain $T(x)T(y)h = T(\psi(x, y))h$. The theorem is proved. □

Appendix B
Pseudodifferential Operators

Pseudodifferential operators, that is, functions of x and $-i\partial/\partial x$, are widely used throughout this book. However, we do not assume that the readers are familiar with the topic, and this appendix is intended to make the book self-contained by presenting the definitions and main properties of pseudodifferential operators.

Pseudodifferential operators, originally known as "singular integral operators" in the theory of elliptic equations, have a long and intricate history. Accordingly, there are quite a few expositions of this theory in the literature, starting with [109] and followed by [84], [175], [178], and others. However, we follow neither of the cited papers here, and after an elementary introduction, where we consider the classical Kohn–Nirenberg pseudodifferential operators (ψDO), we present an exposition in the spirit of noncommutative analysis of the theory of ψDO's which may have rapidly oscillating symbols. Here we mainly follow the paper [103].

1 Elementary Introduction

The aim of the present section is to present an elementary introduction to the theory of pseudodifferential operators (ψDO's in the sequel). Thus, in this section the reader will find the motivation for the appearence of the notion of pseudodifferential operator and the main definitions and theorems of the theory of ψDO's rather than an accurate description of function spaces and precise statements and proofs of the theorems. We hope, however, that this section will be of use for the beginner in the ψDO theory. The reader who is interested in precise statements and proofs can find them in the subsequent sections.

One of the themes that lead to the notion of pseudodifferential operators is the problem of constructing a parametrix for an elliptic differential operator. For simplicity we shall carry out our considerations in the Cartesian space $\mathbb{R}^n$. To recall the statement of this problem let us consider a differential equation of the form

$$\hat{H}u = f \tag{B.30}$$

where

$$\hat{H} = H\left(\overset{2}{x}, -i\overset{1}{\frac{\partial}{\partial x}}\right) = \sum_{|\alpha|\leq m} a_\alpha(x)\left(-i\frac{\partial}{\partial x}\right)^\alpha \tag{B.31}$$

is an elliptic differential operator in $\mathbb{R}^n$ with smooth coefficients $a_\alpha(x)$ and $f = f(x)$ is a (in general nonsmooth[5]) function of variables $x = (x^1, \dots, x^n) \in \mathbb{R}^n$. The upper indices in (B.31) determine the order of action of the operators included in the latter relation.

The ellipticity of the operator (B.31) means that for any nonzero vector

$$p = (p_1, \dots, p_n) \in \mathbb{R}_n$$

of the Cartesian space $\mathbb{R}_n$ (which is dual to the space $\mathbb{R}^n$) the principal symbol of the operator $\hat{H}$

$$H_m(x, p) = \sum_{|\alpha|=m} a_\alpha(x)\, p^\alpha$$

is not equal to zero:

$$H_m(x, p) \neq 0. \tag{B.32}$$

The problem we shall now consider is to solve equation (B.30) up to sufficiently smooth terms. To formulate precisely the meaning of the latter phrase one must have some tools measuring the smoothness of the functions. This can be done, for example, with the help of Sobolev spaces.

The idea of introducing these spaces is based on the fact that, due to the formula

$$f(x) = \left(\frac{1}{2\pi}\right)^{\frac{n}{2}} \int e^{ipx} \tilde{f}(p)\, dp,$$

representing the function $f(x)$ via its Fourier transform $\tilde{f}(p)$, the smoothness of the function $f(x)$ is determined by the decay of $\tilde{f}(p)$ as $|p| \to \infty$. Actually, the faster the function $\tilde{f}(p)$ decreases at infinity, the smoother is the function $f(x)$. Hence, it makes sense to introduce the space $H^s(\mathbb{R}^n)$ with the norm

$$\|f\|_s^2 = \int \left(1 + |p|^2\right)^{\frac{s}{2}} \left|\tilde{f}(p)\right|^2 dp.$$

Clearly, smoother functions belong to the space $H^s(\mathbb{R}^n)$ for larger values of s.

The spaces $H^s(\mathbb{R}^n)$ are called *Sobolev spaces*. We remark that the family of Sobolev spaces parametrized by s forms a decreasing scale of spaces, that is that

$$H^s(\mathbb{R}^n) \subset H^{s'}(\mathbb{R}^n)$$

for any $s' < s$.

[5]The reader familiar with the theory of L. Schwartz's distributions can view $f(x)$ as a distribution, that is, as an element of the space $S'(\mathbb{R}^n)$.

It can be easily shown that the operator (B.31) acts in Sobolev spaces as follows:

$$\hat{H} : H^s(\mathbb{R}^n) \to H^{s-m}(\mathbb{R}^n)$$

for any real value of s. Thus, the problem of finding a solution to equation (B.30) up to sufficiently smooth terms can be formulated as follows:

Find an operator

$$\hat{R}_N : H^{s-m}(\mathbb{R}^n) \to H^s(\mathbb{R}^n)$$

such that the following relation holds:

$$\hat{H}\,\hat{R}_N = 1 + \hat{Q}_{-N} \tag{B.33}$$

where $\hat{Q}_{-N}$ is a smoothing operator of order $-N$ for sufficiently large values of N, that is, the operator

$$\hat{Q}_{-N} : H^s(\mathbb{R}^n) \to H^{s+N}(\mathbb{R}^n)$$

is continuous.

The operator $\hat{R}_N$ satisfying relation (B.33) is called *a parametrix* of the operator $\hat{H}$. If the parametrix for the operator $\hat{H}$ is constructed, then the solution to equation (B.30) up to elements of the space $H^{s+N}(\mathbb{R}^n)$ is given by $u = \hat{R}_N f$; we note that the larger values of N we choose, the smoother are the functions from $H^{s+N}(\mathbb{R}^n)$.

Let us begin with the construction of a parametrix. To do this, we represent equation (B.30) in the form

$$\hat{H}u(x) = \left(\frac{1}{2\pi}\right)^{\frac{n}{2}} \int e^{ipx} \tilde{f}(p)\,dp.$$

The solution of this equation (parametrix) can be found in the form

$$u(x) = \int G(x,p)\tilde{f}(p)\,dp \tag{B.34}$$

if the function $G(x, p)$ satisfies the equation

$$\hat{H}G(x,p) = \left(\frac{1}{2\pi}\right)^{\frac{n}{2}} e^{ipx}. \tag{B.35}$$

Moreover, to construct a solution to equation (B.30) up to sufficiently smooth terms one has to solve equation (B.35) up to functions with sufficiently strong decay at infinity. Thus, we have introduced in our problem a large numerical parameter $|p|$ which corresponds to asymptotic expansions with respect to smoothness. As we shall see below, the asymptotic expansions for solutions to equation (B.35) can be constructed in terms of homogeneous functions in the variables p of decreasing order of homogeneity.

Let us try to find a solution to equation (B.35) in the form

$$G(x,p) = e^{ipx} \sum_{j=k_0}^{-\infty} g_j(x,p) \tag{B.36}$$

where $g_j(x, p)$ are homogeneous functions of order j with respect to the variables p (we do not yet know the value of k_0, that is, the order of homogeneity of the principal term of the expansion (B.36)). Substituting the latter equality in equation (B.35) we obtain

$$\hat{H}\left[e^{ipx}\sum_{j=k_0}^{-\infty} g_j(x, p)\right] = \left(\frac{1}{2\pi}\right)^{\frac{n}{2}} e^{ipx}. \tag{B.37}$$

Now we must commute the factor e^{ipx} with the differential operator $\hat{H}$. To do this, we note that, due to the relation

$$-i\frac{\partial}{\partial x^k}\left[e^{ipx}g(x, p)\right] = e^{ipx}\left(-i\frac{\partial}{\partial x^k} + p_k\right)g(x, p)$$

the following commutation formula is valid:

$$H\left(\overset{2}{x}, -i\overset{1}{\frac{\partial}{\partial x}}\right)\left[e^{ipx}g(x, p)\right] = e^{ipx}H\left(\overset{2}{x}, p - i\overset{1}{\frac{\partial}{\partial x}}\right)g(x, p).$$

Using this relation, we rewrite (B.37) in the form

$$e^{ipx}H\left(\overset{2}{x}, p - i\overset{1}{\frac{\partial}{\partial x}}\right)\sum_{j=k_0}^{-\infty} g_j(x, p) = \left(\frac{1}{2\pi}\right)^{\frac{n}{2}} e^{ipx}.$$

Cancelling out the factor e^{ipx} we arrive at the equality

$$\left[H_m(x, p) + \sum_{k=0}^{m-1} H_k\left(\overset{2}{x}, p, -i\overset{1}{\frac{\partial}{\partial x}}\right)\right]\sum_{j=k_0}^{-\infty} g_j(x, p) = \left(\frac{1}{2\pi}\right)^{\frac{n}{2}} \tag{B.38}$$

where $H_k\left(\overset{2}{x}, p, -i\overset{1}{\frac{\partial}{\partial x}}\right)$ are differential operators of order $m - k$ with coefficients homogeneous in p of order k. Equating terms of equal order of homogeneity in the right- and left-hand sides of equality (B.38) we arrive at the following recurrence system for the unknown functions $g_j(x, p)$ (this equality also shows that the order k_0 of the leading term of the expansion (B.36) equals $-m$):

$$H_m(x, p)g_{-m}(x, p) = \left(\frac{1}{2\pi}\right)^{\frac{n}{2}}, \tag{B.39}$$

$$H_m(x, p)g_j(x, p) = -\sum_{l=1-j}^{m-j} H_{m-l-j}\left(\overset{2}{x}, p, -i\overset{1}{\frac{\partial}{\partial x}}\right)g_l(x, p),\ j \le -m - 1.$$

In the latter formula we set $g_j(x, p) = 0$ for $j \ge -m$.

Due to the ellipticity condition (B.32) the latter system of equations has the unique solution

$$g_{-m}(x, p) = \left(\frac{1}{2\pi}\right)^{\frac{n}{2}} \frac{1}{H_m(x, p)}$$

$$g_{-m-1}(x, p) = -H_{m-1}\left(\overset{2}{x}, p, -i\overset{1}{\frac{\partial}{\partial x}}\right) g_{-m}(x, p) \tag{B.40}$$

$$\vdots$$

The functions $g_j(x, p)$ are homogeneous functions in p of order $-j$. Thus, to construct a parametrix for equation (B.30) one should truncate the expansion (B.36) at a sufficiently high level. Actually, substituting the truncated expansion in (B.34) one has:

$$u(x) = R_N\, f(x) = \sum_{j=-m}^{-N-m} \int e^{ipx} g_j(x, p) \tilde{f}(p)\, dp, \tag{B.41}$$

where the $g_j(x, p)$ are determined from (B.40) and N is a sufficiently large number. Unfortunately, the functions $g_j(x, p)$, being homogeneous functions of large negative order, may have a nonintegrable singularity at the origin and, hence, the integrals involved in the right-hand side of (B.41) are, in general, divergent. To overcome this difficulty, we use a smooth cut-off function $\chi(p)$ which is equal to zero in some neighbourhood of the origin and equals unity outside a compact set of $\mathbb{R}_n$. Thus, we obtain the formula for the required parametrix

$$R_N\, f(x) = \sum_{j=-m}^{-N-m} \int e^{ipx} g_j(x, p) \tilde{f}(p) \chi(p)\, dp. \tag{B.42}$$

It is easy to see that the latter expression does not depend on the choice of the cut-off function $\chi(p)$ up to arbitrary smooth terms, that is, up to terms from the Sobolev space $H^s(\mathbb{R}^n)$ for arbitrary values of s.

Remark B.1 In the case when the homogeneity order of the function $g_j(x, p)$ in (B.42) is positive, then the use of the cut-off function in this formula is not necessary.

Remark B.2 Any differential operator $H(\overset{2}{x}, -i\overset{1}{\frac{\partial}{\partial x}})$ can be written in a form similar to the right-hand side of (B.42):

$$H\left(\overset{2}{x}, -i\overset{1}{\frac{\partial}{\partial x}}\right) = \sum_{j=0}^{m} \int e^{ipx} H_j(x, p) \tilde{f}(p)\, dp \tag{B.43}$$

where

$$H_j(x, p) = \sum_{|\alpha|=j} a_\alpha(x)\, p^\alpha, \; j = 0, \ldots, m$$

are the components of the complete symbol $H(x, p)$ of the operator $\hat{H}$ of order j. Here we have omitted the cut-off function $\chi(p)$ in accordance with the previous remark.

Thus, we arrive at the notion of pseudodifferential operator using the operator on the right in (B.42) as a model.

Definition B.1 The operator

$$P\left(\overset{2}{x}, -i\overset{1}{\frac{\partial}{\partial x}}\right) = \int e^{ipx} P(x, p) \tilde{f}(p) \chi(p)\, dp \tag{B.44}$$

is called *a pseudodifferential operator with symbol* $P(x, p)$.

The only question which we must discuss to complete the definition of a pseudodifferential operator is the class of symbols used in Definition B.1.

If we consider in (B.44) only polynomial (with respect to the variables p) symbols, then the class of pseudodifferential operators will coincide (due to (B.43)) with the class of differential operators. Certainly, this symbol class is too small; it does not even contain parametrices of elliptic differential operators.

The smallest suitable class of symbols, which was in essence introduced in the original paper by J. J. Kohn and L. Nirenberg [109], is the class of so-called classical symbols. We say that the smooth function $P(x, p)$ is *a classical symbol of order m* if for any multi-indices α, β the estimate

$$\left| \left(\frac{\partial}{\partial x}\right)^\alpha \left(\frac{\partial}{\partial p}\right)^\beta P(x, p) \right| \leq C_{\alpha\beta} \left(1 + |p|^2\right)^{\frac{m-|\beta|}{2}}$$

with some positive constant $C_{\alpha\beta}$. We denote by S^m the set of classical symbols[6] of order m.

The set of classical symbols is suitable for constructing parametrices for elliptic differential operators. However, for other problems this class can turn out to be too small. A rather general symbol class will be considered in the subsequent sections; here we restrict ourselves to consideration of the class S^m.

Let us formulate the two main theorems of the theory of pseudodifferential operators. The first of them is concerned with the action of a pseudodifferential operator in the Sobolev spaces $H^s(\mathbb{R}^n)$, and the second treats the composition of pseudodifferential operators.

---

[6] For a rigorous definition one should impose some conditions on the behavior of symbols as $|x| \to \infty$. Exact conditions of this kind are presented, for example, in the cited paper. For simplicity, the reader can assume that all symbols considered have compact support with respect to the variable x, or that these symbols do not depend on x outside a compact set (which can itself depend on the symbol).

Theorem B.1 *Let $P(x, p)$ be a classical symbol of order m. Then the corresponding ψDO*

$$P\left(\overset{2}{x}, -i\overset{1}{\frac{\partial}{\partial x}}\right) : H^s(\mathbb{R}^n) \to H^{s-m}(\mathbb{R}^n)$$

is a continuous operator for any $s \in \mathbb{R}^n$.

Theorem B.2 *Let $P(x, p)$ and $Q(x, p)$ be classical symbols of orders m_P and m_Q, respectively. Then the relation*

$$P\left(\overset{2}{x}, -i\overset{1}{\frac{\partial}{\partial x}}\right) Q\left(\overset{2}{x}, -i\overset{1}{\frac{\partial}{\partial x}}\right) \equiv R\left(\overset{2}{x}, -i\overset{1}{\frac{\partial}{\partial x}}\right)$$

holds modulo operators of order $-\infty$ in the Sobolev scale. The symbol $R(x, p)$ of the operator $R\left(\overset{2}{x}, -i\overset{1}{\frac{\partial}{\partial x}}\right)$ has the asymptotic expansion

$$R(x, p) \equiv \sum_{|\alpha| \geq 0} \frac{(-i)^{|\alpha|}}{\alpha!} \frac{\partial^{|\alpha|} P(x, p)}{\partial p^\alpha} \frac{\partial^{|\alpha|} Q(x, p)}{\partial x^\alpha}$$

as $|p| \to \infty$.

We shall not present the proofs of these theorems here. The reader can find them in many books and papers concerned with this topic (see, for example, [84], [175] and others). We also remark that we have omitted from the framework of this short presentation such important questions as the behavior of pseudodifferential operators under change of variables, the theory of pseudodifferential operators on smooth manifolds, and so on. These topics can also be found in the literature cited above.

2 Symbol Spaces and Generators

The pseudodifferential operators used in this book usually act in symbol spaces. Technically, various symbol spaces may be considered; however, the most important in applications is $S^\infty(\mathbb{R}^n)$, the poly-Banach algebra of symbols of tempered growth introduced in Chapter III. For this reason, we shall define and study pseudodifferential operators in $S^\infty(\mathbb{R}^n)$. In what follows we use the abbreviation ψDO for "pseudodifferential operator".

Recall that $S^\infty(\mathbb{R}^n)$ is defined as the union of intersections

$$S^\infty(\mathbb{R}^n) = \bigcup_l \bigcap_k S_l^k(\mathbb{R}^n) \tag{B.45}$$

where $S_l^k(\mathbb{R}^n)$ is the space of C^k functions on $\mathbb{R}^n$ with finite norm

$$\|f\|_{S_l^k(\mathbb{R}^n)} = \sup_{\mathbb{R}^n} \left(1+|x|^2\right)^{-\frac{l}{2}} \left(\sum_{|\alpha|\le k} \left|f^{(\alpha)}(x)\right|\right),$$

and is endowed with the corresponding convergence (see Chapter III). The space $S_l^k(\mathbb{R}^n)$ is not a Hilbert space, and in order to make our exposition as elementary as possible (and almost independent of Chapter III) we consider the following representation of $S^\infty(\mathbb{R}^n)$ in terms of Hilbert spaces:

$$S^\infty(\mathbb{R}^n) = \bigcup_l \bigcap_k H_l^k(\mathbb{R}^n), \tag{B.46}$$

where $H_l^k(\mathbb{R}^n)$ is the completion of C_0^∞ with respect to the Hilbert norm

$$\|u\|_{H_l^k(\mathbb{R}^n)} = \left(\int \left(1+|x|^2\right)^{-l} \left|(1-\Delta)^{\frac{k}{2}} u(x)\right|^2 dx\right)^{\frac{1}{2}}.$$

(As an exercise, one can easily check that (B.45) and (B.46) give the same result by using Sobolev's embedding theorems).

The convergence in $S^\infty(\mathbb{R}^n)$ can be described as follows: a generalized sequence $\{f_\alpha\} \subset S^\infty(\mathbb{R}^n)$ is said to *converge to zero* if there exists an l such that for any k we have $f_\alpha \in H_l^k$ for sufficiently large α and $\|f_\alpha\|_{H_l^k} \to 0$.

Pseudodifferential operators are elements of the algebra $\mathcal{L}(S^\infty, S^\infty)$ of all continuous linear operators on $S^\infty(\mathbb{R}^n)$. For Hilbert spaces B_1, B_2 let $\mathcal{L}(B_1, B_2)$ denote the Banach space of all continuous linear operators $A : B_1 \to B_2$ equipped with the operator norm

$$\|A\|_{\mathcal{L}(B_1,B_2)} = \sup_{x\in B_1, x\neq 0} \frac{\|Ax\|_{B_2}}{\|x\|_{B_1}}.$$

The algebra $\mathcal{L}(S^\infty, S^\infty)$ has the form

$$\mathcal{L}(S^\infty, S^\infty) = \bigcap_l \bigcup_r \bigcap_s \bigcup_k \mathcal{L}\left(H_l^k, H_r^s\right), \tag{B.47}$$

that is, each operator $T \in \mathcal{L}(S^\infty, S^\infty)$ has the following property:
$\forall l\, \exists r\, \forall s\, \exists k$ such that T extends by continuity to a bounded linear operator from H_l^k to H_r^s.

The convergence on $\mathcal{L}(S^\infty, S^\infty)$ is introduced as follows: a generalized sequence $\{T_\alpha\} \in \mathcal{L}(S^\infty, S^\infty)$ is said to converge to zero if $\forall l\, \exists r\, \forall s\, \exists k$ such that $T_\alpha \in \mathcal{L}\left(H_l^k, H_r^s\right)$ for sufficiently large α and

$$\|T_\alpha\|_{\mathcal{L}\left(H_l^k, H_r^s\right)} \to 0.$$

Both S^∞ and $\mathcal{L}(S^\infty, S^\infty)$ are poly-Banach algebras with respect to the convergence described above.

We are interested in S^∞-generators in $\mathcal{L}(S^\infty, S^\infty)$. Recall that an operator $A \in \mathcal{L}(S^\infty, S^\infty)$ is an S^∞-*generator* if there exists a (necessarily unique) continuous homomorphism

$$\mu : S^\infty\left(\mathbb{R}^1\right) \to \mathcal{L}\left(S^\infty, S^\infty\right)$$

such that $\mu(y) = A$ where y is a coordinate on $\mathbb{R}^1$.

Lemma B.1 *An operator A is an S^∞-generator if and only if there exists a one-parameter subgroup*

$$\{e_t,\ t \in \mathbb{R}\} \subset \mathcal{L}\left(S^\infty, S^\infty\right)$$

such that

i) *e_t is differentiable with respect to t and*

$$-i\frac{de_t}{dt} = A;$$

ii) *the subgroup $\{e_t\}$ is of tempered growth in $\mathcal{L}(S^\infty, S^\infty)$ as $t \to \infty$, that is,*

$$\forall l\ \exists r\ \forall s\ \exists k\ \exists p \left\{\|e_t\|_{H_l^k \to H_r^s} \le C\,(1+|t|)^p\right\}. \tag{B.48}$$

Proof. We shall not prove the "only if" part of the lemma since it is not used in our subsequent considerations. Let us prove the "if" part, that is, the sufficiency of conditions i) and ii). Let $f \in S^\infty\left(\mathbb{R}^1\right)$. We intend to define $\mu(f) = f(A)$. This can be done as follows. For some l we have

$$f \in \bigcap_k H_l^k(\mathbb{R}^1)$$

whence it follows that the Fourier transform of $\varphi(y) = \left(1+y^2\right)^{-N} f(y)$ is continuous and decays rapidly as $|y| \to \infty$ (it suffices to take $N > l + 1/2$). Consider the integral

$$I = \frac{1}{\sqrt{2\pi i}} \int \tilde{\varphi}(t) e_t\, dt, \tag{B.49}$$

where $\tilde{\varphi}(t)$ is the Fourier transform of $\varphi(y)$. Since $\tilde{\varphi}(t)$ decays faster than any negative power of t as $t \to \infty$, it follows from (B.48) that $\forall l\ \exists r\ \forall s\ \exists k$ such that the integral in (B.49) converges in $\mathcal{L}(H_l^k, H_r^s)$. By the definition of convergence in $\mathcal{L}(S^\infty, S^\infty)$, this integral converges in $\mathcal{L}(S^\infty, S^\infty)$. Hence, we can set

$$f(A) = \left(1+A^2\right)^N \frac{1}{\sqrt{2\pi i}} \int \tilde{\varphi}(t) e_t\, dt.$$

Trivial computations show that the result does not depend on the choice of admissible N and that the mapping $f \mapsto f(A)$ is a homomorphism, obviously continuous.

Furthermore, for $f(y) = y$ we have

$$\begin{aligned} f(A) &= \left(1 + A^2\right) \frac{1}{\sqrt{2\pi i}} \int F_{y \to t} \left(\frac{y}{1+y^2}\right) e_t \, dt \\ &= \frac{1}{\sqrt{2\pi i}} \int F_{y \to t} \left(\frac{y}{1+y^2}\right) \left(1 - \frac{\partial^2}{\partial t^2}\right) e_t \, dt, \end{aligned}$$

since it follows easily from condition i) that

$$-i \partial e_t / \partial t = A e_t$$

for all $t \in \mathbb{R}$. Here $F_{y \to t}$ is the Fourier transform. Continuing the computations, we obtain

$$\begin{aligned} f(A) &= \frac{1}{\sqrt{2\pi i}} \int e_t \left(1 - \frac{\partial^2}{\partial t^2}\right) F_{y \to t} \left(\frac{y}{1+y^2}\right) dt \\ &= \frac{1}{\sqrt{2\pi i}} \int e_t F_{y \to t} (y) \, dt \\ &= -\int e_t \delta'(t) \, dt \\ &= \left. \frac{\partial e_t}{\partial t} \right|_{t=0} = A, \end{aligned}$$

as desired (we have used elementary properties of the Fourier transform). The proof is complete. □

Using the Lemma, we can easily show that the following operators are S^∞-generators in $\mathcal{L}(S^\infty, S^\infty)$:

a) The operator of multiplication by the coordinate x_j for each $j = 1, \ldots, n$ (this operator will be denoted by the same symbol x_j).

b) The differentiation operator $\hat{p}_j = -i\partial/\partial x_j$, $j = 1, \ldots, n$.

c) The operator $\hat{p}_j + \varphi(x)$, where $\varphi(x)$ is a smooth real function on $\mathbb{R}^n$ all of whose derivatives are bounded.

Indeed, we have only to construct the corresponding one-parameter groups and to check that they are of tempered growth. In case a) we have

$$e_t \equiv \exp\left(i x_j t\right) = e^{i x_j t}$$

(the operator of multiplication by $e^{i x_j t}$); we obviously have

$$\left| \frac{\partial^\alpha}{\partial x^\alpha} \left(e^{i x_j t}\right) \right| \leq C \left(1 + |t|\right)^{|\alpha|},$$

whence it follows that

$$\|e_t\|_{H_l^k \to H_l^k} \leq C\,(1+|t|)^{|k|}\,.$$

In case b) we have

$$e_t = \exp(t\frac{\partial}{\partial x_j})$$

and

$$e_t\Psi(x) = \Psi(x_1, \ldots, x_{j-1}, x_j + t, x_{j+1}, \ldots, x_n);$$

it is easy to observe that

$$\|e_t\|_{H_l^k \to H_l^k} \leq C\,(1+|t|)^{|l|}\,.$$

Finally, in case c) we have

$$e_t(\hat{p}_j + \varphi(x)) = \exp\left\{ i \int_0^t \varphi(x_1, \ldots, x_j + h\tau, \ldots, x_n)\, d\tau \right\} \exp\left\{ it\hat{p}_j \right\},$$

which is a one-parameter group of tempered growth in $\mathcal{L}(S^\infty, S^\infty)$.

3 Pseudodifferential Operators

Pseudodifferential operators are functions of the multiplication and differentiation operators

$$x = (x_1, \ldots, x_n) \quad \text{and} \quad -i\frac{\partial}{\partial x} = \left(-i\frac{\partial}{\partial x_1}, \ldots, -i\frac{\partial}{\partial x_n}\right),$$

which satisfy the commutation relations

$$\left[x_k, -i\frac{\partial}{\partial x_j}\right] = i\delta_{kj}.$$

Now let $f \in S^\infty(\mathbb{R}^n_x \times \mathbb{R}^n_p)$.

Definition B.2 The operator

$$f\left(\overset{2}{x}, -i\overset{1}{\frac{\partial}{\partial x}}\right) = f\left(\overset{2}{x}_1, \ldots, \overset{2}{x}_n, -i\overset{1}{\frac{\partial}{\partial x_1}}, \ldots, -i\overset{1}{\frac{\partial}{\partial x_n}}\right) \tag{B.50}$$

is called the *pseudodifferential operator* with symbol f.

It is easy to show that the operator (B.50) acts on an arbitrary function $u(x) \in C_0^\infty(\mathbb{R}^n)$ according to the formula

$$f\left(\overset{2}{x}, -i\overset{1}{\frac{\partial}{\partial x}}\right) u(x) = \tilde{F}_{p\to x} f(x,p) F_{y\to p} u(y), \tag{B.51}$$

where

$$F_{y\to p}\varphi(y) = \left(\frac{1}{2\pi i}\right)^{\frac{n}{2}} \int e^{-ipy}\varphi(y)\,dy$$

and

$$\tilde{F}_{p\to x}\Psi(p) = \left(\frac{1}{2\pi i}\right)^{\frac{n}{2}} \int e^{-ipy}\Psi(p)\,dp$$

are the direct and the inverse Fourier transforms, respectively.

Consider the function $e_p(x) = \exp(ipx)$, which belongs to $S^\infty(\mathbb{R}^n)$ for any $p \in \mathbb{R}^n$. Let $T \in \mathcal{L}(S^\infty, S^\infty)$ be an arbitrary operator. With T we associate a function of the variables p and $x \in \mathbb{R}^n$ by setting

$$\operatorname{smb}\{T\}(x,p) = e_{-p}(x) T e_p(x). \tag{B.52}$$

Definition B.3 The function $\operatorname{smb}\{T\}$ defined in (B.52) is called *the symbol* of the operator T.

This definition is justified by the following lemma.

Lemma B.2 *If*

$$T = f\left(\overset{2}{x}, -i\overset{1}{\frac{\partial}{\partial x}}\right), \quad f \in S^\infty(\mathbb{R}^n_x \times \mathbb{R}^n_p)$$

is a ψDO, then

$$\operatorname{smb}\{T\} = f. \tag{B.53}$$

Proof. To prove (B.53) we use (B.51). We have

$$\begin{aligned} Te_p(x) &= f\left(\overset{2}{x}, -i\overset{1}{\frac{\partial}{\partial x}}\right) e^{ipx} \\ &= \tilde{F}_{\eta\to x} f(x,\eta)\tilde{F}_{y\to\eta} e^{ipy} \\ &= \tilde{F}_{\eta\to x}\left\{f(x,\eta)\,(2\pi i)^{n/2}\,\delta(\eta-p)\right\} \\ &= \tilde{F}_{\eta\to x}\left\{f(x,p)\,(2\pi i)^{n/2}\,\delta(\eta-p)\right\} \\ &= f(x,p)e^{ipx} = f(x,p)e_p(x). \end{aligned} \tag{B.54}$$

On multiplying both sides in (B.54) by $e_{-p}(x)$, we obtain the desired identity. □

Our next step is to widen the class of symbols for which pseudodifferential operators are defined so as to make an arbitrary element of $\mathcal{L}(S^\infty, S^\infty)$ be a "pseudodifferential operator" in the sense that

$$T = \operatorname{smb}\{T\}\left(\overset{2}{x}, -i\overset{1}{\frac{\partial}{\partial x}}\right). \tag{B.55}$$

Let $H^{s_1,s_2}_{r_1,r_2}\left(\mathbb{R}^n_x \times \mathbb{R}^n_p\right)$ denote the completion of $C_0^\infty\left(\mathbb{R}^n_x \times \mathbb{R}^n_p\right)$ with respect to the norm

$$\begin{aligned} \|f\|_{H^{s_1,s_2}_{r_1,r_2}} &= \left\{\int\int \left(1+|p|^2\right)^{-r_2}\left(1+|x|^2\right)^{-r_1}\right. \\ &\times \left.\left|(1-\Delta_p)^{s_2/2}(1-\Delta_x)^{s_1/2} f(x,p)\right|^2 dx\,dp\right\}^{1/2}. \end{aligned}$$

Consider the poly-Banach space

$$\mathcal{L}^\infty\left(\mathbb{R}^n_x \times \mathbb{R}^n_p\right) = \bigcap_{s_2}\bigcup_{r_1}\bigcap_{s_1}\bigcup_{r_2} H^{s_1,s_2}_{r_1,r_2}\left(\mathbb{R}^n_x \times \mathbb{R}^n_p\right).$$

This is a poly-Banach algebra with respect to pointwise multiplication. Obviously, we have

$$S^\infty\left(\mathbb{R}^n_x \times \mathbb{R}^n_p\right) \subset \mathcal{L}^\infty\left(\mathbb{R}^n_x \times \mathbb{R}^n_p\right),$$

and the embedding is continuous.

Let $T \in \mathcal{L}^\infty\left(\mathbb{R}^n_x, \mathbb{R}^n_p\right)$. Then the function $\operatorname{smb}\{T\}$ defined in (B.52) obviously belongs to $\mathcal{L}^\infty\left(\mathbb{R}^n_x \times \mathbb{R}^n_p\right)$.

Theorem B.3 *The mapping*

$$f \mapsto f\left(\overset{2}{x}, -i\overset{1}{\frac{\partial}{\partial x}}\right)$$

can be extended to a linear homeomorphism

$$\mu : \mathcal{L}^\infty\left(\mathbb{R}^n_x \times \mathbb{R}^n_p\right) \to \mathcal{L}^\infty\left(S^\infty, S^\infty\right).$$

The inverse mapping is given by the formula

$$\mu^{-1}(T) = \operatorname{smb}\{T\}.$$

Proof. It is well-known that for any $f \in L^2\left(\mathbb{R}^n_x \times \mathbb{R}^n_p\right)$ the operator

$$\hat{f} = f\left(\overset{2}{x}, -i\overset{1}{\frac{\partial}{\partial x}}\right)$$

is a Hilbert–Schmidt operator in $L^2(\mathbb{R}^n)$, and hence we have

$$\|\hat{f}\|_{L^2(\mathbb{R}^n)\to L^2(\mathbb{R}^n)} \le \left[\operatorname{Tr}\left(\hat{f} * \hat{f}\right)\right]^{1/2} = \frac{1}{(2\pi)^{n/2}} \|f\|_{L^2\left(\mathbb{R}^n_x\times\mathbb{R}^n_p\right)} . \tag{B.56}$$

From the last estimate, by multiplying $\hat{f}$ by $(1-\Delta)^s$ and $\left(1+x^2\right)^k$ with appropriate k and s, we obtain the following estimate

$$\|\hat{f}\|_{H^{r+k}_l(\mathbb{R}^n)\to H^r_{l+s}(\mathbb{R}^n)} \le C_{rlsk} \|f\|_{H^{r,l}_{s,k}\left(\mathbb{R}^n_x\times\mathbb{R}^n_p\right)} \tag{B.57}$$

for any $f \in H^{r,l}_{s,k}\left(\mathbb{R}^n_x \times \mathbb{R}^n_p\right)$.

In $S^\infty\left(\mathbb{R}^n_x \times \mathbb{R}^n_p\right)$ consider the convergence inherited from $\mathcal{L}^\infty\left(\mathbb{R}^n_x \times \mathbb{R}^n_p\right)$. Then $\mathcal{L}^\infty$ is a completion of S^∞ in this convergence; it follows from the estimate (B.57) that the mapping

$$\mu : S^\infty\left(\mathbb{R}^n_x \times \mathbb{R}^n_p\right) \to \mathcal{L}\left(S^\infty, S^\infty\right)$$

takes generalized Cauchy sequences into generalized Cauchy sequences and hence extends to be a continuous linear mapping from $\mathcal{L}^\infty\left(\mathbb{R}^n_x \times \mathbb{R}^n_p\right)$ to $\mathcal{L}\left(S^\infty, S^\infty\right)$.

Let us show that μ is a homomorphism, that is, the inverse mapping

$$\mu^{-1} = \operatorname{smb} : \mathcal{L}\left(S^\infty, S^\infty\right) \to \mathcal{L}^\infty\left(\mathbb{R}^n_x \times \mathbb{R}^n_p\right)$$

is continuous. To this end we establish the following estimate on the symbol of an operator T in $\mathcal{L}\left(S^\infty, S^\infty\right)$: $\forall l \in 2\mathbb{Z}_+ \; \exists s \in \mathbb{R} \; \forall r \in 2\mathbb{Z}_+ \; \exists k \in \mathbb{R}$ such that

$$\|\operatorname{smb}\{T\}\|_{H^{r,l}_{s+l,r-k}\left(\mathbb{R}^n_x\times\mathbb{R}^n_p\right)} \le C^{(1)}_{rlsk} \|T\|_{H^{k+n}_{r+n}(\mathbb{R}^n)\to H^r_s(\mathbb{R}^n)} . \tag{B.58}$$

Here $2\mathbb{Z}_+$ is the set of nonnegative even integers. The estimate (B.58) is equivalent to the continuity of the mapping

$$\operatorname{smb} : \mathcal{L}\left(S^\infty, S^\infty\right) \to \mathcal{L}^\infty\left(\mathbb{R}^n_x \times \mathbb{R}^n_p\right).$$

Let us now prove (B.58). Since $T \in \mathcal{L}(S^\infty, S^\infty)$, it follows that $\forall l \in 2\mathbb{Z}_+ \; \exists s \in \mathbb{R}$ $\forall r \in 2\mathbb{Z}_+ \; \exists k \in \mathbb{R}$ such that

$$\|T\|_{H^{k+n}_{r+n}\to H^r_s} < \infty.$$

Set

$$\operatorname{smb}\{T\} = f;$$

then

$$T = f\left(\overset{2}{x}, -i\overset{1}{\frac{\partial}{\partial x}}\right) = \hat{f}.$$

It follows from (B.56) that

$$\|f\|^2_{H^{r,l}_{s+l,r-k}} = (2\pi)^n \operatorname{Tr}\left(\hat{f}_1 \hat{f}_1^*\right),$$

where

$$f_1(x,p) = \left(1+|x|^2\right)^{-\frac{s+l}{2}} = \left(1+|p|^2\right)^{\frac{k-r}{2}} (1-\Delta_x)^{\frac{r}{2}} \left(1-\Delta_p\right)^{\frac{l}{2}} f(x,p).$$

Consequently, we have

$$\|f\|_{H^{r,l}_{s+l,r-k}} \le C^{(2)}_{lrks} \left\| \hat{f}_1 (1-\Delta)^{\frac{n}{2}} \left(1+|x|^2\right)^{\frac{n}{2}} \right\|_{L^2\to L^2}.$$

in the scale $\{H^k_l\}$. This is easy to carry out by using the formula that relates f_1 to f (we have purposely chosen $l \in 2\mathbb{Z}_+$, so that f_1 is expressed in terms of the derivatives of f of integral order). As a result, we obtain

$$\|f\|_{H^{r,l}_{s+l,r-k}} \le C^1_{lrsk} \|\hat{f}\|_{H^{k+n}_{r+n} \to H^r_s}.$$

The theorem is proved. □

Corollary B.1 *An operator* $f\left(\overset{2}{x}, -i\overset{1}{\frac{\partial}{\partial x}}\right)$ *is continuous in* $S^\infty(\mathbb{R}^n)$ *if and only if* $f \in \mathcal{L}^\infty\left(\mathbb{R}^n_x \times \mathbb{R}^n_p\right)$.

We observe that the linear structure and the convergence structure of the spaces $\mathcal{L}^\infty\left(\mathbb{R}^n_x \times \mathbb{R}^n_p\right)$ and $\mathcal{L}\left(S^\infty(\mathbb{R}^n), S^\infty(\mathbb{R}^n)\right)$ are the same. However, these spaces have different algebraic structures. By using the homomorphism μ we can transfer the multiplication from $\mathcal{L}(S^\infty, S^\infty)$ to $\mathcal{L}^\infty\left(\mathbb{R}^n_x \times \mathbb{R}^n_p\right)$, thus obtaining a noncommutative multiplication in the symbol space $\mathcal{L}^*\left(\mathbb{R}^n_x \times \mathbb{R}^n_p\right)$. This multiplication is given by the formula

$$f * g = \operatorname{smb}\left\{ [\![f\left(\overset{2}{x}, -i\overset{1}{\frac{\partial}{\partial x}}\right)]\!] [\![g\left(\overset{2}{x}, -i\overset{1}{\frac{\partial}{\partial x}}\right)]\!] \right\} \tag{B.59}$$

where $[\![\ ]\!]$ are autonomous brackets (see Chapter I). An explicit formula for the symbol of the product in (B.59) is given by the following theorem.

Theorem B.4 *The following relation holds:*

$$f * g = f\left(\overset{2}{x}, p - i\overset{1}{\frac{\partial}{\partial x}}\right) g(x, p)$$

whenever $g \in S^{\infty}\left(\mathbb{R}^n_x \times \mathbb{R}^n_p\right)$.

The proof follows from the results of Chapter II for the case in which both f and g lie in $S^{\infty}\left(\mathbb{R}^n_x \times \mathbb{R}^n_p\right)$. The general result ($f \in \mathcal{L}^{\infty}\left(\mathbb{R}^n_x \times \mathbb{R}^n_p\right)$, $g \in S^{\infty}\left(\mathbb{R}^n_x \times \mathbb{R}^n_p\right)$) then follows by continuity.

Glossary

Ado's theorem A theorem stating that each finite-dimensional Lie algebra can be represented as a subalgebra of a matrix Lie algebra.

asymptotic expansion If H is a linear space equipped with a decreasing filtration,

$$H = H_0 \supset H_1 \supset H_2 \supset \cdots,$$

then an *asymptotic expansion* of an element $h \in H$ with respect to the filtration is a sequence $\{h_n\}$ of elements of H such that

$$h - h_n \in H_n, \quad n = 1, 2, 3, \ldots$$

Depending on the choice of the space and the filtration, various particular types of asymptotic expansions can be obtained.

asymptotic expansion with respect to parameter A type of *asymptotic expansion* in which H is a space of bounded mappings $f : (0, \infty) \to W, \lambda \mapsto f(\lambda)$, where W is a Banach space, and the filtration is determined by the condition

$$f \in H_n \Leftrightarrow \lambda^n f \in H \quad \text{(asymptotic expansion as } \lambda \to \infty)$$

or

$$f \in H_n \Leftrightarrow \lambda^{-n} f \in H \quad \text{(asymptotic expansion as } \lambda \to 0).$$

asymptotic expansion with respect to smoothness An asymptotic expansion in which H is a function space and the filtration is determined by the condition

$$f \in H_n \quad \Leftrightarrow \quad \text{the derivatives of } f \text{ up to the } n\text{th order exist and belong to } H.$$

asymptotic expansion with respect to growth at infinity An expansion in which H is a space of functions of variable(s) x and the greater is n, the less rapidly do $f \in H_n$ grow at infinity (and, for n large, the more rapidly do they decay at infinity).

autonomous brackets A kind of brackets $[\![\ ]\!]$ used to limit the scope of Feynman indices in operator expressions. A subexpression in autonomous brackets has to be computed separately and used in subsequent computations as a simple operator; Feynman indices in the subexpression are valid only within the brackets. Examples:

$$(\overset{1}{A}+\overset{2}{C})^2 = A^2 + 2CA + C^2,$$

but

$$[\![\overset{1}{A}+\overset{2}{C}]\!]^2 = (A+C)^2 = A^2 + AC + CA + C^2;$$

$$e^B e^A = e^{\overset{1}{A}+\overset{2}{B}} \neq e^{[\![\overset{1}{A}+\overset{2}{B}]\!]} = e^{A+B} \quad \text{if} \quad [A, B] \neq 0.$$

The autonomous brackets themselves may bear Feynman indices, which is to be assigned to the operator resulting from the computation of the expression in brackets. Thus,

$$\left(\overset{1}{[\![}\overset{1}{A}\overset{3}{B}]\!] + \overset{2}{C}\right)^2 = BABA + C^2 + 2CBA,$$

whereas

$$(\overset{1}{A}\overset{3}{B} + \overset{2}{C})^2 = B^2A^2 + C^2 + 2BCA.$$

If we use Feynman indices only in a part of an operator expression, it is preferable to enclose parts of operator expressions in autonomous brackets so as to avoid misunderstanding, e.g. for

$$Ce^{\overset{1}{A}+\overset{2}{B}} \quad (e^{\overset{1}{A}+\overset{2}{B}} \text{ multiplied by } C \text{ on the left})$$

it is better to write $\overset{3}{C}e^{\overset{1}{A}+\overset{2}{B}}$ or $C[\![e^{\overset{1}{A}+\overset{2}{B}}]\!]$.

Often autonomous brackets prove useful even if there are no Feynman indices at all, or at least inside the bracket, as in the formula

$$e^{-iS(x)}\left[H\left(\overset{2}{x}, -i\overset{1}{\frac{\partial}{\partial x}}\right)\right]e^{iS(x)} = H\left(\overset{2}{x}, \overset{1}{[\![}\frac{\partial S}{\partial x} - i\frac{\partial}{\partial x}]\!]\right);$$

we could also write $H\left(\overset{2}{x}, \overset{1}{\overline{\partial S/\partial x - i\partial/\partial x}}\right)$ for the right-hand side.

Banach scale A collection of Banach spaces $\{B_\alpha\}_{\alpha\in I}$, indexed by a poset I, such that there is a continuous embedding $B_\alpha \subset B_\beta$ for $\alpha < \beta$. Given a Banach scale, one can obtain various poly-Banach spaces by choosing an arbitrary filter Λ of sections of I and by setting

$$\mathcal{B}_\Lambda = \bigcup_{A\in\Lambda}\bigcap_{\alpha\in A}\mathcal{B}_\alpha,$$

with the convergence defined as follows: a generalized sequence $\{x_\mu\}$ is convergent if it is convergent, for some $A \in \Lambda$, in all B_α with $\alpha \in A$.

Campbell–Hausdorff–Dynkin formula An important formula of Lie theory expressing $\ln(e^B e^A)$ via A, B, and their commutators:

$$\ln(e^B e^A) = A + B + \sum_{k=1}^{\infty} \frac{(-1)^k}{k+1} \sum_{l_i+m_i>0, l_i\geq 0, m_i \geq 0} \frac{[B^{l_1} A^{m_1} \dots B^{l_k} A^{m_k} B]}{l_1! m_1! \dots l_k! m_k!},$$

where the bracket in the numerator stands for the commutator

$$[B^{l_1} A^{m_1} \dots B^{l_k} A^{m_k} B] = \frac{(\mathrm{ad}_B)^{l_1} (\mathrm{ad}_A)^{m_1} \dots (\mathrm{ad}_B)^{l_k} (\mathrm{ad}_A)^{m_k} (B)}{l_1 + \dots + l_k + 1}.$$

Can be obtained by developing into a usual Taylor series in powers of t from the closed formula

$$\ln(e^{tB} e^A) = A + \int_0^t \frac{\ln(e^{t\,\mathrm{ad}_B} e^{\mathrm{ad}_A})}{e^{t\,\mathrm{ad}_B} e^{\mathrm{ad}_A} - 1} (B)\, dt$$

provided by noncommutative analysis.

Campbell–Hausdorff theorem A theorem stating that in a local Lie group G the multiplication law is completely defined by the Lie bracket in its Lie algebra $\mathcal{G}$. In the exponential coordinates of first genus,

$$A \cdot B = A + B + \frac{1}{2}[A, B] + \cdots$$

where $A, B \in \mathcal{G}$, the neighborhood of zero in $\mathcal{G}$ is identified with that in G via the exponential mapping, and the dots stand for commutators of order ≥ 2.

characteristics Let $\mathcal{A}$ be an operator algebra, and let $A = (\overset{1}{A_1}, \dots, \overset{n}{A_n})$ be a Feynman tuple possessing a left ordered representation $l_1, \dots, l_n$. If $B = f(\overset{1}{A_1}, \dots, \overset{n}{A_n})$ is a function of A, then one may consider the pseudodifferential operator

$$l_B = f((\overset{1}{l_1}, \dots, \overset{n}{l_n}) \equiv H\left(y, -i\frac{\partial}{\partial y}\right).$$

The zeroes of its principal symbol (Hamiltonian) are called the *characteristics* of B. In fact, the definition depends on the type of homogeneity considered, which, in turn, is closely related to which asymptotic expansions are to be obtained for solutions to the equation $Bu = v$.

commutation formula The formula

$$[A, f(B)] = \overset{2}{[A, B]}\frac{\delta f}{\delta y}(\overset{1}{B}, \overset{3}{B}).$$

commutator The commutator $[A, B]$ of two elements A, B of an associative algebra $\mathcal{A}$ is defined as

$$[A, B] = AB - BA.$$

commutation relations Let $\mathcal{A}$ be an algebra determined by a set of generators $A_1, \ldots, A_n$ and relations

$$\{\omega(\overset{j_1}{A_{i_1}}, \ldots, \overset{j_k}{A_{i_k}}) = 0\}_{\omega\in\Sigma}.$$

The system of relations Σ is called a system of commutation relations if it possesses a left ordered representation, that is, there exist operators

$$L_1, \ldots, L_n$$

on the space of n-ary symbols such that

$$(L_j f)(\overset{1}{A_1}, \ldots, \overset{n}{A_n}) = A_j[\![f(\overset{1}{A_1}, \ldots, \overset{n}{A_n})]\!]$$

for any n-ary symbol f. In other words, the system of relations is rich enough to guarantee that in any product we can rearrange the operators $A_1, \ldots, A_n$ in the order prescribed by their indices.

composite function formula A formula expressing

$$f([\![g(\overset{1}{A}, \overset{2}{B})]\!]) \overset{\text{def}}{=} f(C),$$

where $C = g(\overset{1}{A}, \overset{2}{B})$. It reads

$$f([\![(g(\overset{1}{A}, \overset{2}{B})]\!]) = f(g(\overset{1}{A}, \overset{2}{B}))$$

$$+\overset{5}{[A, B]}\frac{\delta g}{\delta y_2}(\overset{3}{A}, \overset{4}{B}, \overset{6}{B})\frac{\delta g}{\delta y_1}(\overset{2}{A}, \overset{7}{A}, \overset{8}{B})\frac{\delta^2 f}{\delta y^2}(\overset{1}{[\![}g(\overset{1}{A}, \overset{2}{B})]\!], g(\overset{2}{A}, \overset{8}{B}), g(\overset{7}{A}, \overset{8}{B})).$$

For the particular case in which $g(y_1, y_2) = y_1 + y_2$ one has, say,

$$f([\![A + B]\!]) = f(\overset{1}{A}, \overset{2}{B}) + \overset{1}{[A, B]}\frac{\delta^2 f}{\delta y^2}(\overset{1}{[\![}A + B]\!], \overset{1}{A} + \overset{2}{B}, \overset{1}{A} + \overset{4}{B}),$$

and there is a variety of similar formulas.

Daletskii–Krein formula A formula expressing the derivative with respect to a parameter of a function of an operator depending on this parameter. It reads

$$\frac{d}{dt} f(A(t)) = \overset{2}{\overline{A'(t)}} \frac{\delta f}{\delta x}(\overset{1}{A(t)}, \overset{3}{A(t)}),$$

where

$$\frac{\delta f}{\delta x}(x, y) = \frac{f(x) - f(y)}{x - y}$$

is the difference derivative of $f(x)$. It can also be arranged as

$$df(A) = \frac{\delta f}{\delta x}(R_A, L_A).$$

Here $df(A)$ is the differential of the function f, considered as a mapping

$$\begin{aligned} f : \mathcal{A} &\to \mathcal{A} \\ A &\mapsto f(A) \end{aligned}$$

at the point A, and R_A and L_A are the operators of right and left multiplication by A, respectively. However, the latter formula seems less justified in case f is only partially defined on $\mathcal{A}$ (recall that $f(A)$ is meaningless if A is not a generator).

derivation formula The formula

$$D(f(A)) = \overset{2}{\overline{DA}} \frac{\delta f}{\delta y}(\overset{1}{A}, \overset{3}{A}),$$

where D is any derivation of the operator algebra.

difference derivatives The difference derivative of a function $f(y)$, $y \in \mathbb{R}$, is a function of two variables defined by

$$\frac{\delta f}{\delta y}(y_1, y_2) = \begin{cases} \dfrac{f(y_1) - f(y_2)}{y_1 - y_2}, & y_1 \neq y_2, \\ f'(y_1), & y_1 = y_2. \end{cases}$$

Difference derivatives of higher order are defined inductively; to find $\delta^k f/\delta y^k$ one should fix all but one argument of $\delta^{k-1} f/\delta y^{k-1}$ and apply $\delta/\delta y$ with respect to the remaining argument.

Here are some useful properties of difference derivatives.

a) $\delta^k f/\delta y^k$ is a symmetric function of its $k + 1$ arguments; if f is smooth, then so is $\delta^k f/\delta y^k$ for any k.

b)

$$\frac{\delta^k f}{\delta y^k}(y_1, \ldots, y_{k+1}) = \int_0^1 d\mu_1 \int_0^{1-\mu_1} d\mu_2 \ldots \int_0^{1-\mu_1-\cdots-\mu_{k-1}} d\mu_k f^{(k)}\Big(\mu_1 y_1 + \cdots + \mu_k y_k + \Big(1 - \sum_j \mu_j\Big) y_{k+1}\Big)$$

(here $f^{(k)}$ is the kth derivative of f);

c)

$$\frac{\delta^k f}{\delta y^k}(y_1, \ldots, y_{k+1}) = \sum_{j=1}^{k+1} f(y_j) \prod_{i=1, i\neq j}^{k} (y_j - y_i)^{-1}$$

if $y_i \neq y_j$ for $i \neq j$;

d)

$$\frac{\delta^k f}{\delta y^k}(y, \ldots, y) = \frac{1}{k!} f^{(k)}(y);$$

e)

$$\frac{\delta^k f}{\delta y^k}(y, x, \ldots, x) = \frac{1}{(k-1)!} \int_0^1 (1-\tau)^{k-1} f^{(k)}(\tau y + (1-\tau)x)\, d\tau;$$

f)

$$\left[\left(\frac{\partial}{\partial y_1}\right)^{\alpha_1} \cdots \left(\frac{\partial}{\partial y_{k+1}}\right)^{\alpha_{k+1}} \frac{\delta^k f}{\delta y^k}(y_1, \ldots, y_{k+1})\right]\Bigg|_{y_1=\cdots=y_{k+1}=y} = \frac{\alpha_1! \ldots \alpha_{k+1}!}{(k+|\alpha|)!} f^{(k+|\alpha|)}(y).$$

difference-differential equations An equation containing both derivatives and finite differences. Such an equation can always be considered as a pseudodifferential equation since the difference derivatives can be expressed via the derivation operator; e.g.

$$\delta_h : f(x) \mapsto \frac{f(x+h) - f(x)}{h}$$

is expressed as

$$\delta_h = \frac{e^{h\partial/\partial x} - 1}{h},$$

etc.

differential of an operator function The linear part of the increment

$$f(A+B)-f(A)$$

considered as a function of B. The differential is the element of $\mathcal{L}(\mathcal{A})$, the algebra of linear operators in the operator algebra $\mathcal{A}$, and is given by the formula

$$d[f(A)]=\frac{\delta f}{\delta y}(L_A,R_A),$$

where L_A and R_A are the operators of left and right multiplication by A, respectively.

intertwining operators Let $A=(A_1,\dots,A_n)$ and $B=(B_1,\dots,B_n)$ be linear operators on the spaces E and F, respectively, and let $\mu : E\to F$ be a linear operator. Then say that μ is an intertwining operator for the tuples A and B, or merely that μ intertwines A and B, if the diagram

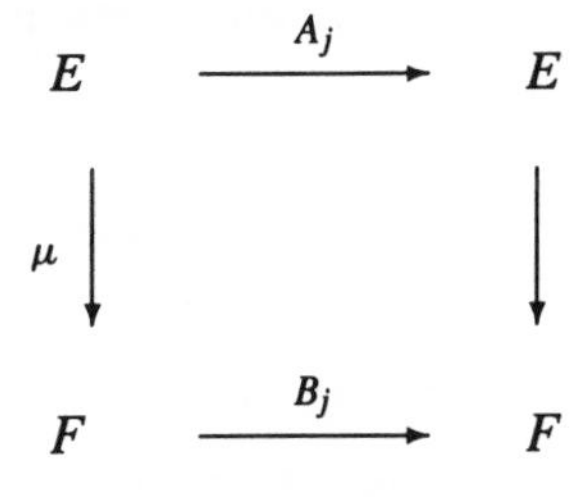

commutes for any $j=1,\dots,n$.

factor extracting rule The rule of noncommutative calculus stating that in the operator expression

$$f(\overset{j_1}{A_1},\dots,\overset{j_n}{A_n})\,g(\overset{k_1}{B_1},\dots,\overset{k_m}{B_m})$$

the factor $f(\overset{j_1}{A_1},\dots,\overset{j_n}{A_n})$ can be isolated by enclosing it into autonomous provided that the indices $k_1,\dots,k_m$ all fall within two categories: either

$$k_i>\max(j_1,\dots,j_n),$$

or

$$k_i<\min(j_1,\dots,j_n),$$

$i=1,\dots,m$. This being true, we have

$$f(\overset{j_1}{A_1},\dots,\overset{j_n}{A_n})\,g(\overset{k_1}{B_1},\dots,\overset{k_m}{B_m})=\overset{j_1}{[\![}f(\overset{j_1}{A_1},\dots,\overset{j_n}{A_n})]\!]\,g(\overset{k_1}{B_1},\dots,\overset{k_m}{B_m}).$$

Feynman indices Indices in operator expressions showing the order in which operators act (the arrangement of operators), the greater is an index, the closer to the left the corresponding operator stands. In writing Feynman indices they are placed just over the corresponding operator argument or over the left autonomous bracket, e.g.,

$$\overset{1}{(}A + \overset{2}{B})\,\overset{1}{(}A - \overset{2}{B}) = \overset{2}{A} - BA + BA + B^2 = A^2 + B^2,$$

$$\overset{1}{[\![}\overset{1}{A}\overset{3}{B} - C\overset{7}{]\!]}\overset{2}{D} = D(BA - C),$$

etc.

The value of an operator expressions depends on the order relation between Feynman indices rather than on the indices themselves. Also indices over operator arguments commuting with all the other arguments in an operator expression are irrelevant and may be omitted by convention.

Feynman quantization Here *quantization* is understood as a rule taking each symbol $f(y_1, \dots, y_n)$ into the operator $f(A_1, \dots, A_n)$ (for a prescribed tuple $A_1, \dots, A_n$). Different quantizations use different ways to define

$$f(A_1, \dots, A_n).$$

The term "Feynman quantization" refers to the case in which one sets

$$f(y_1, \dots, y_n) = f(\overset{1}{A_1}, \dots, \overset{n}{A_n}),$$

i.e., Feynman indices are used to remove the ambiguity in the definition of $f(A_1, \dots, A_n)$.

Feynman tuple A collection $A = (\overset{j_1}{A_1}, \dots, \overset{j_n}{A_n})$ of *generators* in an algebra $\mathcal{A}$, equipped with *Feynman indices* $j_1, \dots, j_n$. If $f(y_1, \dots, y_n)$ is an n-ary symbol then we can define the function of A by setting

$$f(A) \overset{\text{def}}{=} f(\overset{j_1}{A_1}, \dots, \overset{j_n}{A_n}).$$

Feynman's extraction formula The formula

$$T\text{-}\exp\Big[\int_0^t (A(\tau) + B(\tau))\,d\tau\Big] = T\text{-}\exp\Big(\int_0^t A(\tau)\,d\tau\Big)\,T\text{-}\exp\Big(\int_0^t C(\tau)\,d\tau\Big),$$

where

$$C(t) = \Big[T\text{-}\exp\Big(\int_0^t A(\tau)\,d\tau\Big)\Big]^{-1} B(t)\,T\text{-}\exp\Big(\int_0^t A(\tau)\,d\tau\Big),$$

allowing one to extract the factor $T\text{-}\exp\left(\int_0^t A(\tau)\,d\tau\right)$ from the T-exponential of $A(\tau) + B(\tau)$.

Fourier integral operators Functions of $(\overset{2}{x}, -i\overset{1}{\partial/\partial x})$ with symbols given by oscillatory integrals and naturally occurring as parametrices of the Cauchy problem (pseudo)differential equations of hyperbolic type.

generators A *generator* is an element of an *operator algebra* on which functions can be defined. Specifically, let $\mathcal{F}$ be a symbol space. An element A of an operator algebra $\mathcal{A}$ is called an $\mathcal{F}$-generator (or simply a generator, if $\mathcal{F}$ is clear from the context) if there exists a continuous algebra morphism

$$\mu_A : \mathcal{F} \to \mathcal{A}$$

such that $\mu_A(y) = A$. The morphism μ_A, if it exists at all, is always unique. Notation: $\mu_A(f) = f(A)$. The element $f(A)$ of the algebra $\mathcal{A}$ is called the *function* of A with *symbol* $f(y)$.

graded Lie algebras Let G be an abelian group, and let

$$\chi : G \times G \to \mathbb{C} \setminus \{0\}$$

be a mapping such that

i) $\chi(g, \cdot)$ is a homomorphism of groups for each fixed $g \in G$;

ii) $\chi(g, h)\chi(h, g) \equiv 1$.

A G-graded Lie algebra of colour χ is a g-graded vector space $L = \oplus_{g\in G} L_g$ equipped with the bilinear graded Lie bracket $[\cdot, \cdot]$ such that

$$[A, B] = -\chi(|A|, |B|)[B, A]$$

(graded antisymmetry)

$$[A, [B, C]] = [[A, B], C] + \chi(|A|, |B|)[B, [A, C]]$$

(graded Jacobi identity) (here $|A|$ is the gradation of A, etc.).

Any G-graded associative algebra $\mathcal{A} = \oplus_{g\in G}\mathcal{A}_g$ can be made into a graded Lie algebra by setting

$$[A, B] = AB - \chi(|A|, |B|)BA.$$

In applications to particle physics (supersymmetries) one mostly deals with the case

$$\chi(x, y) = (-1)^{xy}.$$

Hamiltonian See characteristics.

Hamilton–Jacobi equation The equation

$$H\left(x, \frac{\partial S(x)}{\partial x}\right) = 0,$$

satisfied by the phase $S(x)$ in the WKB-approximation

$$\psi(x) = e^{(i/h)S(x)}\varphi(x)$$

to the solution of the Schrödinger equation

$$H\left(\overset{2}{x}, -ih\overset{1}{\frac{\partial}{\partial x}}\right)\psi = 0$$

with Hamiltonian $H(x, p)$, and various similar equations arising in the context of Maslov's canonical operator, Fourier integral operators, etc.

Heisenberg algebra The Lie algebra

$$\mathbb{R}^{2n+1} \ni (x, y, \xi) = (x_1, \dots, x_n, y_1, \dots, y_n, \xi)$$

with the bracket (commutation relations)

$$[x_j, y_k] = i\xi\delta_{jk}, \quad [x_j, x_k] = [y_j, y_k] = [x_j, \xi] = [y_j, \xi] = 0,$$

where δ_{jk} is the Kronecker delta. It has the representation

$$\begin{aligned} x_j &\mapsto x_j \quad \text{(multiplication operators)}, \\ y_j &\mapsto -i\frac{\partial}{\partial x_j}, \\ \xi &\mapsto 1 \end{aligned}$$

in $L_2(\mathbb{R}^n)$, widely used in quantum mechanics and the theory of differential, pseudodifferential, and Fourier integral operators. The left ordered representation for the tuple $(\overset{2}{x}_1, \dots, \overset{2}{x}_n, \overset{1}{y}_1, \dots, \overset{1}{y}_n, \overset{3}{\xi})$ is

$$l_\xi = \xi, \quad l_{x_j} = x_j, \quad l_{y_j} = y_j - i\xi\frac{\partial}{\partial x_j}.$$

Heisenberg commutation relations See Heisenberg algebra.

index permutation formula Any of the formulas permitting one to pass from $f(\overset{1}{A}, \overset{2}{B})$ to $f(\overset{2}{A}, \overset{1}{B})$, such as

$$f(\overset{1}{A}, \overset{2}{B}) = f(\overset{1}{A}, \overset{2}{B}) + \overset{3}{[B, A]}\frac{\delta^2 f}{\delta x \delta y}(\overset{1}{A}, \overset{5}{A}, \overset{2}{B}, \overset{4}{B}).$$

Jacobi condition The Jacobi condition is a condition on the operators of the left ordered representation. Let $(\overset{1}{A_1}, \dots, \overset{n}{A_n})$ be a Feynman tuple, and suppose that the operators $A_1, \dots, A_n$ satisfy a system of alglebraic relations, say,

$$\left\{\omega(\overset{j_1}{A_{i_1}}, \dots, \overset{j_s}{A_{i_s}}) = 0\right\}_{\omega\in\Omega},$$

where the sequences $i_1, \dots, i_s$ and $j_1, \dots, j_s$ are, in general, chosen differently for different $\omega \in \Omega$.

Suppose also that there exists a left ordered representation $(l_1, \dots, l_n)$ of the tuple $(A_1, \dots, A_n)$.

Then the Jacobi condition requires that

$$\omega(\overset{j_1}{l}_{i_1}, \dots, \overset{j_s}{l}_{i_s}) = 0$$

for all $\omega \in \Omega$.

In fact, the Jacobi condition guarantees monomorphy of the mapping

$$f(x_1, \dots, x_n) \to f(\overset{1}{A}, \dots, \overset{n}{A})$$

provided that the system Ω and its consequences exhaust all possible relations between the operators $A_1, \dots, A_n$. In the context of Lie algebras, this leads directly to the Poincaré–Birkhoff–Witt theorem stating that the ordered monomials $A_n^{\alpha_n}, \dots, A_1^{\alpha_1}$ form a basis in the enveloping algebra of the Lie algebra with basis $A_1, \dots, A_n$.

The system of equations given by the Jacobi condition can be viewed as a system for finding $l_1, \dots, l_n$. Being equipped with the additional regularity condition

$$f(\overset{1}{l_1}, \dots, \overset{n}{l_n})(1) = f(y_1, \dots, y_n),$$

it may well serve as such.

Jacobi identity The identity

$$[A, [B, C]] = [[A, B], C] + [B, [A, C]]$$

for the commutation in a Lie algebra, or the identity

$$[A, [B, C]] = [[A, B], C] + \chi(|A|, |B|)[B, [A, C]]$$

for the graded χ-commutator on a graded Lie algebra.

Leibniz rule The identity

$$D(AB) = D(A)B + AD(B)$$

to be satisfied by any derivation D of an algebra.

Nelson's condition A condition on the generators $A_1, \dots, A_n$ of a strongly continuous unitary representation of an n-dimensional Lie group G. Nelson's condition states that $A_1, \dots, A_n$ must be essentially self-adjoint on some dense subset D in the representation space H together with the operator

$$\Delta = -(A_1^2 + A_2^2 + \dots + A_n^2).$$

Newton formula The expansion of $f(C) - f(A)$

$$f(C) - f(A) = \sum_{k=1}^{N_1} \overset{2}{[\![C - A]\!]} \dots \overset{2k}{[\![C - A]\!]} \frac{\delta^k f}{\delta x^k}(\overset{1}{A}, \overset{3}{A}, \dots, \overset{2k+1}{A}) + R_N$$

with the remainder

$$R_N = \overset{2}{[\![C - A]\!]} \dots \overset{2N}{[\![C - A]\!]} \frac{\delta^k f}{\delta x^k}(\overset{1}{C}, \overset{3}{A}, \dots, \overset{2k+1}{A}).$$

Lie algebra A Lie algebra L is a linear space L equipped with a bilinear operation $[\cdot, \cdot]$ (Lie bracket) such that

$$[A, B] + [B, A] = 0 \quad \text{for any} \quad A, B \in L, \quad \text{(skew-symmetry)}$$
$$[[A, B], C] + [[B, C], A] + [[C, A], B] = 0 \quad \text{(Jacobi identity)}.$$

normal form Let $\mathcal{A}$ be an operator algebra and $A = (\overset{1}{A_1}, \dots, \overset{n}{A_n})$ a Feynman tuple of $\mathcal{F}$-generators on $\mathcal{A}$ (here $\mathcal{F}$ is some fixed symbol class). A normal form of an operator $B \in \mathcal{A}$ is its representation as

$$B = f(\overset{1}{A_1}, \dots, \overset{n}{A_n})$$

with some $f \in \mathcal{F}_n$. The normal form neither exists nor is unique in general. However, if the tuple A has the left ordered representation $(l_1, \dots, l_n)$ consisting of $\mathcal{F}$-generators in $\mathcal{F}$, then any operator of the form

$$B = \varphi(\overset{1}{A_{j_1}}, \dots, \overset{s}{A_{j_s}}), \quad \varphi \in \mathcal{F}_s$$

can be reduced to a normal form by setting

$$f(y_1, \dots, y_n) = \varphi(\overset{1}{l_{j_1}}, \dots, \overset{s}{l_{j_s}})(1).$$

The normal form is unique if the left ordered representation operators satisfy the generalized Jacobi condition.

operator algebras In this book, by an "operator algebra" we mean an associative algebra with identity element equipped with an appropriate *convergence* (see Chapter IV for details on this point). That is, an operator algebra $\mathcal{A}$ is a linear space $\mathcal{A}$ over $\mathbb{C}$ on which a bilinear operation (multiplication) $A, B \mapsto AB$ is defined such that

(i) $(AB)C = A(BC)$ for any $A, B, C \in \mathcal{A}$;

(ii) there exists an element $1 \in \mathcal{A}$ (the two-sided identity element) such that

$A1 = 1A = A$ for any $A \in \mathcal{A}$;

(iii) the linear operations and the multiplication are continuous, e.g., if A_α and B_α are two (generalized) sequences in $\mathcal{A}$ converging, respectively, to A and B, then $A_\alpha B_\alpha$ is convergent to AB.

Typical examples of operator algebras are supplied by algebras of continuous linear operators in some linear space. What is more, any operator algebra can be realized in this way by using the actual embedding $\mathcal{A} \to \mathcal{L}(\mathcal{A})$, where $\mathcal{L}(\mathcal{A})$ is the algebra of linear operators in $\mathcal{A}$ (see regular representations).

operator-valued symbol Let W and V be operator algebras, let $\mathcal{F}$ be a symbol class, and let f be a W-valued function of $(y_1, \dots, y_n)$ such that $f \in \mathcal{F}_n \hat{\otimes} W$. If $A_1, \dots, A_n$ are $\mathcal{F}$-generators in V, then the operator

$$B = f(\overset{1}{A_1}, \dots, \overset{n}{A_n}) \in V \hat{\otimes} W$$

is well-defined (the mapping $f \mapsto f(\overset{1}{A_1}, \dots, \overset{n}{A_n})$ extends by continuity from $\mathcal{F}_n \otimes W$, where its definition is obvious, to $\mathcal{F}_n \hat{\otimes} W$). The operator B is called a function of $\overset{1}{A_1}, \dots, \overset{n}{A_n}$ with *operator-valued symbol* f.

A slightly more complicated situation occurs if f takes values in the algebra itself where $A_1, \dots, A_n$ lie, $f \in \mathcal{F}_n \hat{\otimes} V$. The values of f need not commute with $A_1, \dots, A_n$, so that one must be careful. We define

$$B = \overset{k}{f}(\overset{1}{A_1}, \dots, \overset{n}{A_n}),$$

where k, the *Feynman index of the symbol*, is distinct from $1, \dots, n$, as follows. Let

$$f = \sum f_i \otimes v_i \in \mathcal{F}_n \otimes V.$$

We set

$$\overset{k}{f}(\overset{1}{A_1}, \dots, \overset{n}{A_n}) = \sum f_i(\overset{1}{A_1}, \dots, \overset{n}{A_n}) \overset{k}{v_i}$$

(this is well-defined since v_i occur only linearly, and we need not assume that v_i is a generator). This definition extends to the entire $\mathcal{F}_n \hat{\otimes} V$ by continuity.

ordered representations Let $\mathcal{F}$ be a fixed symbol space, $A = (\overset{1}{A_1}, \dots, \overset{n}{A_n})$ a Feynman tuple in an operator algebra $\mathcal{A}$. The *left ordered representation* of A on $\mathcal{F}_n$ is the Feynman tuple $l = (\overset{1}{l_1}, \dots, \overset{n}{l_n})$ of operators determined by the following properties:

(i) $(l_i f)(A) = A_i [\![f(A)]\!]$ for any $i = 1, \dots, n$ and $f \in \mathcal{F}_n$;

(ii) $(l_i f)(y) = y_i f(y)$ whenever f is independent of $y_{i+1}, \dots, y_n$.

Under the assumption that $l_1, \dots, l_n$ are $\mathcal{F}$-generators in $\mathcal{F}_n$, the following important property holds for these operators:

$$[\![f(A)]\!] [\![g(A)]\!] = [f(l)g](A)$$

for any $f, g \in \mathcal{F}_n$.

The right ordered representation is defined similarly; we replace the conditions (i) and (ii) by the conditions

(i′) $(r_i f)(A) = [\![f(A)]\!] A_i$ for any $i = 1, \dots, n$ and $f \in \mathcal{F}_n$;

(ii′) $(r_i f)(y) = y_i f(y)$ whenever f is independent of $y_1, \dots, y_{i-1}$.

Then

$$[\![f(A)]\!] [\![g(A)]\!] = [g(r)f](A)$$

for any $f, g \in \mathcal{F}_n$.

path integral Also Feynman's path integral. The expression

$$\begin{aligned} \psi(x,t) &= \int\!\!\int [dx] \left[\frac{d\dot{x}}{2\pi}\right] \exp\left\{ i \int_0^t (p\dot{q} - H)\, d\tau \right\} \psi_0(x_0) dx_0 \\ &\stackrel{\text{def}}{=} \lim_{N\to\infty,\ \max \Delta t_i \to 0} \int_{\mathbb{R}^{2N}} e^{i\left(\sum_{k=0}^{N_1} p_k(x_{k+1}-x_k) - H(x_{k+1}, p_k, \tau_k)\Delta t_k\right)} \\ &\quad \times \psi_0(x_0) \prod_{k=0}^{N-1} \frac{dx_k dp_k}{2\pi} \end{aligned}$$

for the solution to the Schrödinger equation

$$-i\frac{\partial \psi}{\partial t} + H\left(\overset{2}{x}, -i\overset{1}{\frac{\partial}{\partial x}}\right)\psi = 0$$

with the initial data

$$\psi(x,0) = \psi_0(x).$$

Poincaré–Birkhoff–Witt theorem A theorem stating that in the enveloping algelbra $U(L)$ of a given Lie algebra L the elements $a_n^{\alpha_1}, \dots, a_1^{\alpha_n}$ form a basis, where $(a_1, \dots, a_n)$ is a basis of L.

Poisson algebra A space $\underset{\sim}{\Phi}$ of smooth functions on a manifold M, equipped with a *Poisson bracket*. A Poisson bracket is a bilinear operation $\{\cdot, \cdot\}$ on $\underset{\sim}{\Phi}$ such that

$$\{f, g\} = \sum_{j,k} \omega_{jk}(x) \frac{\partial f}{\partial x_j} \frac{\partial g}{\partial x_k}$$

in local coordinates. For this operation the skew-symmetry property

$$\{f, g\} = -\{g, f\},$$

and the Jacobi identity

$$\{f, \{g, h\}\} + \{h, \{f, g\}\} + \{g, \{h, f\}\} = 0$$

are satisfied.

poly-Banach space A linear space $\mathcal{B}$ equipped with a polynorm

$$p : \mathcal{B} \times I \to \mathbb{R}_+ \cup \{\infty\}$$

satisfying the axioms given below and with convergence defined as follows: a generalized sequence $\{x_\mu\}$ is convergent in $\mathcal{B}$ if for some $A \in \Lambda$ (where Λ is a given filter of sections of the bidirected set I) and for any $\alpha \in \Lambda$ $p(x_\mu, \alpha) < \infty$ and x_μ is convergent in the seminorm $p(\cdot, \alpha)$.

Axioms of polynorm:

1) For any $x \in \mathcal{B}$ there exists an α such that $p(x, \alpha) < \infty$.

2) $p(\cdot, \alpha)$ is a seminorm on $\mathcal{B}_\alpha = \{x \in \mathcal{B} \mid p(x, \alpha) < \infty\}$.

3) $p(x, \alpha) \le p(x, \beta)$ whenever $\alpha \ge \beta$.

product theorem A theorem stating that if a Feynman tuple $A = (\overset{1}{A_1}, \dots, \overset{n}{A_n})$ possesses a left ordered representation, then for any two symbols $f, g \in \mathcal{F}_n$ the product $[\![f(\overset{1}{A_1}, \dots, \overset{n}{A_n})]\!][\![g(\overset{1}{A_1}, \dots, \overset{n}{A_n})]\!]$ can be reduced to a normal form,

$$[\![f(\overset{1}{A_1}, \dots, \overset{n}{A_n})]\!][\![g(\overset{1}{A_1}, \dots, \overset{n}{A_n})]\!] = [f(\overset{1}{l_1}, \dots, \overset{n}{l_n})g](\overset{1}{A_1}, \dots, \overset{n}{A_n})$$

provided that $l_1, \dots, l_n$ can be substituted into f (e.g., f is a polynomial or $l_1, \dots, l_n$ are $\mathcal{F}$-generators in $\mathcal{F}_n$). Similarly,

$$[\![f(\overset{1}{A_1}, \dots, \overset{n}{A_n})]\!][\![g(\overset{1}{A_1}, \dots, \overset{n}{A_n})]\!] = [g(\overset{-1}{r_1}, \dots, \overset{-n}{r_n})f](\overset{1}{A_1}, \dots, \overset{n}{A_n}),$$

where $r_1, \dots, r_n$ are the operators of the right ordered representation. We use here the negative indices over the operators $r_1, \dots, r_n$.

pseudodifferential operators Functions of the operators

$$(\overset{2}{x}, -i\overset{1}{\partial/\partial x})$$

with symbols $f(x, p)$ that are sums of homogeneous functions with respect to p (the classical pseudodifferential operators) or with symbol $f(x, p)$ satisfying the estimates

$$\left|\frac{\partial^{\alpha+\beta} f}{\partial x^{\alpha} \partial p^{\beta}}(x, p)\right| \leq C(1+|p|)^{m-\rho|\alpha|+\delta|\beta|}, \quad |\alpha|, |\beta| = 0, 1, 2, \ldots$$

when x lies in a compact set; here $0 \leq \delta < \rho \leq 1$ (Hörmander's classes of pseudodifferential operators).In the broad sense, *any* function of $(\overset{2}{x}, -i\overset{1}{\partial/\partial x})$ may be referred to as a pseudodifferential operator.

quantization This word has quite a few different meanings. In this book it refers to the choice of ordering of operators in substituting them for (commuting) numerical arguments of a function. See Feynman quantization, Weyl quantization.

quantum oscillator The quantum system described by the Schrödinger equation

$$H\left(x, -ih\frac{\partial}{\partial x}\right)\psi = E\psi$$

with quadratic Hamiltonian

$$H(x, p) = \frac{p^2}{2m} + \omega^2 x^2.$$

quantum groups Quantum groups were introduced in [114], [166], and [47] as a formal object useful in the quantum method of the inverse scattering problem. The theory of quantum groups has been developing intensively since then; we refer the reader to [38], [170], [32], and the paper cited therein for detailed information (one should also mention the very important paper [152]). Here we only give some brief remark, following [38] and [170].

First, let us recall the notion of a Poisson algebra. Let M be a smooth manifold and $A = F(M)$ a space of functions on M. Suppose that A ia equipped with a bilinear operation

$$(a, b) \mapsto \{a, b\}$$

that is antisymmetric and satisfies the Jacobi identity. Then we say that A is a Poisson algebra and refer to $\{\,,\,\}$ as the Poisson bracket. A *quantization* of a Poisson algebra A on M is an algebra A_h over the algebra $\mathbb{C}[[h]]$ of formal power series such that

i) there is an isomorphism $A_h/hA_h \cong A$;

ii) $\frac{1}{h}[a, b] \bmod h = \{a, b\}$ under this isomorphism;

iii) A_h is a topologically free $\mathbb{C}[[h]]$-module.

Now let G be a Lie group and $H = F(G)$ a space of functions on G. Suppose that H is equipped with a Poisson bracket such that the multiplication map

$$\mu : G \times G \to G$$

is a morphism of Poisson manifolds.

Note that H is a Hopf algebra; namely, the mappings

$$\Delta : H \to H \otimes H \quad \text{(comultiplication)},$$
$$S : H \to H \quad \text{(antipode), and}$$
$$\varepsilon : H \to \mathbb{C} \quad \text{(counit)}$$

are defined by duality to the multiplication, inversion $x \mapsto x^{-1}$, and unit $\{e\} \to G$, respectively, and these mappings satisfy the axioms of a Hopf algebra.

A *quantum group* is a quantization H_q of H in the class of Poisson Hopf algebras (here q is the tuple of quantization parameters). Thus, in the spirit of algebraic geometry, we consider a ring H of functions on G as the primary object and quantize H rather that G itself. Note that G can be reconstructed from H by applying the functor Spec; in contrast, H_q is usually not commutative, so applying Spec to H_q would make little sense. Here by Spec H we denote the space of maximal ideals in the ring H.

It is convenient to describe quantum groups in dual terms. Let H be the Hopf algebra of regular functions on G. Then the universal enveloping algebra $U(\mathcal{G})$ of the corresponding Lie algebra $\mathcal{G}$ is contained in H^*, and H is known to be isomorphic to a Hopf algebra consisting of matrix elements of finite-dimentional representations of $U(\mathcal{G})$. The quantization $U_q(\mathcal{G})$ of $U(\mathcal{G})$ is easy to describe (see the example below). Then H_q can be defined as the Hopf algebra generated by matrix elements of finite-dimensional representations of $U_q(\mathcal{G})$.

Consider a simple example (quantization of $SL_2(\mathbb{C})$). Let

$$e = \begin{pmatrix} 0 & 1 \\ 1 & 0 \end{pmatrix}, \quad h = \begin{pmatrix} 1 & 0 \\ 0 & -1 \end{pmatrix}, \quad f = \begin{pmatrix} 0 & 0 \\ 1 & 0 \end{pmatrix}$$

be the standard Chevalley basis of the Lie algebra $sl_2 = sl_2(\mathbb{C})$. We have the commutation relations

$$[e, f] = h, \quad [h, e] = 2e, \quad [h, f] = -2f. \tag{B.60}$$

The universal enveloping algebra $U(sl_2)$ is the algebra with generators e, f, h, determined by the relations (B.60).

Now let $q \neq 0, \pm 1$ be a complex number. Consider the associative algebra $U_q(sl_2)$ with generators e, f, a, a^{-1} and defining relations

$$\begin{aligned} aa^{-1} &= a^{-1}a = 1, \\ ae &= q^2 ea, \quad af = q^{-2} fa, \\ [e, f] &= (a - a^{-1})/(q - q^{-1}). \end{aligned}$$

The algebra $U(sl_2)$ can be obtained as the limit of $U_q(sl_2)$ as $q \to 1$ in the following way. Let

$$q = 1 + \varepsilon, \quad a = 1 + \varepsilon h + \cdots, \quad \varepsilon \to 0.$$

Then we obtain

$$e + \varepsilon he = (1 + 2\varepsilon)(e + \varepsilon eh) + \cdots,$$

or

$$\varepsilon[h, e] = 2\varepsilon e + \cdots,$$

and $[h, e] \to 2e$ as $\varepsilon \to 0$. Similarly, $[h, f] \to -2f$ and $[e, f] \to h$. We omit the verification of the fact that $U_q(sl_2)$ has the required Hopf algebra structure.

Hence, we have described the quantization of the enveloping algebra $U(sl_2)$.

regular representations Representations of an algebra on itself by left and right multiplications. Let $\mathcal{A}$ be an associative algebra with identity element. For any $A \in \mathcal{A}$ denote by L_A and R_A two linear operators acting on the linear space $\mathcal{A}$ by the formula

$$L_A B = AB, \quad R_A B = BA, \quad B \in \mathcal{A}.$$

The linear mappings

$$\begin{aligned} L : \mathcal{A} &\to \mathcal{L}(\mathcal{A}) & & & R : \mathcal{A} &\to \mathcal{A}(\mathcal{A}) \\ & & \text{and} & & & \\ A &\mapsto L_A & & & A &\to R_A, \end{aligned}$$

where $\mathcal{L}(\mathcal{A})$ is the algebra of linear operators on $\mathcal{A}$, are called the *left* and the *right* regular representations of $\mathcal{A}$. In fact, L is a representation, but R is an *antirepresentation* since it reverses the order of factors,

$$R_A R_B = R_{BA}.$$

Both L and R are faithful representations, and they commute with each other,

$$L_A R_B = L_B R_A$$

for any $A, B \in \mathcal{A}$.

representations A representation of an associative algebra $\mathcal{A}$ on a linear space E is a homomorphism of $\mathcal{A}$ into the algebra $\mathrm{End}(()E)$ of linear operators on E. A representation of a Lie algebra is defined in the same way, but the algebra $\mathrm{End}(()E)$ should be considered as a Lie algebra with Lie bracket

$$[A, B] \stackrel{\mathrm{def}}{=} AB - BA, \quad A, B \in \mathrm{End}(E).$$

Sometimes a representation is understood as a morphism into an arbitrary algebra. A representation $\varphi : \mathcal{A} \to \mathrm{End}(E)$ is said to be faithful if $\operatorname{Ker} \varphi = 0$, i.e. $\varphi(a) = 0$ implies $a = 0$.

resolvent formulas The identites for the resolvent of a linear operator A:

$$R_\lambda(A) - R_\mu(A) = R_\lambda(A) R_\mu(A)\,(\mu - \lambda)$$

(first resolvent identity);

$$R_\lambda(A) - R_\lambda(B) = R_\lambda(A)(A - B)R_\lambda(B)$$

(second resolvent identity).

semigroup property A family of linear operators $T(t)$, $t \in \mathbb{R}_+$, is said to satisfy the *semigroup property* if

$$T(t + \tau) = T(t)T(\tau), \quad t, \tau \in \mathbb{R}_+.$$

stationary phase method A method to obtain the asymptotic expansion as $h \to 0$ of the integral

$$I(h) = \left(\frac{i}{2\pi h}\right)^{n/2} \int\limits_{\mathbb{R}^n} e^{iS(x)/h} \varphi(x)\, dx,$$

where $h \in (0, 1]$, $S(x)$ is smooth and real-valued, and $\varphi(x)$ is smooth and compactly supported. Suppose that $S(x)$ has a unique stationary point x_0 on $\operatorname{supp} \varphi(x)$, that is,

$$\frac{\partial S}{\partial x}(x_0) = 0, \quad \frac{\partial S}{\partial x}(x) \neq 0 \quad \text{for} \quad x \neq x_0, \quad x \in \operatorname{supp} \varphi.$$

Furthermore, suppose that the stationary point is nondegenerate in the sense that

$$J = \det\left(-\frac{\partial^2 S(x_0)}{\partial x_i \partial x_j}\right) \neq 0.$$

Then the integral $I(h)$ can be expanded into the asymptotic series

$$I(h) = I_0 + hI_1 + h^2 I_2 + \cdots,$$

where

$$I_0 = \frac{e^{iS(x_0)/h}\varphi(x_0)}{\sqrt{J}},$$

and the argument of J is determined by

$$\arg J = \sum_{k=1}^{n} \arg \lambda_k,$$

where λ_k are the eigenvalues of the matrix

$$\left\| \frac{\partial^2 S(x_0)}{\partial x_i \, \partial x_j} \right\|,$$

and $\arg \lambda_k$ are chosen in the interval

$$-\frac{3\pi}{2} \leq \arg \lambda_k < \frac{\pi}{2}.$$

There is also a version of the stationary phase method for the case in which the function $S(x)$ is complex-valued.

symbol spaces Intuitively, symbols are functions whose arguments we can replace by operators, thus obtaining functions of operators and operator expressions.

In order to obtain a calculus, some assumptions should however be made. They are listed below, and throughout the book we require them to be valid for all symbol spaces considered.

a) *Unary* symbols. A space $\mathcal{F}$ of unary sumbols is a space consisting of functions $f(y)$ depending on a simple variable y (complex or real). We require that:

(a1) The functions $f \in \mathcal{F}$ are defined either on some domain in $\mathbb{R}$ or $\mathbb{C}$ or in the neighborhood of some closed subset in $\mathbb{R}$ or $\mathbb{C}$.

(a2) $\mathcal{F}$ is an algebra with respect to pointwise multiplication of fuctions. $\mathcal{F}$ contains the constant functions and the coordinate function $f(y) = y$.

(a3) $\mathcal{F}$ is equipped with some convergence (cf. Chapter IV), and the algebraic operations are continuous.

(a4) The difference derivative $\delta/\delta y$ acts continuously in the spaces

$$\frac{\delta}{\delta y} : \mathcal{F} \to \mathcal{F}\hat{\otimes}\mathcal{F},$$

where $\mathcal{F}\hat{\otimes}\mathcal{F}$ is the completed tensor product as defined in Chapter IV.

(b) *Multivariate* symbols. The space $\mathcal{F}_n$ of n-ary symbols is defined as the product

$$\mathcal{F}_n = \underbrace{\mathcal{F}\hat{\otimes} \cdots \hat{\otimes}\mathcal{F}}_{n \text{ copies}} \overset{\text{def}}{\equiv} \mathcal{F}^{\hat{\otimes} n},$$

where $\mathcal{F}$ is the given space of unary symbols. Sometimes a more complicated construction is used, in which different spaces of unary symbols are allowed for different variables. Specifically,

$$\mathcal{F}_n = \mathcal{F}_{(1)} \hat{\otimes} \cdots \hat{\otimes} \mathcal{F}_{(n)},$$

where $\mathcal{F}_{(1)} \dots \mathcal{F}_{(n)}$ are the given spaces of unary symbols.

The latter construction is mostly used for the case in which some of

$$\mathcal{F}_{(1)}, \dots, \mathcal{F}_{(n)}$$

are the algebra of polynomials, and the other are equal to some $\mathcal{F}$.

synchronous asymptotic expansion A particular type of asymptotic expansion in which the filtration chosen provides two or more simple types of asymptotic expansions simultaneously.

For example, if

$$H = H_0 \supset H_1 \supset H_2 \supset \cdots$$

is the filtration associated with asymptotic expansions with respect to $h \to 0$, and

$$H = G_0 \supset G_1 \supset G_2 \supset \cdots$$

is the filtration (in the same space) associated with asymptotic expansions with respect to smoothness, then the filtration

$$H = H_0 \cap G_0 \supset H_1 \cap G_1 \supset H_2 \cap G_2 \supset \cdots$$

is said to provide *simultaneous* asymptotic expansions with respect to the parameter h and differentiability.

Taylor formula The formula expressing $f(C) - f(A)$ in the form

$$f(C) - f(A) = \sum_{k=1}^{N} \frac{1}{k} f^{(k)}(A)[\![(\overset{1}{C} - \overset{2}{A})^k]\!] + Q_N$$

with the remainder

$$Q_N = \overset{2}{[\![}(\overset{1}{C} - \overset{2}{A})^k]\!] \frac{\delta^N f}{\delta x^N}(\overset{1}{C}, \overset{3}{A}, \dots, \overset{3}{A})$$

and having the disadvantage of that it cannot provide the expression of $f(C) - f(A)$ in powers of ε for $C = A + \varepsilon B$, unless the commutators of A and B satisfy some additional conditions such as nilpotency, etc.

tensor product The *algebraic tensor product* of two linear spaces A and B over $\mathbb{C}$ is the space generated by finite formal sums

$$\sum \alpha_i a_i \otimes b_i, \quad a_i \in A, \quad b_i \in B, \quad \alpha_i \in \mathbb{C}$$

and factorized by the relations

$$\begin{aligned} a \otimes (b+c) &= a \otimes b + a \otimes c, \\ (a+b) \otimes c &= a \otimes c + b \otimes c, \\ \lambda(a \otimes b) &= \lambda a \otimes b = a \otimes \lambda b. \end{aligned}$$

The tensor product of A and B is denoted by $A \otimes B$.

The *projective tensor product* of Banach spaces A and B is the completion of $A \otimes B$ with respect to the *projective norm*

$$\|f\| = \inf \sum \|a_i\| \, \|b_i\|,$$

where the infimum is taken over all (finite) representations

$$f = \sum_i a_i \otimes b_i.$$

The projective tensor product is denoted by $A \hat{\otimes} B$. The projective tensor product of poly-Banach spaces is defined as the completion of $A \otimes B$ with respect to a certain projective polynorm on $A \otimes B$.

In any case, the tensor product satisfies the following universal mapping property: if $\psi : A \times B \to C$ is a (continuous) bilinear mapping, then there exists a unique (continuous) linear mapping $\tilde{\psi} : A \otimes B \to C$ such that $\psi = \tilde{\psi} \circ i$, where

$$\begin{aligned} i : A \times B &\to A \otimes B, \\ (a, b) &\mapsto a \otimes b \end{aligned}$$

is the natural bilinear mapping A.

***T*-exponentials** Let $A(t)$ be a family of operators with parameter $t \in \mathbb{R}$. Consider the Cauchy problem

$$\frac{dU(t)}{dt} = A(t)U, \quad U(0) = I \quad \text{(identity operator).}$$

Were the operators $U(t)$ commuting with one another, we would have

$$U(t) = \exp\left(\int_0^t A(t)\, dt \right).$$

In lack of commutativity, we denote

$$U(t) = T\text{-}\exp\left(\int_0^t A(t)\,dt\right).$$

The prefix T indicates that the operators $A(t)$ in that expression are ordered by their arguments t. Using Feynman indices, we can write

$$U(t) = \exp\left(\int_0^t \overset{t}{A}(t)\,dt\right),$$

i.e., $U(t)$ is a function of the "infinite Feynman tuple" $\left\{\overset{t}{A}(t)\right\}_{t\in\mathbb{R}}$. Accordingly, the symbol also depends on infinitely many arguments, i.e., is a functional: $U(t) = f(\{\overset{t}{A}(t)\})$, where

$$f(\{y(t)\}) = \exp\left(\int_0^t y(t)\,dt\right).$$

Either of these expressions is referred to as the T-exponential.

Trotter formula Any of the formulas like

$$e^{[\![A+B]\!]} = \lim_{n\to\infty} \underbrace{e^{A/n}e^{B/n}\dots e^{A/n}e^{B/n}}_{2n \text{ factors}}$$

expressing the exponential of a sum via the exponentials of the summands.

Trotter-type formulas Any of the formulas like

$$T\text{-}\exp\Big(\int_0^t \overset{\tau}{[\![} f(\overset{1}{A}, \overset{2}{B}, \tau)]\!]\,d\tau\Big) = T\text{-}\exp\Big(\int_0^t f(\overset{\tau}{A}, \overset{\tau+0}{B}, \tau)\,d\tau\Big)$$

which allows one to remove the autonomous brackets in a T-exponential and further to pass from the T-exponential to the *Feynman path integral*.

twisted product The product $*$ in the symbol space introduced via the operators of the left ordered representation:

$$f * g = f(l)(g).$$

With this product, the mapping symbol $\to$ operator is a homomorphism of algebras.

The associativity of the $*$-product is generated by the Jacobi condition.

Weyl quantization A method of *quantization* in which all operators are assumed to act "simultaneously." More precisely, if $f(y_1, \dots, y_n)$ is a power of a linear function and $A = (A_1, \dots, A_n)$ is a tuple of operators, then the Weyl-quantized function $f_w(A_1, \dots, A_n)$ is defined as the power of the given linear function of $A_1, \dots, A_n$: if

$$f(y_1, \dots, y_n) = (\alpha_1 y_1 + \dots + \alpha_n y_n)^m,$$

then

$$f_w(A_1, \dots, A_n) = [\![\alpha_1 A_1 + \dots + \alpha_n A_n]\!]^m.$$

By linearity, this definition extends to arbitrary polynomials. As for wider symbol classes, the definition for, say, Fourier-representable functions can be given as follows:

$$f_w(A_1, \dots, A_n) = \left(\frac{1}{2\pi i}\right)^n \int\limits_{\mathbb{R}^n} \tilde{f}(t_1, \dots, t_n) \exp(i[\![t_1 A_1 + \dots + t_n A_n]\!])\, dt,$$

where $\tilde{f}(t)$ is the Fourier transform of $f(y)$.

The Weyl quantization is a more delicate issue then the Feynman quantization both in that it imposes stronger analytic requirecounts and in that the calculus is less evident with the Weyl quantization.

Wick normal form The representation of an operator A in the quantum state space as a function of the creation a^+ and annihilation a^- operators in which all a^- act first and all a^+ act last:

$$A = f(\overset{2}{a^+}, \overset{1}{a^-}) = \sum_{j,k=0}^{\infty} f_{jk} (a^+)^j (a^-)^k.$$

Bibliographical Remarks

Chapter I

Ordering operators by indices, one of the main ideas in noncommutative analysis, was first introduced by Feynman [54], but its systematic use in conjunction with other tools such as autonomous brackets was inspired only by Maslov's book [129]. Attempts to define functions of several noncommuting operators necessarily involve some kind of ordering. The Feynman ordering is apparently the most convenient one and is the only ordering used in this book. However, there exist several other kinds of ordering, of which the most widespread and familiar is undoubtedly the Weyl ordering (Subsection 2.6). It originally arose in connexion with problems of quantum mechanics. We refer the reader to [191], [2], [3], [4], [151] and [101] for details concerning the Weyl ordering. Almost not touched upon in this book are problems pertaining to functions of several *commuting* operators. Not that we consider them too simple; they merely lie off the mainstream of our exposition, and we give only the information absolutely necessary to maintain the desired level of rigor in dealing with the noncommutative case. The reader interested in problems specific to commutative functional calculus may refer to either textbooks on functional analysis such as [78], [105], and [197] or to more special treatments such as [14] and [174].

T-exponentials (Subsection 1.1) were introduced by Feynman [54], who also discovered a striking relationship between T-exponentials and path integrals (cf. [55], [150]). We describe this relationship in Subsection 1.7 following the paper [135]. Various functional-analytic subtleties, which are beyond the framework of an elementary textbook, arise in connexion with the T-exponentials; e.g., see [105], [135], [179], and [180].

The idea of representing operators of quantum mechanics as functions of creation and annihilation operators (Subsection 1.2) is very old. Perhaps the most rigorous treatment of the subject from the mathematical viewpoint can be found in [10] and [11].

The fact that (pseudo)differential operators are functions of operators of multiplication and differentiation by the coordinates was clearly understood very long ago. For example, the treatment of pseudodifferential operators in [109] in fact uses the Feynman ordering (however, this use is implicit since Feynman indices never occur in the paper).

From the very beginning of the systematic development of noncommutative analysis it was realized that the emphasis should be put on common algebraic properties rather than on specific details of the definition of functions of noncommuting operators. At first glance, the most attractive idea is to present these properties in the form of a set of axioms; then for each specific definition one will only have to check these axioms. This was the approach adopted in [129]. However, there is the disadvantage that serious technical difficulties occur whenever one tries to consider morphisms of operator algebras and find what happens to functions of noncommuting operators under these morphisms. That is why in the subsequent publications [99] and [101] the set of axioms was replaced by another set, which deals with the properties of the symbol classes and the mapping symbol $\to$ operator rather than with the purely algebraic properties and their behavior under transformations. We hope that in the present book the definitions (see Section 2) have assumed their final form. In any case they have become simple, universal, and more convenient than before.

The fundamental formulas of noncommutative differential calculus presented in Section 3 come from a number of sources. The Daletskii–Krein formula expressing the derivative with respect to a parameter of a function of an operator depending on the parameter was first obtained in [29] and [30]. It was extended to a general derivation formula in [96] and [97]. Permutation formulas (Subsection 3.4) and the composite function formula (Subsection 3.5) can be found in [96], [97], [98], and [129]. We should point out that only a small part of a broad variety of formulas of noncommutative differential calculus is presented in the book. A great many of these formulas were obtained by Karasev. Here we cite only a few of his main papers; the complete list of references can be found in [98] and [101].

The Campbell–Hausdorff theorem and Dynkin's formula (Section 4) have a long and intricate history. Surprisingly, they have a number of important practical applications not limited to the reconstruction of the multiplication law in a Lie group [19], [53], [104], [117], [190]. There have been different approaches to the proof of Dynkin's formula [43]; e.g., see [21], [34], [53], [56], and [190]. The proof presented in Section 4 was obtained in [139], and the generalization to T-exponentials in [102].

Chapter II

The method of ordered representation is the most powerful tool in special noncommutative analysis (that is, noncommutative analysis that deals with functions of a fixed Feynman tuple of operators). There is a substantial body of literature devoted, in essence, to functions of a tuple of operators satisfying a special class of relations. Primarily, these are Lie relations or graded Lie relations (see [12], [26], [28], [61], [62], [110], [119], [140], etc).

The ordered representation (Section 1) was introduced in [129], and in [130] the essential role of the generalized Jacobi condition (Section 4) (which consists in the requirement that the operators of ordered representation should satisfy the same relations as the original operators) was announced, and was later studied in [133], [141], and

[134]. The ordered representation permits one to compute products in terms of symbols and leads (under the condition that the generalized Jacobi condition be satisfied) to the notion of a hypergroup (see [130] and cf. [122], [119]). The Jacobi condition for the ordered representation leads immediately to the Poincaré–Birkhoff–Witt (PBW) theorem (see [86], [162]), which states the PBW property for enveloping algebras of Lie algebras. Far more interesting results are obtained if the Jacobi condition is applied to non-Lie commutation relation involving parameters. This results in a certain finite number of equations in these parameters. (Numerous examples of this sort were considered in [130], and some of them are included in Section 2 and Section 4.) These equations are closely related to the Yang–Baxter equations [7], [196] heavily used in the theory of quantum groups and quadratic algebras, which has *been recently* intensively under development (e.g., see [32], [35], [36], [37], [38], [47], [46], [70], [71], [76], [90], [91], [92], [113], [114], [115], [116], [121], [122], [152], [158], [161], [166], [167], [168], [169], [170], [181], [182], [192], [193], [194], [195], and [198]). The situation is as follows: the Yang–Baxter equations are easier to verify but only guarantee that the PBW property is valid on polynomials of degree ≤ 3, whereas the generalized Jacobi condition is much harder to check but ensures the PBW property for polynomials of arbitrary degree. The difficulty lies, of course, in how to compute the ordered representation operators. Some methods for this computation for certain classes of relations not restricted to Lie relations were proposed in [133], [134], [141], and [101] (see also references cited therein).

Note that there is a very important branch of special noncommutative analysis not included in the present textbook, since it goes far beyond the elementary level. We mean *asymptotic noncommutative analysis*, which deals with commutation relations involving a small parameter and provides ordered representations, composition formulas, etc., in the form of asymptotic expansions with respect to this parameter. The small parameter being interpreted as Planck's constant, we arrive at the old, familiar quantization problem. Its solution leads to rich geometric constructions involving symplectic geometry (cf. [75]). We refer the reader to [99], [100], and [101], where a relatively complete bibliography on the topic can also be found.

However, there is a case in which asymptotic and exact noncommutative analysis coincide, i.e., they give the same results. We mean the case in which the Feynman tuple in question satisfies Lie commutation relations. In this case the noncommutative analysis of functions of this tuple is closely related to the analysis on the corresponding Lie group. The relationship is considered in Section 6. For the reader's convenience, simple basic facts concerning Lie algebras, Lie groups, and their representations are collected in the Appendix at the end of the book. They can also be found in any standard textbook on the topic, e.g. [86], [162], [184] and others.

The functional-analytic conditions on the generators of a representation of a Lie group such as the Krein–Shikhvatov and Nelson theorems are discussed in numerous works, e.g., [42], [58], [69], [78], [93], [95], [105], [107], [138], [149], [164], and [197]. Analysis in Hilbert (or Banach) scales (e.g., see [111]) is a natural framework for our considerations in Section 6. The symbol classes introduced there are suitable for solving asymptotic problems arising in the theory of differential equations.

Chapter III

This chapter comprises several examples in which noncommutative analysis is applied to obtain some kind of asymptotic solution to a problem for a differential equation. For standard settings of asymptotic problems for differential equations and related topics we refer the reader to [41], [79], [80], [81], [82], [84], [108], [109], [137], [142], [143], [165], [175], and [178]. However, the applications we cover also include nonstandard settings of asymptotic problems and our exposition chiefly follows [131], [132], and [134].

The approaches to difference and differential-difference equations in Section 2 are taken from [129] and [131].

The problem on propagation of electromagnetic waves in plasma (Section 3) first considered by Lewis [120]. The rigorous mathematical treatment of this problem was carried out in [145], [146], [147], and [148] (cf. also [134]).

The problem of obtaining synchronous asymptotics for equations with coefficients growing at infinity (Section 4) has been considered by many authors (let us mention, e.g., the papers [52], [163], and the literature cited therein). Here we include the results concerning coefficients of polynomial growth and obtained in [141]; these results were published in [133] and [134].

In Section 5 we consider a model example based on the geostrophic wind equations introduced in [153].

In Section 6 we demonstrate the method of re-expansions using the exact solution of some standard operator [129] on the example of a degenerate equation taken from [141]. Here the standard operator is just the operator of the quantum oscillator.

Finally, in Section 7 we consider a problem involving an operator with double characteristics. This problem is well-studied (e.g., see [15], [16], [74], [80], [154], [155], and [157]) and we present no new results. Our aim is merely to demonstrate the use of operator-valued symbols [129] in problems of such sort in the context of noncommutative analysis. To this end, we first reduce the problem to a simple normal form in the spirit of [172], [124], and [1].

Chapter IV

The general definitions of noncommutative analysis given in Section 2 require an appropriate functional-analytic framework. This is provided by the language of polynormed spaces. The theory of polynormed spaces (and, more generally, of spaces with convergence) has been developed intensively for a long time, see [13], [22], [23], [24], [31], [68], [72], [77], [87], [88], [89], [98], [126], and [138]. We reproduce the results necessary for our purposes in Section 1; the results of Section 2 are apparently new; at least we could not find them elsewhere in the literature. The exposition in Section 3 chiefly follows [134] and [141].

Bibliography

[1] S. Alinhac, On the reduction of pseudo-differential operators to canonical forms. *J. Differential Equations* **31**, No. 2, 1979, 165–182.

[2] R. F. V. Anderson, The Weyl functional calculus. *J. Funct. Anal.* **4**, No. 2, 1969, 240–267.

[3] —, On the Weyl functional calculus. *J. Funct. Anal.* **6**, No. 1, 1970, 110–115.

[4] —, Laplace transform methods in multivariate spectral theory. *Pacific J. Math.* **51**, No. 2, 1974, 339–348.

[5] V. I. Arnol'd, *Mathematical Methods of Classical Mechanics*, 2nd ed., Graduate Texts in Math. 60, Springer-Verlag, Berlin–Heidelberg–New York 1989.

[6] —, and A. B. Givental', Symplectic geometry. In: *Itogi Nauki i Tekhniki*, Sovrem. Probl. Mat., No. 4, pp. 7–139. VINITI, Moscow 1985. [Russian].

[7] R. J. Baxter, Partition function of the eightvertex lattice model. *Ann. Physics* **70**, 1972, 193–228.

[8] R. Beals, A general calculus of pseudodifferential operators. *Duke Math. J.* **42**, 1979, 1–42.

[9] —, and C. Fefferman, Spatially inhomogeneous pseudodifferential operators. *Comm. Pure Appl. Math.* **27**, 1974, 1–24.

[10] F. A. Berezin, *Methods of Second Quantization*. Pure and Applied Physics 24, Academic Press, New York 1966.

[11] —, Wick and anti-Wick operator symbols. *Math. USSR-Sb.* **15**, No. 4, 1971, 577–606.

[12] —, and G. I. Kac, Lie groups with commuting and anticommuting parameters. *Math. USSR-Sb.* **11**, No. 3, 1970, 311–325.

[13] G. Birkhoff, Moore–Smith convergence in general topology. *Ann. of Math.* **38**, 1937, 39–56.

[14] N. Bourbaki, *Théories Spectrales*. Hermann, Paris 1967.

[15] L. Boutet de Monvel, Hypoelliptic operators with double characteristics and related pseudodifferential operators. *Comm. Pure Appl. Math.* **27**, 1974, 585–639.

[16] —, and F. Trèves, On a class of pseudodifferential operators with double characteristics. *Invent. Math.* **24**, 1974, 1–34.

[17] V. S. Buslaev, The generating integral and the canonical Maslov operator in the WKB method. *Functional Anal. Appl.* **3**, 1969, 181–193.

[18] —, Quantization and the WKB method. In *Proc. Steklov Inst. Math.* **110**, 1972, 1–27.

[19] J. E. Campbell, *Introductory Treatise on Lie's Theory of Finite Continuous Transformation Groups*. Reprint. Chelsea, New York 1966.

[20] W. B. Campbell, P. Finkler, C. E. Jones, and M. N. Misheloff, Path integrals with arbitrary generators and the eigenfunction problem. *Ann. Physics* **96**, 1976, 286–302.

[21] P. Cartier, Démonstration algébrique de la formula de Hausdorff. *Bull. Soc. Math. France* **84**, 1956, 241–249.

[22] C. H. Cook, Compact pseudo-convergences. *Math. Ann.* **202**, No. 3, 1973, 193–202.

[23] —, and H. R. Fisher, On equicontinuity and continuous convergence. *Math. Ann.* **159**, 1965, 94–104.

[24] —, and H. R. Fisher, Uniform convergence structures. *Math. Ann.* **173**, 1967, 290–306.

[25] A. Cordoba and C. Fefferman, Wave packets and Fourier integral operators. *Comm. Partial Differential Equations* **3**, No. 11, 1978, 979–1005.

[26] L. Corwin, Y. Ne'eman, and S. Sternberg, Graded Lie algebras in mathematics and physics. (Bose–Fermi symmetry). *Rev. Modern Phys.* **47**, No. 3, 1975, 573–603.

[27] —, and L. Rothschild, Nesessary conditions for local solvability of homogeneous left-invariant differential operators on nilpotent Lie groups. *Acta Math.* **147**, 1981, 265–288.

[28] Yu. L. Daletskii, Lie superalgebras in a Hamiltonian operator theory. In: *Nonlinear and turbulent processes in physics*, Vol. 3 (Kiev 1983), pp. 1289–1295, Harwood Academic Publ., Chur 1984.

[29] —, and S. G. Krein, A formula for differentiating functions of Hermitian operators with respect to a parameter. *Dokl. Akad. Nauk SSSR* **76**, No. 1, 1951, 13–66. [Russian].

[30] —, and S. G. Krein, Integration and differentiation of operators depending on a parameter. *Uspekhi Mat. Nauk* **12**, No. 1, 1957, 182–186. [Russian].

[31] G. F. C. De Bruyn, Concepts in vector spaces with convergence structures. *Canad. Math. Bull.* **18**, No. 4, 1975, 499–502.

[32] E. E. Demidov, On some aspects of the theory of quantum groups. *Uspekhi Mat. Nauk* **48**, No. 6, 1993, 39–74. [Russian].

[33] P. A. M. Dirac, *Lectures on quantum mechanics*. Belfer Graduate School of Science. Yeshiva Univ., New York 1964.

[34] D. Ž. Djokovi´c, An elementary proof of the Baker–Campbell–Hausdorff–Dynkin formula, *Math. Z.* **143**, 1975, 209–211.

[35] V. G. Drinfeld, Hamiltonian structures on Lie groups, Lie bialgebras and the geometric meaning of the classical Yang–Baxter equations. *Soviet Math. Dokl.* **27**, No. 1, 1983, 68–71.

[36] —, On constant, quasiclassical solutions of the Yang–Baxter quantum equation. *Soviet Math. Dokl.* **28**, No. 3, 1983, 667–671.

[37] —, On quadratic commutation relations in the quasiclassical case. In *Mat. Fiz. i Funkts. Anal.*, pp. 25–34. Naukova Dumka, Kiev 1986 [Russian].

[38] —, Quantum groups. *J. Soviet Math.* **41**, No. 2, 1988, 898–915.

[39] B. A. Dubrovin, A. T. Fomenko, and S. P. Novikov, *Modern Geometry – Methods and Applications, Part I, II, III*. Graduate Texts in Math. 93, 14, 124 , Springer-Verlag, Berlin–New York 1984 – 1990].

[40] J. Duistermaat and V. Guillemin, The spectrum of positive elliptic operators and periodic bicharacteristics. *Invent. Math.* **29**, 1975, 39–79.

[41] —, and L. Hörmander, Fourier integral operators II. *Acta Math.* **128**, 1972, 183–269.

[42] N. Dunford and J. T. Schwartz, *Linear Operators*, Vol. I, II. Wiley Interscience 1963.

[43] E. B. Dynkin, On a representation of the series $\log(e^x e^y)$ in noncommuting x and y via their commutators. *Mat. Sb.* **25 (67)**, No. 1, 1949, 155–162. [Russian].

[44] Yu. V. Egorov, Canonical transformations and pseudodifferential operators. *Trans. Moscow Math. Soc.* **24**, 1974, 1–28.

[45] —, *Linear Differential Equations of Principal Type*. Plenum Publishing Corp., New York–London 1987.

[46] L. D. Faddeev, N. Yu. Reshetikhin, and L. A. Takhtajan, Quantization of Lie groups and Lie algebras. In: *Algebraic analysis. Papers dedicated to Prof. Mikio Sato on the occasion of his sixtieth birthday*, Vol. 1, pp. 129–139, Academic Press, Boston 1989.

[47] —, and L. A. Takhtajan, A Liouville model on the lattice. In: *Field Theory, quantum gravitiy and strings (Meudon/Paris, 1984/1985)*, pp. 166–179, Lecture Notes in Phys. **246**, Springer-Verlag, Berlin–Heidelberg-New York 1986.

[48] M. V. Fedoryuk, *The Saddle-Point Method*. Nauka, Moscow 1977. [Russian].

[49] C. Fefferman and D. Phong, On positivity of pseudodifferential operators. *Proc. Nat. Acad. Sci. USA* **75**, 1978, 4673–4674.

[50] —, —, On the eigenvalue distribution of a pseudodifferential operator. *Proc. Nat. Acad. Sci. USA* **77**, 1980, 5622–5625.

[51] —, —, The uncertainty principle and sharp Gårding inequalities. *Comm. Pure Appl. Math.* **34**, 1981, 285–331.

[52] V. I. Feigin, New classes of pseudodifferential operators in $\mathbb{R}^n$ and some applications. *Trans. Moscow Math. Soc.* **1979**, issue 2, 153–195.

[53] F. Fer, Résolution de l'équation matricielle $\frac{dU}{dt} = pU$ par produit infini d'exponentielles matricielles. *Acad. Roy. Belg. Bull. Cl. Sci. (5)* **44**, 1958, 818–829.

[54] R. P. Feynman, An operator calculus having applications in quantum electrodynamics. *Phys. Rev.* **84**, No. 2, 1951, 108–128.

[55] —, and A. R. Hibbs, *Quantum Mechanics and Path Integrals*. McGraw-Hill, New York 1965.

[56] D. Finkelstein, On relations between commutators. *Comm. Pure Appl. Math.* **8**, 1955, 245–250.

[57] H. R. Fisher, Limesräume. *Math. Ann.* **137**, 1959, 269–303.

[58] M. Flato, J. Simon, H. Snellman, and D. Sternheimer, Simple facts about analytic vectors and integrability. *Ann. Sci. École Norm. Sup.* **5**, 1972, 423–434.

[59] V. A. Fok, On the canonical transformation in classical and quantum mechanics. *Vestn. Leningrad Univ. Math.* **16**, 1959, 67. [Russian].

[60] —, *Fundamentals of Quantum Mechanics*, 2nd ed., Mir, Moscow 1978.

[61] G. B. Folland, Subelliptic estimates and function spaces on nilpotent Lie groups. *Ark. Mat.* **13**, No. 2, 1975, 161–207.

[62] —, and E. M. Stein, Estimates for $\bar{\partial}_b$ complex and analysis on the Heisenberg group. *Comm. Pure Appl. Math.* **27**, No. 4, 1974, 429.

[63] A. T. Fomenko, *Differential Geometry and Topology*. Plenum Publ. Corporation, New York–London 1987.

[64] —, *Integrability and Nonintegrability in Geometry and Mechanics*. Kluwer Acad. Publ., Dordrecht–Boston–London 1988.

[65] —, *Symplectic Geometry*. Gordon & Breach, London 1989.

[66] —, *Visual Geometry and Topology*. Springer-Verlag, Berlin–Heidelberg–New York 1993.

[67] —, and V. V. Trofimov, *Integrable System of Lie Algebras and Symmetric Spaces*. Gordon & Breach, London 1987.

[68] A. Frölicher and W. Bucher, *Calculus in Vector Spaces without Norm*. Lecture Notes in Math. 30, Springer-Verlag, Berlin–New York–Heidelberg 1966.

[69] I. M. Gel'fand, One-parameter groups of operators in a normed space. *Dokl. Akad. Nauk SSSR* **25**, No. 9, 1939, 713–718. [Russian].

[70] —, and I. V. Cherednik, The abstract Hamiltonian formalism for the classical Yang–Baxter bundles. *Russian Math. Surveys* **38 : 3**, 1983, 1–22.

[71] —, and I. Ya. Dorfman, Hamiltonian operators and the classical Yang–Baxter equation. *Functional Anal. Appl.* **16**, No. 4, 1982, 241–248.

[72] K. K. Golovkin, Parametric-normed Spaces and Normed Massives. *Proc. Steklov Inst. Math.* **106** (1969), American Math. Society, Providence 1972.

[73] R. W. Goodman, Nilpotent Lie Groups: Structure and Applications to Analysis. Lecture Notes in Math. 562. Springer-Verlag, Berlin–Heidelberg–New York 1976.

[74] V. V. Grushin, On a class of elliptic pseudodifferential operators degenerate on a submanifold. *Math. USSR-Sb.* **13**, No. 2, 1971, 155–185.

[75] V. Guillemin and S. Sternberg, *Geometric Asymptotics*. Math. Surveys Monographs 14, revised edition. Amer. Math. Soc., Providence, Rhode Island 1990.

[76] D. I. Gurevich, Poisson brackets associated with the classical Yang–Baxter equation. *Functional Anal. Appl.* **23**, No. 1, 1989, 57–59.

[77] A. Ya. Helemskii, *Banach and Locally Convex Algebras.* Clarendon Press, Oxford 1993.

[78] E. Hille and R. S. Phillips, *Functional Analysis and Semigroups*, Amer. Math. Soc. Colloq. Publ. 31, Amer. Math. Soc., Providence, Rhode Island 1957

[79] L. Hörmander, Pseudo-differential operators. *Commun. Pure Appl. Math.* **18**, 1965, 501–517.

[80] —, Hypoelliptic second order differential equations. *Acta Math.* **119**, 1967, 147–171.

[81] —, The calculus of Fourier integral operators. In: *Prospects in mathematics* (Proc. Symp., Princeton 1970), pp. 33–57, Ann. of Math. Stud. **70**, Princeton Univ. Press, Princeton 1971.

[82] —, Fourier integral operators I. *Acta Math.* **127**, 1971, 79–183.

[83] —, The Weil calculus of pseudodifferential operators. *Comm. Pure and Appl. Math.* **32**, 1979, 355–443.

[84] —, *The Analysis of Linear Partial Differential Operators I, II, III, IV.* Grundlehren Math. Wiss. 256, 257, 274, 275, Springer-Verlag, Berlin–Heidelberg–New York 1983–1985.

[85] R. Howe, Quantum mechanics and partial differential equations. *J. Funct. Anal.* **38** 1980, 188–254.

[86] N. Jacobson, *Lie Algebras.* Wiley Interscience 1962.

[87] H. Jarchow, Dualität und Marinescu-Räume. *Math. Ann.* **182**, No. 2, 1969, 134–144.

[88] —, Marinescu-Räume. *Comment. Math. Helv.* **44**, No. 2, 1969, 138–163.

[89] —, On tensor algebras of Marinescu spaces. *Math. Ann.* **187**, No. 3, 1970, 163–174.

[90] J. Jimbo, A q-difference analogue of $u(g)$ and the Yang–Baxter equation. *Lett. Math. Phys.* **10**, 1985, 63–69.

[91] —, A q-analoque of $u(gl(n+1))$, Hecke algebras and the Yang–Baxter equation. *Lett. Math. Phys.* **11**, 1986, 247–252.

[92] —, Quantum r-matrix for the generalized Toda system. *Comm. Math. Phys.* **102**, 1986, 537–547.

[93] P. T. Jorgensen, Perturbation and analytic continuation of group representation. *Bull. Amer. Math. Soc.* **82**, 1976, 921–928.

[94] V. G. Kac, *Infinite-Dimensional Lie Algebras*, 3rd ed., Cambridge University Press, Cambridge 1990.

[95] S. D. Karakozov, *Representations of Lie semigroups in a locally compact space*. PhD Dissertation, Mathematical Institute, Siberian Division of the USSR Academy of Sciences, Novosibirsk 1985. [Russian].

[96] M. V. Karasev, Expansions of functions of noncommuting operators. *Soviet Math. Dokl.* **15**, No. 1, 1974, 346–350.

[97] —, Formulas for functions of ordered operators. *Math. Notes* **18**, 1975, 746–751.

[98] —, *Exercises in Operator Calculus*. Moscow Inst. Electr. Mach., Moscow 1979. [Russian].

[99] —, and V. P. Maslov, Algebras with general commutation relations and their applications. II. Unitary-nonlinear operator equations. *J. Soviet Math.* **15**, No. 3, 1981, 273–368.

[100] —, and V. P. Maslov, Global asymptotic operators of the regular representation. *Soviet Math. Dokl.* **23**, No. 2, 1981, 228–232.

[101] —, and V. P. Maslov, *Nonlinear Poisson Brackets. Geometry and Quantization*, Transl. Math. Monographs 119, American Mathematical Society, Providence, Rhode Island 1993.

[102] —, and M. V. Mosolova, Infinite products and T-products of exponentials. *Theoret. and Math. Phys.* **28**, No. 2, 1976, 721–729.

[103] —, and V. E. Nazaikinskii, On the quantization of rapidly oscillating symbols. *Math. USSR-Sb.* **34**, No. 6, 1978, 737 – 764.

[104] M. Kashiwara and M. Vergne, The Campbell–Hausdorff formula and invariant hyperfunctions. *Invent. Math.* **47**, No. 3, 1978, 249–272.

[105] T. Kato, *Perturbation Theory for Linear Operators*, 2nd ed., Grundlehren Math. Wiss. 132, Springer-Verlag, Berlin–Heidelberg–New York 1976.

[106] A. A. Kirillov, The characters of unitary representations of Lie groups. *Functional. Anal. Appl.* **2**, No. 2, 1968, 133–146.

[107] —, *Elements of the Theory of Representations*. Grundlehren Math. Wiss. 220, Springer-Verlag, Berlin–Heidelberg–New York 1976.

[108] J. Kohn, Pseudodifferential operators and hypoellipticity. In: *Partial Differential Equations*, pp. 61–69, Proc. Symp. Pure Math. 23, American Math. Society, Providence 1973.

[109] J. J. Kohn and L. Nirenberg, An algebra of pseudo-differential operators. *Comm. Pure Appl. Math.* **18**, 1965, 269–305.

[110] B. Kostant, Graded manifolds, graded Lie theory, and prequantization. In: *Differential Geometrical Methods in Mathematical Physics*, pp. 177–306, Lect. Notes in Math. **570**, Springer-Verlag, Berlin–Heidelberg–New York 1975.

[111] S. G. Krein and Yu. I. Petunin, Scales of Banach spaces. *Russian Math. Surveys* **21**, No. 2, 1966, 85–159.

[112] —, and A. M. Shikhvatov, Linear differential equations on a Lie group. *Functional. Anal. Appl.* **4**, No. 1, 1970, 46–54.

[113] I. M. Krichever, Baxter's equations and algebraic geometry. *Functional Anal. Appl.* **15**, No. 2, 1981, 92–103.

[114] P. P. Kulish and N. Yu. Reshetikhin, Quantum linear problem for the Sine–Gordon equation and higher representations. *Zap. Nauch. Sem. LOMI* **101**, 1981, 101–110. [Russian].

[115] —, and E. K. Sklyanin, Solutions of the Yang–Baxter equation. *J. Soviet Math.* **19**, 1982, 1596–1620.

[116] —, N. Yu. Reshetikhin, and E. K. Sklyanin, Yang–Baxter equations and representation theory. I. *Lett. Math. Phys.* **5**, 1981, 393–403.

[117] Chen Kuo-Tsai, Integration of paths, geometric invariants and generalized Baker–Hausdorff formula. *Ann. of Math.* **65**, No. 1, 1957, 163–178.

[118] P. Lax and L. Nirenberg, On stability for difference schemes; a sharp form of Gårding's inequality. *Comm. Pure and Appl. Math.* **19**, 1966, 473–492.

[119] B. M. Levitan, *The Theory of Generalized Shift Operators*. Nauka, Moscow 1973. [Russian].

[120] R. M. Lewis, *Asymptotic Theory of Transients*. Pergamon Press, New York 1967.

[121] G. L. Litvinov, On double topological algebras and Hopf topological algebras. *Trudy Sem. Vektor. Tenzor. Anal.* **18**, 1978, 372–375. [Russian].

[122] —, Hypergroups and hypergroup algebra. In: *Itogi Nauki i Tekhniki*, Sovrem. Probl. Mat. 26, pp. 57–106. VINITI, Moscow 1985. [Russian].

[123] G. Lusztig, Quantum deformations of certain simple modules over enveloping algebras. *Adv. Math.* **70**, 1988, 237–249.

[124] V. V. Lychagin and B. Yu. Sternin, On the microlocal structure of pseudo-differential operators. *Math. USSR-Sb.* **56**, 1987, 515 – 527.

[125] Yu. I. Manin, Some remarks on Koszul algebras and quantum groups. *Ann. Inst. Fourier* **38**, 1988, 191–206.

[126] G. Marinescu, Opérations linéaires dans les espaces vectorials pseudo-topologiques et produits tensoriels. *Bull. Math. Soc. Sci. Math. R. S. Roumanie (N.S.)* **2**, No. 1, 1958, 49–54.

[127] V. P. Maslov, *Théorie des Perturbations et Méthods Asymptotiques*. Dunod, Paris 1972.

[128] —, *Complex Markov Chains and Feynman's Path Integral*. Nauka, Moscow 1976. [Russian].

[129] —, *Operational Methods*. Mir, Moscow 1976.

[130] —, Application of the method of ordered operators to obtain exact solutions. *Theoret. and Math. Phys.* **33**, No. 2, 1977, 960–976. [Russian].

[131] —, Nonstandard characteristics in asymptotic problems. *Russian Math. Surveys* **38 : 6**, 1983, 1–42.

[132] —, *Asymptotic Methods for Solving Pseudodifferential Equations*. Nauka, Moscow 1987. [Russian].

[133] —, and V. E. Nazaikinskii, Algebras with general commutation relations and their applications. I. Pseudodifferential equations with increasing coeffiicients. *J. Soviet Math.* **15**, No. 3, 1981, 176–273.

[134] —, and V. E. Nazaikinskii, *Asymptotics of Operator and Pseudo-Differential Equations*. Plenum Publishing Corp., New York 1988.

[135] —, and I. A. Shishmarev, On the T-product of hypoelliptic operators,. In: *Itogi Nauki i Tekhniki*, Sovrem. Probl. Mat. 8, pp. 137–198. VINITI, Moscow 1977.

[136] A. Melin, Parametrix constructions for right invariant differential operators on nilpotent groups. *Comm. Partial Differential Equations* **6**, 1981, 1363–1405.

[137] A. S. Mishchenko, V. E. Shatalov, and B. Yu. Sternin, *Lagrangian Manifolds and the Maslov Operator*. Springer-Verlag, Berlin–Heidelberg–New York 1990.

[138] R. T. Moore, Exponentiation of operator Lie algebras on Banach spaces. *Bull. Amer. Math. Soc.* **71**, 1965, 903–908.

[139] M. V. Mosolova, New formula for $\log(e^B e^A)$ in terms of commutators of A and B. *Math. Notes* **23**, No. 5–6, 1978, 448–452.

[140] —, Functions of non-commuting operators that generate a graded Lie algebra. *Math. Notes* **29**, No. 1–2, 1981, 17–22.

[141] V. E. Nazaikinskii, *Ordering and Regular Representations of Noncommuting Operators*. PhD Dissertation, MIEM, Moscow 1981. [Russian].

[142] —, V. G. Oshmyan, B. Yu. Sternin, and V. E. Shatalov, Fourier integral operators and the canonical operator. *Russian Math. Surveys* **36 : 2**, 1981, 93–161.

[143] —, V. E. Shatalov, and B. Yu. Sternin, *Contact Geometry and Linear Differential Equations*. De Gruyter Exp. Math. 6, Walter de Gruyter, Berlin–New York 1992.

[144] —, B. Yu. Sternin, and V. E. Shatalov, Contact geometry and linear differential equations. *Russian Math. Surv.* **48**, 1993, 97–134.

[145] —, —, —, Applications of noncommuting operators to diffractional problems. In: *Waves and Diffraction*, pp. 33–40. Nauka, Moscow 1990. [Russian].

[146] —, —, —, On an application of the Maslov operator method to a diffraction problem. *Soviet Math. Dokl.* **43**, No. 2, 1991, 547–549.

[147] —, —, —, Introduction to Maslov's operational method (noncommutative analysis and differential equations). In: *Global Analysis – Studies and Applications V*, pp. 81–91, Lecture Notes in Math. 1520, Springer-Verlag, Berlin–Heidelberg–New York 1992.

[148] —, —, —, Maslov operational calculus and noncommutative analysis. In: *Operator Calculus and Spectral Theory*, pp. 225–243, Operator Theory: Advances and Applications 57, Birkhäuser Verlag, Basel–Boston 1992.

[149] E. Nelson, Analytic vectors. *Ann. of Math.* **70**, No. 2, 1959, 572–615.

[150] —, Feynman integrals and the Schrödinger equation. *J. Math. Phys.* **5**, No. 3, 1964, 332–343.

[151] —, Operants: A functional calculus for non-commuting operators. In: *Functional Analysis and Related Fields*, ed. F. E. Browder, pp. 172–187, Springer-Verlag, Berlin–Heidelberg–New York 1970.

[152] S. P. Novikov, Various doublings of Hopf algebras. Operator algebras on quantum groups, complex cobordisms. *Russian Math. Surveys* **47 : 5**, 1992, 198–199.

[153] M. A. Obukhov, Concerning the geostrophic wind. *Izv. Akad. Nauk SSSR, Ser. Geograf. Geofiz.* **13**, 1949, 281–306. [Russian].

[154] O. A. Oleynik, Linear equations of second order with nonnegative characteristic form. *Mat. Sb.* **69**, No. 1, 1966, 111–140. [Russian].

[155] —, and E. V. Radkevich, *Second order equations with nonnegative characteristic form*. Plenum Press, New York–London 1973.

[156] S. B. Priddy, Koszul resolutions. *Trans. Amer. Math. Soc.* **152**, 1970, 39–60.

[157] E. V. Radkevich, Hypoelliptic operators with multiple characteristics. *Math. USSR-Sb.* **8**, No. 2, 1969, 181–205.

[158] M. Rosso, Comparaison des groupes $SU(2)$ quantiques de Drinfeld et Woronowicz. *C. R. Acad. Sci. Paris Sér. I Math.* **304**, 1987, 323–326.

[159] L. P. Rotschild and E. M. Stein, Hypoelliptic differential operators and nilpotent groups. *Acta Math.* **137**, No. 3–4, 1976, 247–320.

[160] M. Sato, T. Kawai, and M. Kashiwara, Microfunctions and pseudodifferential equations. In: *Hyperfunctions and pseudodifferential equations*, pp. 265–529, Lecture Notes in Math. 287, Springer-Verlag, Berlin–Heidelberg–New York 1973.

[161] M. A. Semenov-Tyan-Shanskii, Classical r-matrices and quantization. *J. Soviet Math.* **31**, 1985, 3411–3431.

[162] J.-P. Serre, *Lie Algebras and Lie Groups*. Benjamin, New York 1965.

[163] M. A. Shubin, *Pseudodifferential Operators and Spectral Theory*. Springer-Verlag, Berlin–Heidelberg–New York 1987.

[164] J. Simon, On the integrability of finite-dimensional real Lie algebras. *Comm. Math. Phys.* **28**, 1972, 39–46.

[165] J. Sjöstrand, Parametrices for pseudodifferential operators with multiple characteristics. *Ark. Mat.* **12**, No. 1, 1974, 85–130.

[166] E. K. Sklyanin, Some algebraic structures connected with the Yang–Baxter equation. *Funct. Anal. Appl.* **16**, No. 4, 1983, 263–270.

[167] —, Some algebraic structures connected with the Yang–Baxter equation. Representations of quantum algebras. *Funct. Anal. Appl.* **17**, No. 4, 1983, 273–284.

[168] —, On an algebra generated by quadratic relations. *Uspekhi Mat. Nauk* **40**, 1985, 214. [Russian].

[169] —, L. A. Takhtadzhyan, and L. D. Faddeev, Quantum inverse problem method. *Theoret. and Math. Phys.* **40**, No. 2, 1979, 688–706.

[170] Ya. S. Soibel'man and L. L. Vaksman, On some problems in the theory of quantum groups. In: *Representation theory and dynamical systems*, ed. A. M. Vershik, pp. 3–55, Adv. in Soviet Math. 9, American Math. Society, Providence 1992.

[171] E. Stein, Invariant pseudo-differential operators on a Lie group. *Ann. Scuola Norm. Sup. Pisa Cl. Sci.* **26**, 1972, 587–611.

[172] B. Yu. Sternin, Differential equations of subprincipal type. *Math. USSR-Sb.* **53**, No. 1, 1986, 37–67.

[173] —, and V. E. Shatalov, On a method of solving equations with simple characteristics. *Math. USSR-Sb.* **44**, No. 1, 1983, 23–59.

[174] M. E. Taylor, Functions of several self-adjoint operators. *Proc. Amer. Math. Soc.* **19**, 1968, 91–98.

[175] —, *Pseudodifferential Operators*. Princeton Univ. Press, Princeton 1981.

[176] —, *Pseudodifferential operators on contact manifolds*. Graduate lecture notes. Stony Brook, Spring 1981.

[177] —, Noncommutative microlocal analysis. *Mem. Amer. Math. Soc.* **52**, No. 313, 1984, 1–182.

[178] F. Trèves, *Introduction to Pseudodifferential and Fourier Integral Operators*, Vol. 1, 2. Plenum Publishing Corp., New York–London 1980.

[179] H. F. Trotter, Approximation of semigroups of operators. *Pacific J. Math.* **8**, No. 4, 1958, 887–920.

[180] —, On the product of semi-groups of operators. *Proc. Amer. Math. Soc.* **10**, 1959, 545–551.

[181] L. L. Vaksman and Ya. S. Soibel'man, Algebra of functions on the quantum group $SU(2)$. *Functional Anal. Appl.* **22**, 1988, 170–181.

[182] A. M. Vershik, Algebras with quadratic relations. In: *Spectral theory of operators and infinite-dimensional analysis*, pp. 32–57, Akad. Nauk Ukrain. SSR, Inst. Mat., Kiev 1984. [Russian].

[183] A. Voros, An algebra of pseudodifferential operators and the asymptotics of quantum mechanics. *J. Funct. Anal.* **29**, 1978, 104–132.

[184] F. W. Warner, *Foundations of Differential Manifolds and Lie Groups*. Graduate Texts in Math. 93, Springer-Verlag, Berlin–Heidelberg-New York 1983.

[185] Alan Weinstein, Symplectic manifolds and their Lagrangian submanifolds. *Adv. Math.* **6**, 1971, 329–346.

[186] —, Lagrangian submanifolds and Hamiltonian systems. *Ann. of Math.* **98**, No. 2, 1973, 377–410.

[187] —, On Maslov's quantization conditions. In: *Fourier Integral Operators and Partial Differential Equations*, Lecture Notes in Math. 459, pp. 341–372, Springer-Verlag, Berlin–Heidelberg–New York 1975.

[188] —, Lectures on symplectic manifolds. *CBMS Regional Conf. Ser. in Math.* 29, American Math. Society, Providence 1979.

[189] —, Noncommutative geometry and geometric quantization. In: *Symplectic Geometry and Mathematical Physics*, pp. 446–462, Progr. Math. 99, Birkhäuser Verlag, Basel–Boston 1991.

[190] G. H. Weiss and K. Maradudin, The Baker–Hausdorff formula and a problem in crystal physics. *J. Math. Phys.* **3**, No. 4, 1962, 771–777.

[191] H. Weyl, *Gruppentheorie und Quantenmechanik.* Leipzig 1928.

[192] S. L. Woronowicz, Compact matrix pseudogroups. *Comm. Math. Phys.* **111**, 1987, 613–655.

[193] —, Twisted $SU(2)$ group. An example of a non-commutative differential calculus. *Publ. Res. Inst. Math. Sci.* **23**, 1987, 117–181.

[194] —, Tannaka–Krein duality for compact matrix pseudogroups. Twisted $SU(n)$ groups. *Invent Math.* **93**, 1988, 35–76.

[195] —, Differential calculus on compact matrix pseudogroups (quantum groups). *Comm. Math. Phys.* **122**, 1989, 125–170.

[196] C. N. Yang, Some exact results for the many-body problem in one dimension with repulsive delta-function inversation. *Phys. Rev. Lett.* **19**, 1967, 1312–1314.

[197] K. Yosida, *Functional Analysis.* Grundlehren Math. Wiss. 123, Springer-Verlag, Berlin–Heidelberg 1965.

[198] A. B. Zamolodchikov and Al. B. Zamolodchikov, Two-dimensional factorizable s-matrices as exact solutions of some quantum field theory models. *Ann. Physics* **120**, 1979, 253–291.

Index